SILICON EARTH

Introduction to the Microelectronics and Nanotechnology Revolution

We are in the swirling center of the most life-changing technological revolution the Earth has ever known. In only 60 years, a blink of the eye in human history, a single technological invention has launched the mythical thousand ships, producing the most sweeping and pervasive set of changes ever to wash over humankind; changes that are reshaping the very core of human existence, on a global scale, and at a relentlessly accelerating pace. And we are just at the very beginning. *Silicon Earth* introduces readers with little or no technical background to the many marvels of microelectronics and nanotechnology, using easy, nonintimidating language and an intuitive approach with minimal math. The general scientific and engineering underpinnings of microelectronics and nanotechnology are addressed, as well as how this new technological revolution is transforming a broad array of interdisciplinary fields, and civilization as a whole. Special "widget deconstruction" chapters address the inner workings of ubiquitous micro- or nano-enabled pieces of technology such as cell phones, flash drives, GPS, DVDs, and digital cameras.

John D. Cressler is the Ken Byers Professor of Electrical and Computer Engineering at The Georgia Institute of Technology. He received his Ph.D. in applied physics from Columbia University in 1990. After working on the research staff at the IBM Thomas J. Watson Research Center for eight years, he began his academic career at Auburn University in 1992 and then joined the faculty at Georgia Tech in 2002. His research interests center on developing blindingly fast, next-generation electronic components using silicon-based heterostructure devices and circuits. He and his research team have published more than 450 scientific papers in this area. He has served as associate editor for three IEEE journals and on numerous conference program committees. He is a Fellow of the IEEE and was awarded the C. Holmes MacDonald National Outstanding Teacher Award (Eta Kappa Nu, 1996), the IEEE Third Millennium Medal (2000), the Office of Naval Research Young Investigator Award (1994), the Birdsong Merit Teaching Award (1998), and the Outstanding Faculty Leadership for the Development of Graduate Research Assistants Award (2007). His previous books include *Silicon-Germanium Heterojunction Bipolar Transistors* (2003), *Reinventing Teenagers: The Gentle Art of Instilling Character in Our Young People* (2004), and *Silicon Heterostructure Handbook: Materials, Fabrication, Devices, Circuits, and Applications of SiGe and Si Strained-Layer Epitaxy* (2006). He is an avid hiker, gardener, and wine collector.

SILICON EARTH

Introduction to the Microelectronics and Nanotechnology Revolution

John D. Cressler
Georgia Institute of Technology

CAMBRIDGE UNIVERSITY PRESS
Cambridge, New York, Melbourne, Madrid, Cape Town, Singapore, São Paulo, Delhi

Cambridge University Press
32 Avenue of the Americas, New York, NY 10013-2473, USA

www.cambridge.org
Information on this title: www.cambridge.org/9780521705059

First published 2009

Printed in the United States of America

A catalog record for this publication is available from the British Library.

Library of Congress Cataloging in Publication data

Cressler, John D.
Silicon Earth : introduction to the microelectronics and nanotechnology revolution / John D. Cressler.
    p.   cm.
Includes bibliographical references and index.
ISBN 978-0-521-87939-2 (hardback)
ISBN 978-0-521-70505-9 (pbk.)
1. Microelectronics – Popular works.   2. Nanotechnology – Popular works.   3. Telecommunication –
Popular works.   4. Technological innovations – Social aspects – Popular works.   I. Title.
TK7874.C746 2009
621.381–dc22   2009006712

Description of the Cover Art
Curious about the slick cover art? Starting from NASA's famous "Earth at Night" image, we played some games with Adobe Photoshop to introduce curvature and atmosphere to the Earth and then added silicon integrated circuit oceans (read: Silicon Earth). The original concept of showing the Earth with an integrated circuit overlay was mine, but Peter Gordon led me to NASA's night view, and Poornima Ozarkar, the artistic wife of one of my Ph.D. students, executed a large number of design iterations to get it right. Final result? Pretty slick!

The "Earth at Night" (courtesy of NASA, 2000) is actually a composite of images created from data gathered by satellites that are a part of the U.S. Air Force's Defense Meteorological Satellite Program (DMSP), and that reside at about 800 km above the Earth's surface, in Sun-synchronous, near-polar orbits (99 degree inclination). The night view of the Earth clearly shows evidence of human civilization, speaking volumes about our reliance on electricity, and also suggests why it's awfully tough to get a great view of the Milky Way when you live in big cities! The brightest areas are obviously not the most populated, but rather the most urbanized (read: light pollution?!). The integrated circuit spanning the oceans of the Silicon Earth is actually a SiGe radar chip designed by my Georgia Tech research team.

To the many aspiring students
of microlectronics and nanotechnology:
May your imaginings be bold, your vision keen,
And your re-creation of a new world full of hope and promise.

And . . .

For Maria:
My beautiful wife, best friend, and soul mate for these 26 years.
For Matthew John, Christina Elizabeth, and Joanna Marie:
God's awesome creations, and our precious gifts.
May your journey of discovery never end.

The most beautiful thing
We can experience
Is the mysterious.
It is the source
Of all true art and science.
He to whom this emotion
Is a stranger,
Who can no longer pause
To wonder
And stand rapt in awe,
Is as good as dead:
His eyes are closed.

– Albert Einstein

Contents

Preface

We are in the swirling center of the most life-changing technological revolution the Earth has ever known. In only 60 years, a blink of the eye in human history, a single technological discovery has launched the mythical thousand ships, producing the most sweeping and pervasive set of changes ever to wash over humankind; changes that are reshaping the very core of human existence, on a global scale, and at a relentlessly accelerating pace. More important, these changes are only in their infancy! *Silicon Earth: Introduction to the Microelectronics and Nanotechnology Revolution* introduces readers with little or no technical background to the marvels of microelectronics and nanotechnology, using friendly, nonintimidating language and an intuitive approach with minimal math. This book introduces the general scientific and engineering underpinnings of microelectronics and nanotechnology and explores how this new technological revolution is transforming the very essence of civilization. To round things out for the technologically curious, special "widget deconstruction" chapters address the inner workings of ubiquitous micro or nano-enabled pieces of technology such as cell phones, flash drives, GPSs, DVDs, and digital cameras. Want to know how that i-Phone works? Here's your chance!

Is this really such a big deal that it warrants plunking down some hard-won bucks and allocating some quiet reading time? You bet it is! The microelectronics and nanotechnology revolution is profoundly reshaping planet Earth as we speak, changing forever the ways we humans communicate, socialize, shop, play games, create art, elect our leaders, practice medicine, teach, conduct business, and, yes, even think. A big deal. Should you care? You'd better, else you're going to be steamrolled by the relentless advance of micro/nanotechnology sweeping across the globe. Think of this cute little book as an investment in your future, a chance for you to be in the know and ahead of the curve. One up on your friends. So, yes, go ahead, use that ATM card to send a few electrons to your checking account . . . and let's get learning!

1. THROWING DOWN THE GAUNTLET

I am throwing down the gauntlet![1] I would like to issue my challenge to the legions of you bright young people of our world, you students on a learning curve to become the next set of technology practitioners, the future movers and shakers of our world. To win the challenge, you must understand what microelectronics and nanotechnology are really all about and then gleefully revel in all the glamor and excitement, appreciating the incredible myriad of future applications awaiting your inventive minds. Don't let it end there! It mustn't end there! I challenge you to take up the gauntlet. Knowing what you soon will discover in these pages, I invite you to then step back, reflect, muse a bit, and take a stand regarding HOW the development of these remarkable microelectronics and nanotechnology inventions you will conceive can best be put to use in serving our global community for the greater good. The final chapter in this book examines the many evolving societal transformations and the numerous issues swirling around the ensuing microelectronics and nanotechnology revolution and is offered for your serious consideration. First things first, though – let's learn some micro/nano stuff!

2. USING THIS BOOK

This book is intended to be used by two very different audiences: (1) As a textbook for an interdisciplinary, introductory course in microelectronics and nanotechnology and (2) as a pick-up-and-read-cover-to-cover book for those curious about what all this micro/nanobabble is all about. More important, this book assumes no special technical background in the subject matter, and thus should be accessible to your typical university freshman from virtually any discipline. Have you been out of school for a while? No worries, you should do just fine. Some basic physics and electrical engineering refresher appendices are included for those that might need them.

This book serves as the textbook for a course (CoE 3002) I have introduced into the curriculum at Georgia Tech (Fall 2008), titled "Introduction to the Microelectronics and Nanotechnology Revolution." It is intended for freshmen and sophomores in the Georgia Tech Honors' Program and for juniors in Georgia Tech's joint College of Management and College of Engineering's new Technology and Management Program. The students taking this course come from many disciplines (engineering, management, science, social science, etc.), at varying educational levels, and, important for the reluctant among you, with no real background in electrical engineering. That's the intent, so the book is pitched to the level of this audience. My course consists of a mixture of lecture, several tours to real micro/nanotechnology research labs on campus, and roundtable discussions based on the philosophical and social topics addressed in the last chapter of the

---

[1] You know, the Middle Ages, armored knights, swordplay, chain mail, chivalry, duels! A gauntlet is a type of glove with an extended cuff protecting part of the forearm against a sword blow. To throw down the gauntlet is to issue a formal challenge. A gauntlet-wearing knight would challenge a fellow knight to a duel by throwing one of his gauntlets on the ground. His opponent would then pick up the gauntlet to formally accept the challenge. Let the games begin!

book. For these discussions, I form different "debate teams" who first research and then adopt pro–con positions (by draw of the card) on the topic in question to heighten the energy level of the experience. Students also engage in a collaborative capstone research experience in which five-person teams do their own widget deconstructions and present those findings to the class. I make it a competition. Works very well. For those who are interested, my course Web site (*http://users.ece.gatech.edu/~cressler/courses/courses.html*) contains more information on debate topics and class deconstruction projects.

My sincere hope is that this type of come-one-come-all cross-disciplinary university entry-level micro/nano course becomes a groundswell across campuses (hopefully globally). Encouraging signs can already be gleaned at U.S. universities. I believe this material, if placed in the right hands, could also be effectively introduced to select seniors at the high school level.

3. SOME SPECIAL THANKS

I am truly grateful to my editor, Peter Gordon, of Cambridge University Press, for his unwavering support throughout this project. Peter came highly recommended to me by Dave Irwin, a trusted mentor. True, it took a little time for Peter to embrace what has evolved into a decidedly nontraditional approach to authoring a textbook, especially with regard to my intentionally, shall we say, "free-spirited" writing style. From minute one I wanted this book to be fun, not stuffy; a new take on what an engaging textbook could and should be – and something that could simultaneously be embraced by students as well as a general readership. Peter bought into my vision; helped me hone my book into a feasible, cost-effective package; and even agreed not to unleash the English-usage police on me! If the final product gives you any pause for thought, for any reason, please blame me, not Peter!

I'd like to thank Dustin Schisler, Glendaliz Camacho, and their colleagues at Cambridge University Press for their expert handling of the figures and layout. Thanks also to Larry Fox of Aptara for his expertise in production, and Vicki Danahy for her interest and skillful copyediting. I am also grateful to Poornima Ozarkar for her help with the book cover and, expertly exercising the minimiracles of Adobe Photoshop.

I'd like to thank Mark Ferguson, Linda Oldham, Monica Halka, Greg Nobles, Gary May, Joe Hughes, Doug Williams, and Larry Jacobs of Georgia Tech, for their support in getting my new micro/nano course off the ground.

My fundamental approach to writing this book originates from my deeply held belief that ANY subject can be effectively taught to ANY audience if you work hard enough as a teacher–writer. Call me naive! This can be challenging, to be sure, but it has worked here. Does it take audacity to teach a subject so deeply technical as microelectronics and nanotechnology to folks with no real electrical engineering background? You bet! Can it work well? It can. I've had some help along the way, clearly. I'd like to thank my many students, both undergraduate and graduate, over the years for helping me hone my teaching skills.

Yea verily, I am shameless when it comes to subjecting my friends, my colleagues, my graduate students, my nieces and nephews, my brothers- and sisters-in-law, and even my wife and kids to preliminary drafts of several of the chapters in this book for their feedback and a sanity check. Does it make sense? Can you understand it? Am I out in left field? Thanks to all for their indulgence.

In the spirit of the burgeoning social media landscape enabled by the Internet, I have frequently drawn inspiration from various Wikipedia entries on a number of subjects contained in this book. My apologies to any of my fellow professors who may find a wiki citation bloodcurdling! In the present case, given my target (nonexpert) audience, who are all VERY familiar with Wikipedia, it seemed to me silly to not employ the extensive information out on the Web. However, let me amplify what students frequently hear from us professors: Wikipedia should be used as a starting point, not as an end in itself. And, as with all things Web-based, a healthy nonzero level of skepticism regarding the accuracy of what you find on the Web is always warranted. Let me issue a special thanks to all you wiki authors out there who devote your time and energy to improving our global knowledge base.

My earliest draft of the table of Contents for this book dates to January of 2005 – sigh . . . over four years back. Yes, it has been a long road, and a TON of work, but I have to admit, I really enjoyed writing this book! If my treatment remains too technical for you, or if it seems overly obtuse in certain areas, or if I've managed to botch some facts along the way, or if you think my discussion could be improved here and there with a little tweaking, or if heaven forbid, if you really did enjoy my approach . . . please, by all means let me know; I'd love to hear from you!

I now gently lay my labor of love at your feet. I hope my efforts please you, and maybe even make you smile here and there! Enjoy.

John D. Cressler
School of Electrical and Computer Engineering
Georgia Tech
July 2009

About the Author

John D. Cressler is Ken Byers Professor of Electrical and Computer Engineering at The Georgia Institute of Technology (Georgia Tech), Atlanta, GA. He received his Ph.D. from Columbia University, New York, NY, in 1990; was on the research staff of the IBM Thomas J. Watson Research Center, Yorktown Heights, NY, from 1984 to 1992; and was on the engineering faculty of Auburn University, Auburn, AL, from 1992 to 2002, when he left to join Georgia Tech. His research focuses on developing next-generation, high-speed electronic systems and is currently centered on the fundamental understanding, technological development, and clever new uses of nanoscale-engineered, silicon-based, heterostructure transistors and circuits (admittedly a mouthful!). He and his research team have published more than 450 scientific papers in this field. He is the co-author of *Silicon-Germanium Heterojunction Bipolar Transistors* (Artech House, 2003); the editor of *Silicon Heterostructure Handbook: Materials, Fabrication, Devices, Circuits, and Applications of SiGe and Si Strained-Layer Epitaxy* (CRC Press, 2006); and the author of *Reinventing Teenagers: The Gentle Art of Instilling Character in Our Young People* (Xlibris, 2004). He has served as editor for three scientific journals, has played various roles on numerous technical conference committees, and has received a number of awards for both his teaching and his research. To date, he has graduated 27 Ph.D. students and 25 M.S. students during his 17-year academic career. He was elected Fellow of the Institute of Electrical and Electronics Engineers (IEEE) in 2001. On a personal note, John's hobbies include hiking, gardening, bonsai, all things Italian, collecting (and drinking!) fine wines, cooking, history, and carving walking sticks, not necessarily in that order. He considers teaching to be his primary vocation. John has been married to Maria, his best friend and soul mate, for 26 years, and is the proud father of three: Matt (24), Christina (22), and Jo-Jo (19).

Dr. Cressler can be reached at:
School of Electrical and Computer Engineering
777 Atlantic Drive, N.W.
Georgia Institute of Technology
Atlanta, GA 30332-0250 U.S.A.
or
cressler@ece.gatech.edu
http://users.ece.gatech.edu/~cressler/

1 The Communications Revolution

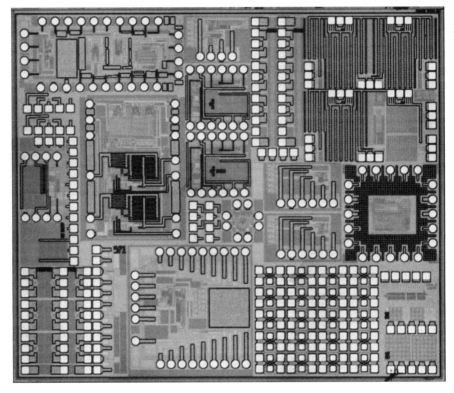

An integrated circuit designed at Georgia Tech and fabricated at IBM. (See color plate 1.)

> *We Predict the Future . . .*
> *By Inventing It.*
> Motto of Xerox PARC

1.1 The big picture

As I intimated in the Preface, we earthlings are smack in the middle of history's most sweeping technological revolution. Period. Amazingly, this global revolution is the play-out of a single pivotal discovery, a discovery unearthed by a trio of

1

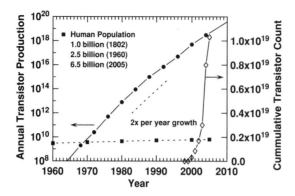

Figure 1.1 Evolution of annual global production of transistors and cumulative transistor count (after G. Moore). Also shown is the human population growth over the same period.

clever scientists. A discovery so profound in its consequences for everyday life that it merits attention by even grade-schoolers and grandmas. That discovery? The "transistor." Skeptical of my bold claim on the importance of this revolution? Clueless as to what a transistor actually is? Well, buck up and read on.

The mighty transistor: A microbial-sized, speed-of-light fast, dirt cheap, on–off switch . . . with a twist. Pervasive. Whether you appreciate it or not, the world is literally awash in transistors. The transistor was discovered–invented by John Bardeen, Walter Brattain, and William Shockley at Bell Laboratories on December 23, 1947, at about 5:00 P.M. EST! A fresh white snow was falling in New Jersey. In 1947, one transistor. Today, there are about 10,000,000,000,000,000,000 (10-billion billion) transistors on Earth, and this number is roughly doubling every 18 months (Fig. 1.1). More on growth trends in a few moments. Given our global population of over 6.5 billion souls [2], this means 1,538,461,538 (1.5 billion) transistors for every man, woman, and child. Imagine – from zero to 10,000,000,000,000,000,000 transistors in only a little over 60 years! For modern humans, growth from zero to 6.5 billion took nearly 30,000 years. But let me not oversimplify the fascinating field of paleoanthropology. Roughly 100,000 years ago, the planet was occupied by a morphologically diverse group of hominids. In Africa and the Near East there were *homo sapiens*; in Asia, *homo erectus*; and in Europe *homo neanderthalensis*. By 30,000 years ago, however, this taxonomic diversity had vanished, and humans everywhere had evolved into a single anatomically and behaviorally modern form near and dear to all of us. Sixty years for 10-billion billion transistors; 30,000 years for 6.5 billion people – quite a contrast in the historical record. Amazingly, those 1,538,461,538 transistors per person would fit comfortably into the palm of your hand. Something most unusual is definitely afoot.

In 1947 the freshly minted transistor was a thumbnail-sized object (Fig. 1.2). By 1997, only 50 years later, we could put 9,300,000 transistors onto a single piece of silicon crystal, 1 cm on a side, to build a microprocessor integrated circuit (in common parlance, a "microchip") for a computer capable of executing

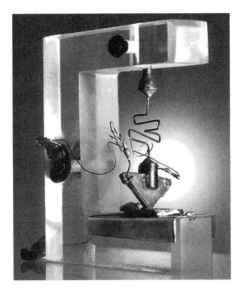

Figure 1.2 A replica of the first transistor, a germanium point-contact device, demonstrated on December 23, 1947. The transistor is roughly the size of your thumbnail (courtesy of Science and Society Picture Library). (See color plate 2.)

1,000,000,000 (1 billion!) numerical computations per second (Fig. 1.3). Talk about progress!

What are we doing with all of these 10,000,000,000,000,000,000 transistors? Gulp... Starting a Communications Revolution – capital C, capital R. This may come as a surprise. Revolution? Why a revolution? Well, you first need

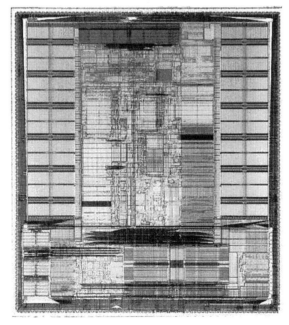

Figure 1.3 DEC ALPHA 21164 (1997), the first billion-instructions-per-second (BIPS) microprocessor. This microprocessor contains over 9,300,000 transistors and is roughly 1.0 cm on a side in size (courtesy of Digital Equipment Corporation).

to open your eyes wide and examine the technology that impacts your day-to-day existence, and then, as with most things, put those observations squarely in the context of history. And then reconsider carefully your world. Here I am using the word "communications" in an intentionally general way: Communications embody the totality of generating, manipulating, storing, transmitting, receiving, and hopefully constructively utilizing information. But what exactly is information? Well, knowledge in its broadest possible sense. Communications also connote movement of information – information flow – that might be as simple as calling your friend on a cell phone, or reading the newspaper, or listening to a Bach concerto – but it also might be as complex as a banking transaction of $5B beamed over a satellite link to other side of the world in 1.7 s (talk about moving money around!), or computing the path of an approaching hurricane in the Gulf of Mexico and then disseminating that warning to 100 million people via television and the Internet – in real time. Whatever the venue, at present, communications, inevitably, at some level – sometimes obvious, often not – trigger the use of Mr. Transistor. Our world, as we know and experience it, literally stands on the broad shoulders of our countless billions of microbial-sized transistors. Pivotal invention indeed. Skeptical? Prepare to be convinced!

Okay, kick back in your chair and prop your feet up so that you are comfortable, and consider: 10,000,000,000,000,000,000 transistors are out there on the streets, and they are ready, willing, and able to forever change the way you talk to your friends, study, socialize, shop, and play games. In short, transistors are radically altering human civilization – read: a BIG deal. This should be a sobering thought. If not, stop texting on your cell phone, go back and reread the last two pages, and then reconsider. Got it? Excellent!

This transistor-changes-the-world fact is something all of us, not just technically literate geeks, should be aware of; MUST be aware of. Hence this book. In this first chapter I set the stage. I first probe the nature of electronic communications using transistors ("digital" information – you know, "1s" and "0s"); attempt to scare you with some doomsday scenarios, in which I imagine life without transistors; introduce Moore's law; explore the magic of the semiconductor silicon, bricks-and-mortar for building Mr. Transistor; and finish by showing you visually just what he looks like. Relax, it's going to be fun. By the way, please turn your cell phone off!

1.2   The evolution of human communications

Okay, first things first. How exactly did we get to electronic communications by using transistors? Well, Fig. 1.4 depicts the (semipersonal, yes, but highly defendable!) milestones in the history of human communications, intentionally spaced in five-century intervals, 1/2 a millenium per tic! From 30,000 B.C.E. to 2000 B.C.E. (28,000 years!), human communication had no written basis. Modern alphabets began to emerge in the Near East circa 2000 B.C.E., marking

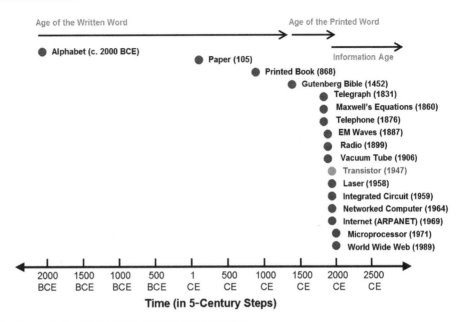

Figure 1.4   Some milestones in the history of human communications.

the beginning of what we will call here the "Age of the Written Word." Early clay tablet documents were hand-lettered, requiring enormous effort to produce, and clearly were reserved for the privileged few. AND, they tended to break when you dropped them on the floor! Enter paper. Although paper was actually invented by the Chinese in the second century C.E., our "Age of the Printed Word" did not begin in force until the development of movable type and the resultant efficient printing press, in the mid-15th century (see sidebar discussion).

Geek Trivia: Exactly How Old Are Books?

The earliest printed book with a date inscription is the "Diamond Sutra," printed in China in 868 C.E. It is believed, however, that book printing may have occurred well before this date. In 1041 C.E., movable clay type was first invented in China. Goldsmith Johannes Gutenberg invented (independently, obviously) his printing press with movable metal type in 1436. His "new" printing process was fine-tuned by 1440, and ready for application to the mass printing of primarily religious texts, the most famous of which was the Gutenberg Bible, first available on September 30, 1452 (Fig. 1.5).

Movable type led to a rapid proliferation of book production, beginning, of course, with the justifiably famous Gutenberg Bible in 1452 (only one tic to the left of present day). Mass-produced books, and soon after all manner of newspapers and pamphlets, were suddenly readily produced, enabling the rapid dissemination of ideas and knowledge. Think universities. Why should you care? Well, this book-based knowledge explosion ultimately spawned the Enlightenment, and with it modern science, together with its practical-minded cohort – engineering – in

Figure 1.5   Two pages of an original Gutenberg Bible (courtesy of the U.S. Library of Congress).

tow, which so marvelously exploits scientific discoveries in such interesting ways, many of which will be revealed within the covers of the book you hold.

It is no coincidence that the milestones in human communications (refer to Fig. 1.4) are very densely packed within the past 100 years. Science (and engineering) had to evolve some sophistication first, of course, before the technological explosion took place. In many ways, the telegraph (you know, dots and dashes – Morse code), which at its central core proved that (rapid) communications at a distance were possible using electrons, set into motion events that set the stage for the invention of the transistor. More on this fascinating story in Chap. 3.

Alas, ubiquitous Mr. Transistor. December 23, 1947, marks the birth of what many term the "Information Age," to my mind, the third principal age in the evolution of human communications.[1] By the way, for the trivia-minded (count me in!), the term "Information Age" was coined by Wilson Dizard in his 1982 book *The Coming Information Age: An Overview of Technology, Economics, and Politics.* After that bright lightbulb of human ingenuity in 1947, the rest seemed to almost naturally unfold, with a lot of hard work and cleverness by the pioneers of course – the integrated circuit, the microprocessor, memory chips, computers, networks, cell phones, the Internet; you name it. All of these downstream technological marvels inevitably trace their roots to the humble transistor, and none could work without it today. Zero, zilch! Transistors, whether you appreciate them or not (don't worry, you soon will!), exist by the millions in

---

[1] For an interesting discussion of the parallels of Gutenberg's invention and the transistor as enablers of their respective communications revolutions, see J. T. Last's discussion in [1].

every piece of technology you own: your car, your cell phone, your i-Pod, your dishwasher, your TV, your camera, probably even your credit card!

Imagine – a new age of human civilization beginning only 60-ish years ago. We live in exciting times. The Information Age was born, and the Communications Revolution that is currently playing out in this new millennium represents the means by which this new age of human civilization will unfold. Shouldn't you know something about this? Shouldn't everybody? Read on.

Geek Trivia: Stump Your Friends!

Care for some relevant trivia to stump your friends? Sure! Question: How many personal computers were there on Earth? Answer: 593,085,000 (all of these answers are 2005 numbers). Question: How many Internet users were there worldwide? Answer: 1,018,057,389. Question: How many cells phones were in use globally? Answer: 2,168,433,600! Question: What country had the highest per capita number of Internet users? Answer: Iceland, with 6,747 per 10,000 inhabitants (surfing the Web, a great way to spend those long, dark winters!). Geek challenge: Update these numbers to present day!

1.3 Doomsday scenarios

Okay, fine . . . I get it. You're wondering . . . "Are transistors really THAT big a deal? Perhaps this guy has overplayed his hand." Hey, a little skepticism is healthy! If you're game, let's engage in a slightly diabolical thought experiment to test my claim regarding the importance of transistors in the grand scheme of things. It will also help you appreciate just how deeply embedded transistors are in everyday life. Here goes. Imagine for a moment that I could suddenly, with the snap of my fingers, erase transistors from the face of the Earth; blot them from the historical record. How? I don't know, maybe a meteor containing "Transistorite" lands in the Gobi Desert (you know, Superman's Kryptonite, but for transistors). Simple question. One second later, only one, how would your life change? For the sake of argument, let's assume you are living, like me, in Atlanta, Georgia, in the United States. Let's now consider the implications of erasing transistors from human civilization, in each of the following life arenas: communications, transportation, health care, and economics. Hold on tight – it gets pretty scary.

Communications: This one is simple. It's back to drums and smoke signals! Your cell phone goes instantly dead mid text message. Hint: There is no such thing as a telephone that does not have a transistor lurking under the covers somewhere. Even a call made from that antique transistorless rotary phone in the attic has to be routed electronically on a landline, so the electronic "clicks" your rotary pulse dialer outputs vaporize into thin air. Ditto for radio – there is nothing to transmit or receive that WREK 91.1 FM broadcast (Georgia Tech's student radio station). Your radio now sits dumb, an ornament on the shelf. Global radio stations go silent. Same story for television. TV screens go blank, worldwide, now a *Poltergeist* snow land. CNN no longer connects you in real time to events unfolding around the world. Oprah is cut off midsentence. How about the Internet? No transistors, no computers; no computers, no networks; no networks, no Internet. No cable modem, DSL line, or T1 fiber line works anyway. So forget all those on-line communications you depend on. Your computer screen goes blank; that IM note (all lowercase, grrrr!) you just typed remains forever unanswered. There are no satellite uplinks or downlinks or optical-fiber backbones, or functional transoceanic cables for moving information around the globe. All instantly silenced. Country-to-country links are dead. There are inevitable political implications. Without communications systems, president-to-president "hot lines" are down, all diplomatic ties severed. The world is profoundly and instantly opaque, akin to days of old when it took 2 months to say "hi" across the Atlantic by boat. Generals with itchy trigger fingers get understandably nervous, hopefully not panicky – not to worry, the nukes can't be launched even if they push the dreaded "red button." All in one second. No transistors, no global communications infrastructure.

Transportation: I'm afraid this one is pretty simple too. It's back to hoofing it! If you are lucky enough to live on a farm, a horse, or maybe a horse and buggy, will reemerge as your preferred means of transport. Bikes are fine for around town – unless you live on Telegraph Hill in San Francisco! I wouldn't really want to bike from Atlanta to New York for a business meeting though. Here's the scoop. Unless your car predates 1980 (if it does you need a new one anyway!), your engine immediately dies, and you coast to a dead stop. Electronic fuel injection (yep, uses transistors) exists in virtually all automobiles today (more trivia to stump your friends – the last passenger car to have a carburetor was the 1990 Subaru Justy). You quip – "but cars were around before transistors!" Correct, but you will not find one today without transistors, lots of electronics, and probably several dozen on-board computers (fuel injectors, engine controls, sensors; the list is long). Unless you are a highly skilled mechanic with a junkyard conveniently located next door, you are unlikely ever to convert your now dead-as-a-doornail Beamer into a functioning car. It would take the car industry years to convert plants to make such a beast, even if they remembered how. Because I'm not a sadist, let's assume you are not on a jet crossing the country. It shouldn't surprise you that a modern passenger jet has a ton of control electronics and literally hundreds of on-board computers, not to mention the satellite-based GPS that flies the plane for 95% of the way. Ever wonder why the flight attendants are so insistent that you turn off all your electronic gadgets during takeoff and landing? They are afraid that spurious transmissions from your gadget (electronic leakage) may inadvertently interfere with their much more expensive gadget! In short, all planes are grounded. Permanently. Ditto my comments on planes before transistors; unless you are an employee of the Smithsonian's Air and Space Museum, and you can fly a Messerschmidt 109, you're out of luck. All in one second. No transistors, no global transportation infrastructure.

Health care: Here things are slightly better, but only if you are healthy and intend to stay that way for a good while. Alas, life is rarely so convenient. Last year, Americans suffered about 1.5 million heart attacks; about 18 million currently have diabetes; about 173,000 were diagnosed with lung cancer. The miracle that is modern medicine is something we all too often take for granted. Yes, it is frighteningly expensive; yes, health care costs are skyrocketing out of control; yes, not enough people have adequate (or any) health insurance – still, when you are very ill, stop to consider what it would be like to go to the hospital for treatment, not today, but say 200 years ago, or even 100 years ago (read: preantibiotics); not at Saint Joseph's Hospital in Atlanta, but to the village barber who would promptly bleed you (using a rusty knife) to improve your health. I think you will readily concede that we are indeed blessed in our access to modern medicine. It may seem less obvious that transistors play a major role in health care, but . . . sigh . . . it's true. Your medical records are stored electronically in a massive computer database. Poof, they are now gone. The drugs that are at the doctor's disposal require development, testing, and manufacturing on a scale unthinkable without computers. Subtract those once the pharmacy runs out. Even medical school training would be awfully tough to contemplate without computers. Diagnostic tests inevitably include computer-based analysis: blood series, urine analysis, EKG, EEG, CT scan, you name it. Your doctor will now have to treat you blind. (A reminder that the quaint notion of the ambling old country doctor who draws only on intuition and experience for treatment is just that – quaint and old. Country doctors will now be forced back into vogue, however.) Any hospital surgical team is effectively shut down now. Bypass, brain tumor, C-section, appendix? Nope. The autoclave that sterilizes surgical instruments is, surprise, computer controlled; not to mention the anesthesia dispensers, and the myriad of real-time health monitors that comfortingly sound their "beep-beep" that all is well. Your insurance processing and records are likely all electronic also. Gone. Importantly, all of the miraculous treatment options available today come to a grinding halt. No CT scan. No PET scan. No Gamma Knife. No chemotherapy. All those remarkable new discoveries in genetics associated with mapping the human genome, and the vast potential for health care they are likely to embody, are no more – EXTREMELY computer-intensive research. The list could go on. When the lights go out on transistors, so to speak, you may perhaps have a little bit more time to coast if you have no health problems, but not likely for very long. We all age, after all. All in one second. No transistors, no global health care infrastructure.

Economics: Remember those bartering-based medieval economies? You'll need to – the industrialized world is going to quickly look very agrarian. It should not surprise you that computers have completely restructured the global economy. In the richest countries, only 5% or so of financial transactions occur with paper money (i.e., "hard" currency). The other 95%, at least at some level, are processed electronically – yep, using transistors. For consumers, for instance, electronic banking [also known as electronic fund transfer (EFT)] gives near-instantaneous 24-hour access, world wide, to our cash through automated teller machines (ATMs), and direct deposit of paychecks into our checking accounts gives us the cash to draw on. No more. No ATMs work. They will soon be smashed open for the remaining cash inside, reluctantly (and mutely) surrendering a rather large, one-time payout. Because you have direct deposit, you now effectively work for free. Your company has no way – without computers – to generate a paycheck for your toils! Not that you could cash it anyway; the bank can't open its electronically controlled vault. Your life savings instantly vanish. Banks are closed, permanently. Forget Jimmy Stewart's *Wonderful Life* triumph over money panic – Potter wins this one. No more 401k. Global bank-to-bank monetary transfer instantly halts. The notion of currency exchange is meaningless; there is no way to move it around. The world's Stock Exchange boards are now blank, the mouths of the suddenly silent commodities traders at the New York Stock Exchange circular and aghast. Stock? How? It can't be bought or sold. Without any form of payment, the movement of goods rapidly grinds to a halt. Fuel production instantly ceases (hopefully not via a "kaboom"). Without goods moving around, the very notion of "stores" and consumerism is suddenly irrelevant. Within a week there will be nothing to buy on the shelves, and no cash to buy it with anyway. If you live on a farm and can grow some food, and perhaps barter for the rest of your needs, you should be in at least decent shape; for a while. Newsflash – not many of the 300+ million Americans live on farms! The rest of us are going to go hungry. Grocery stores will not receive any resupply, and yep, there are computer controls on all the refrigeration systems. Everything is now starting to melt. Cash registers, with their laser-based bar-code scanners, don't work either – there is no such thing as a computerless crank-handle "cha-ching" cash register except in the Smithsonian, and I doubt you can use an abacus. You get the idea. All in one second. No transistors, no global economic infrastructure.

An identical thought process could be applied to virtually any venue of human activity. Consider these activities: (1) *entertainment*: (How will you play a DVD without a DVD player? Forget your i-POD and your i-Phone! How will you watch your favorite TV program? How will you IM or text your friends?); (2) *publishing*: (How will you typeset a magazine or a book without a computer? How do you plan to print it without a printer? How will you get it to the stores?); (3) *sports*: (Will anyone watch the game or listen to it without radio, TV or the Internet? How will players or fans get to the field without a car or a plane? How will you buy your tickets without an ATM machine for cash or on-line ticketing? Will the athletes really play for free if they can't access their multimillion dollar bank accounts? Back to playing just for the love of the game – nice thought!) (4) *education*: (How will you type up that paper without your computer? How will you do on-line searching for your research paper? How will you look up a book at the library? Photocopy it? How will you give a powerpoint presentation?)

This list could stretch on for miles. We're talking utter chaos. Globally. Dark Ages as we have never known them, even in medieval Europe. Anarchy. Mass suffering and loss of life. As I think you must agree by now, my bold claims have in fact not been overstated: Life as we know it would simply cease to exist if we removed transistors from the planet. Again – we are at only the very beginning of the story. Imagine replaying the preceding doomsday scenarios 20 years from now when we are even more dependent on the ubiquitous transistor. A sobering thought indeed.

1.4 Global information flow

Okay, enough of the gloom-and-doom. Relax, it was only a thought experiment to get your attention! A Communications Revolution. THE Communications Revolution. An interesting question presents itself: How much information flow, knowledge transfer if you will, actually exists in this new age of human civilization? To arrive at a reasonable answer, we must first momentarily digress to define a few key concepts to get us rolling.

First we must delineate the "analog" versus "digital" realms. The world we *homo sapiens* inhabit is decidedly analog in nature. By this, I mean that the inputs that our magically delicate and nuanced sense organs detect and our brains then interpret, no matter the type or form [e.g., pressure (P), temperature (T), sound, light, smell, etc.], are continuous in time, meaning that they can in principle assume any range of values from one instant to the next. Such time-dependent analog variables – e.g., $P(t)$, $T(t)$ – are called analog "signals." In the world of microelectronics and nanoelectronics, the most prevalent and useful signals are voltages $V(t)$ and currents $I(t)$, primarily because they are, at their core, simply electrons in action and can be cleverly generated, detected, manipulated, transmitted, and converted from one to the other and back again, in a host of amazingly subtle and useful ways. Virtually all of modern technology relies on this basic manipulation of electrons in one way or another, either in its operation, its design, or its manufacture; or, most commonly, all three.[2]

The world of the global Communications Revolution, in contrast, is largely digital in nature, seemingly alien to the world we humans inhabit and intuitively know. Digital signals are discrete in form, and can assume only finite values (typically only two, in fact – a "1" or a "0") as they change from one instant in time to the next. For instance, in the world of digital computers, digital signals may range between the system voltage of the object in question (e.g., the cell phone or digital camera), typically called V_{DD} (e.g., 2.5 V) and electrical ground (i.e., 0 V), and are interpreted as either a logical "1" (2.5 V) or a logical "0" (0 V) at time t. There are very good reasons for this fundamental analog–digital dichotomy of the sensory world versus the communications–computing world, as will be discussed at length later.

Okay. Now, how do we quantify (count) these digital signals? A "bit" (sometimes abbreviated as b), in the context of electronic information flow, is a single numeric value, either a binary "1" or "0," a Boolean logic "true" or "false," or

---

[2] Therein lies the essential heart of the electrical and computer engineering (ECE) discipline. A definition of ECE that I am especially fond of: ECE – "the clever manipulation of electrons for useful ends."

an "on" or an "off" at the transistor level (all are equivalent). A bit embodies the single, most basic unit of digital information in our Communications Revolution. A "byte" (pronounced bite and sometimes abbreviated as B) is a sequence of such bits. Typically, 8 bits = 1 byte of information.

For example, for the Internet, the so-called Internet Protocol (IP) "address" (think of this as the electronic location that the Internet assigns to your computer when it is connected to the Internet via your cable modem) contains 32 bits, or 4 bytes. The IP address is written digitally so that it can then be easily shared on the network with other computers. The bytes divide the bits into groups in order to express a more complicated address.

Consider: The IP address 192.168.0.1, for instance, can be "encoded" (rewritten in an equivalent, but ultimately far more useful, form) in the binary number system with the following bits and bytes:

$$192.168.0.1 = 11000000 \ 10100100 \ 00000000 \ 000000001 \qquad (1.1)$$

Oh yeah? To "read" this encoded 4-byte stream of 32 bits, recall from grade school the notion of a binary number system from mathematics (base-2, not our more common base-10, decimal system). Because the digital world is binary (meaning "two," i.e., either a "1" or a "0," but nothing else), in the electronics world we write numbers in base 2. For an 8-bit binary number, the places are given in powers of 2 (not 10, as with the decimal system). Hence, for a byte (an 8-bit binary number) we have

$$digit\ 7 \quad digit\ 6 \quad digit\ 5 \quad digit\ 4 \quad digit\ 3 \quad digit\ 2 \quad digit\ 1 \quad digit\ 0 \qquad (1.2)$$

interpreted as values per digit and placed according to

$$128 \ 64 \ 32 \ 16 \ 8 \ 4 \ 2 \ 1. \qquad (1.3)$$

That is,

$$2^7 = 2 \times 2 \times 2 \times 2 \times 2 \times 2 \times 2 = 128, \quad 2^6 = 2 \times 2 \times 2 \times 2 \times 2 \times 2 = 64,$$
$$2^5 = 2 \times 2 \times 2 \times 2 \times 2 = 32,$$
$$2^4 = 2 \times 2 \times 2 \times 2 = 16, \quad 2^3 = 2 \times 2 \times 2 = 8, \quad 2^2 = 2 \times 2 = 4,$$
$$2^1 = 2, \quad 2^0 = 1. \qquad (1.4)$$

Thus,

$$192 = 1 \times 128 + 1 \times 64 + 0 \times 32 + 0 \times 16 + 0 \times 8 + 0 \times 4 + 0 \times 2$$
$$+ 0 \times 1 = 11000000 \qquad (1.5)$$

in binary, so that

$$192.168.0.1 = 11000000 \ 10101000 \ 00000000 \ 00000001 \qquad (1.6)$$

(the space between bytes is interpreted as a decimal point).

One can, of course, do a similar encoding process for letters in the alphabet, typographic symbols, etc., so that written (or spoken) language can be transmitted or stored in binary (digital) form. The most common encoding scheme for the Roman alphabet (e.g., English) is called ASCII (Geek Trivia alert: American Standard Code for Information Interchange), and was introduced in 1963 (ASCII is pronounced as-ski). For instance, $01000001 = 65 = A$, $01000010 = 66 = B$, $01110111 = 119 = w$, $00100001 = 33 = !$, and so forth.

In general terms, bits are grouped into bytes to increase the efficiency of computer hardware design, including networking equipment, as well as hard disks and memory, and therefore digital bytes underly all electronic information flow (see sidebar discussion on the choice of binary coding).

Deep Thoughts #1: Why Binary Coding?

You might reasonably wonder: Why use a binary arithmetic for digital encoding of information? Isn't that unnecessarily complicated? Why not instead use base 3, or base 5, or even base 10 (our everyday arithmetic)? At a deep level, the compelling reason for using binary arithmetic is tied to the requisite transistors needed to generate the digital bit streams in a manner that can be robustly "transmitted" (what is sent from point A on its way to point B) and "received" (what eventually appears at point B) at very high speed over a variety of diverse media (e.g., on a coaxial cable, a fiber-optic line, or even wirelessly through space), and has much to do with the so-called "noise immunity" of the electronics circuits required for sending and receiving the information. In essence, noise immunity is needed to maintain the integrity of the digital signal (is it a "1" or is it a "0") in the presence of spurious environmental noise (think static) that inevitably finds its way into the transmission–reception process. Turns out binary coding and transistors were destined for each other.

We crafty humans love our shorthand notations, and the counting of bytes of information is no exception (see Table 1.1). For instance, a movie DVD can hold 4,700,000,000 bytes of information, and to be "in the know," you would say this is 4.7 gigabytes (in geek slang, "4 point 7 gig"), and written 4.7 GB for short. Similarly, the entire print collection of the U.S. Library of Congress (the largest library in the world) contains some 10,000,000,000,000 (10 trillion) bytes of information, or 10 terabytes (10 TB).

Okay, back to the task at hand. So how much electronic information flow surrounds us? To proceed, we can imagine encoding all available information (spoken, written, visual, etc.) into a digital format, and then counting the bytes that result. This might seem like a hopeless task, given the seemingly endless varieties of information out there, but in fact it can be meaningfully estimated. The result? In 2002, the global information flow by electronic means totaled some 17,905,340 TB, 17.9 exabytes (EB) (see Table 1.2 [3]). The breakdown by medium of that global electronic information flow is enlightening. Believe it or not, far and away the dominant source of information exchange on Earth happens on the telephone (96.6%!) and increasingly by cell phone (see sidebar discussions). The Internet, which might have naively seemed to be the largest fraction of information

Table 1.1 How big is a byte?

Unit	Definition and reference points
Kilobyte (kB)	**1 kB = 1,000 bytes** 2 kB = a typewritten page 100 kB = a low-resolution photograph
Megabyte (MB)	**1 MB = 1,000 kB = 1,000,000 bytes** 1 MB = a small novel 2 MB = a high-resolution photograph 500 MB = storage capacity of a CD-ROM
Gigabyte (GB)	**1 GB = 1,000 MB = = 1,000,000,000 bytes** 1 GB = a pickup truck filled with books 4.7 GB = storage capacity of a DVD 20 GB = a collection of the complete works of Beethoven
Terabyte (TB)	**1 TB = 1,000 GB = 1,000,000,000,000 bytes** 1 TB = 50,000 trees made into paper and printed 10 TB = the print collection of the Library of Congress
Petabyte (PB)	**1 PB = 1,000 TB = 1,000,000,000,000,000 bytes** 2 PB = sum of all U.S. academic research libraries 200 PB = all printed material on Earth
Exabyte (EB)	**1 EB = 1,000 PB = 1,000,000,000,000,000,000 bytes** 2 EB = total volume of information generated on Earth in 1999 5 EB = all words ever spoken by human beings 17.9 EB = total global electronic flow of information in 2002

Source: [2].

Table 1.2 Electronic information flow in 2002 (in terabytes)

Medium	Information flow
Radio	3,488
Television	68,955
Telephone	17,300,000
Internet	532,897
Total	17,905,340

Source: [3].

flow, represented only a measly 3%. Do appreciate, however, that these are 2002 numbers. The estimate for global information flow in 2006? A whopping 161 EB! And growing. You get the idea. These current numbers strongly highlight the rapid growth in global Internet-based services, now hogging a far larger percentage of global information traffic.

Life Digression #1: Cell Phones

As my family is quick to point out, I have an intense dislike of cell phones, despite the fact that my own research in electronics is commonly applied to state-of-the-art cell phones – an ironic twist of fate they beat me up on with now-predictable frequency! Still, I am a proud holdout, and to this day do not carry a cell phone. Cell phones present some unique new challenges to us teachers – in-class calls and text messaging. Grrrrr. On day one of my classes at Georgia Tech, I establish the following cell phone policy – the first call during lecture is free; but the second one I get to answer for you! I have yet to have a second call during lecture.

Historical Anecdote #1: The Telephone

You may recall the famous quotation from a Western Union (think telegraph) internal executive memo speculating on the fate of the freshly invented telephone – "This 'telephone' has too many shortcomings to be seriously considered as a means of communication. The device is inherently of no value to us." This is a fascinating and ironic commentary on our innate human inability to correctly foresee the future. There is an important lesson here for the play-out of the Communications Revolution. More on this later.

1.5 Evolutionary trends: Moore's law

Back to an earlier fact: In 62 short years the world has gone from zero transistors (1947) to 10,000,000,000,000,000,000 transistors. Some growth! Underlying the mechanics of transistor growth trends are some deeper concepts that need to be explored and digested, especially if we are to understand how things came so far so fast, and what the future is likely to entail. Sad to say, a proper appreciation of any growth trend in nature requires a tiny amount of mathematics. No worries – there is a simple point to be made – and only a tiny, little bit of math.

Okay. Relaxed? Cell phone still off? Good. Recall from algebra that a logarithm is simply an exponent. Formally, if b is a positive number other than 1, then $\log_b x$ (read, "the logarithm to the base b of x") represents the power to which b must be raised to produce x. For example, $\log_{10} 100 = 2$, because $10^2 = 10 \times 10 = 100$. In general, $y = \log_b x$ and $x = b^y$ are equivalent statements. Fine. This latter statement of $x = b^y$ is more precisely referred to as "x is increasing exponentially (with y)." If y happens to be time (t) such that $x = b^t$, then we call this process "exponential growth" (x could in principle either be increasing in time if we have $x = b^y$ or decreasing in time if we have $x = b^{-y} = 1/b^y$, but it is still exponential growth).

Now, let's choose an exponential growth pattern that is relevant in the context of transistors: $2\times$/year exponential growth, or simply $2\times$/year growth. It is key to note that saying you have "$2\times$/year growth" is *not* the same as saying $y = 2t$, which is a "linear" growth dependence (y depends linearly on t; that is, y is proportional to t), not an exponential one. Rather, $2\times$/year exponential growth is mathematically written as an exponential function (hence the name): $y = A \times 10^{0.30103t}$, where t is the total time elapsed (say, in years), and the base of the logarithm is 10 (this is called the "common" logarithm – see sidebar discussion). Here, A is the initial starting value of y – that is, if $t = 0$, $y = A \times 10^0 = A \times 1 = A$.

Deep Thoughts #2: Logarithms

Those with some science and engineering background may realize that the "natural" logarithm (base e with $e = 2.8718281828$ – an irrational number – defined by

(continued)

Deep Thoughts #2: Logarithms *(continued)*

Euler in 1728), not the common logarithm (base 10), is more physically relevant to many problems. That is, $y = A \times e^{kt}$. Note that because the base of the natural logarithm (e) is smaller than for the common logarithm (10), it grows less rapidly with time, albeit still VERY rapidly. You might wonder why the natural logarithm is often more physically relevant to the natural world (this is where it gets its name). It turns out that there are a very large number of cases in nature in which the rate of change in time of a given variable (call it the time derivative from calculus, dy/dt) is proportional to that variable, such that $dy/dt = ky$, where k is a constant of proportionality. This is an example of a simple differential equation nature commonly works with. Its solution is simple and involves dividing both sides by y, integrating both sides, and then exponentiating and simplifying, to obtain $y = A \times e^{kt}$, where A is a constant. Examples of such behavior in nature are far ranging (for deep-seated, quantum statistical mechanical reasons) and include bacterial growth, electron physics, chemical reactions, transistor theory, star formation, etc. The natural logarithm is also a very friendly function to use to perform many calculus operations on (e.g., the derivative of e^t is just e^t – it returns itself!), making it a friend to most students!

The value of 0.30103 in this exponential function comes from taking log 2, for 2×/year growth. Let's check it. If we use $t = 1$ (1 year – e.g., from 2009 to 2010), and plug that in (use a calculator!), we find that $y = 2A$; that is, y has doubled in 1 year. Over 10 years at this 2×/year growth rate ($t = 10$), we would have $y = 1{,}024A$; y is now 1,024 × larger than we started with – huge growth! Over 20 years, this is 1,048,576 × (a factor of 1 million increase!). Over 30 years, 1,073,742,146 × (a factor of 1 billion increase!). As illustrated in Fig. 1.6, even though $y = 10^{0.30103t}$ behaves initially like $y = 2t$, this picture very rapidly changes, and after 10 time units (e.g., years) there is a difference of a factor of 51.2× between an exponential growth of 2×/year and linear growth ($y = 2t$); ergo, they have profoundly different physical behaviors as time elapses. Observe

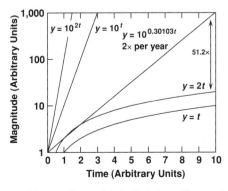

Figure 1.6   The differences in growth behavior between linear and exponential functions over time.

also that if I plot an exponential growth pattern as logarithmic on the y axis versus linear time on the x axis (a "semilog" plot), I get a straight-line behavior (this is very convenient for detecting if you indeed have exponential growth occurring). It begins to get easier to see now how exponential growth can take us from zero transistors in 1947 to 10,000,000,000,000,000,000 transistors today. In fact, it is the only way.

Who was the first person to notice this exponential growth trend in the microelectronics industry? Gordon Moore, former CEO of Intel, in 1965 [7,8] (see Fig. 1.7 and sidebar discussion).

Figure 1.7   A recent photograph of Gordon Moore (courtesy of Intel Corporation).

Historical Anecdote #2: Gordon Moore

Moore's original 1965 prophetic statement reads [7], "The complexity for minimum component costs has increased at a rate of roughly a factor of two per year... Certainly over the short term this rate can be expected to continue, if not to increase. Over the longer term, the rate of increase is a bit more uncertain, although there is no reason to believe it will not remain nearly constant for at least 10 years. That means by 1975, the number of components per integrated circuit for minimum cost will be 65,000. I believe that such a large circuit can be built on a single wafer."

Figure 1.8 shows the four data points Moore had to go on (a tiny number, given the long-term impact of his projection), together with trends in transistor count per

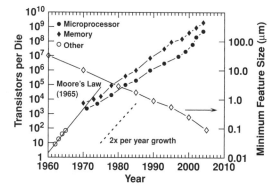

Figure 1.8   Evolution of the number of transistors per die for both microprocessors and memory chips, as well as the minimum feature size used to fabricate those die. Moore's 1965 extrapolation is also shown (after G. Moore).

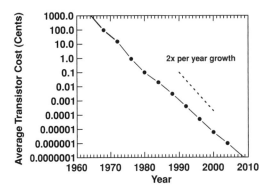

Evolution of annual global transistor cost (in cents) (after G. Moore).

"die" (the roughly centimeter-sized square piece of semiconductor material upon which the transistors are built) for both microprocessor and memory products from 1970 until present. Also shown is the minimum feature size of the transistors [in micrometers (microns for short, or µm), millionths of a meter!]. Moore's earliest observation was roughly a $2\times$/year exponential growth in transistors/die, which settled eventually to more like $2\times$ per 18 months. Not coincidentally, these data are plotted on semilog scales (exponential growth for any resultant straight-line behavior), and to the eye they follow a nice straight-line trend over the past 35 years, a truly remarkable result. Many other metrics of the microelectronics industry follow a similar exponential growth pattern: transistor speed, transistor size, transistor cost. This latter cost behavior is captured in Fig. 1.9. What was $10 per transistor in 1965 is 0.0000001 cents per transistor in 2008. Imagine, single transistors are essentially free! This fact alone belies the fundamental reason why we can build a microprocessor today for your laptop with 100 million transistors on it, and still sell it for only a few bucks. These various exponential growth trends in the microelectronics industry are universally referred to as "Moore's law" (more of an empirical trend, actually).

It is truly difficult to appreciate just how unique Moore's law growth is in the grand scheme of life. A bold statement can be made (I'm getting good at this!): Never before in human history has any industry, at any time, come even close to matching the Moore's law growth trends enjoyed by the microelectronics industry. Skeptical? Consider the automotive industry as a test case. An interesting thought experiment suggests itself. No worries – this will not be nearly so scary as the transistorite thought experiment! Question. What if the car industry had followed the same $2\times$ per 18-month growth patterns as the microelectronics industry? What would a car look like today? As you hopefully know, the world's first mass-produced automobile was Henry Ford's 1913 Model-T Ford Runabout (Fig. 1.10). Considering some key metrics for cars – speed, fuel efficiency, mass, size, cost – after 90+ years of Moore's law growth, what would the new Ford Runabout "Super Deluxe" look like? The shocking results are summarized in Table 1.3.[3] What originally got 20 miles per gallon fuel efficiency now gets

---

[3] Okay, okay – the projected speed is clearly nonphysical as it is faster than the speed of light!

Figure 1.10   A 1913 Model-T Ford Runabout (courtesy of agefotostock.com).

58,144,681,470,000,000,000 miles per gallon and is essentially free. Imagine! In short, if cars had done what microelectronics did, it would make for a really fun way to get around! You could make similar comparisons with any industry you choose – the results would be the same. Never before has the world known this type of truly magical growth – exponentially more performance, while getting exponentially cheaper to achieve that performance. It is a relevant question, of course, as to how long these trends can be extended into the future, and this is the subject of a later chapter.

Ironically, what started as a passive observation on growth trends has in recent years become a driving force for the industry. That is, industry health is to large measure judged today by how well it can maintain this 2× per 18-month Moore's law growth. This can most definitely be a double-edged sword and has at times created serious financial woes for the entire microchip industry. Still, the magnificent achievements of the Communications Revolution are directly linked to it. Long ago, the microelectronics powerhouses of the world, Intel, IBM, Texas Instruments, Motorola, etc., got together to concoct a "roadmap" for where the industry is headed, and to identify (and then attack) early on any technological impediments to sustaining business-as-usual Moore's law growth

Table 1.3 Evolution of a Model-T Ford following a Moore's law growth trend

Metric	Model-T (1913)	Today
Speed (miles per hour)	50	145,361,703,700,000,000,000
Efficiency (miles per gallon)	20	58,144,681,470,000,000,000
Mass (kg)	1,000	0.000,000,000,000,000,343
Luggage Space (ft^3)	18	52,330,213,320,000,000,000
Cost ($)	20,000 (in 1993 $)	0.000,000,000,000,006,88

Table 1.4 2001 SIA roadmap for semiconductors

Year	1995	1999	2001	2003	2005	2008	2011	2014	2016
Feature size (nm)	350	180	130	100	80	70	50	34	22
DRAM size (bits/chip)	64M	256M	512M	1G	2G	6G	16G	48G	–
SRAM size (bits/chip)	4M	16M	64M	256M	–	–	–	–	–
Processor speed (Hz)	300M	750M	1.68G	2.31G	5.17G	6.74G	11.5G	19.3G	28.7G
Transistors per chip	12.0M	23.8M	47.6M	95.2M	190M	539M	1.5G	4.3G	–
Supply voltage (V)	3.3	2.5	1.2	1.0	0.9	0.7	0.6	0.5	0.4

M = Mega = 1,000,000; G = Giga = 1,000,000,000.

for the entire industry. That organization is called the Semiconductor Industry Association (SIA) [9], and they work to annually produce (among other things) the Internal Roadmap for Semiconductors (ITRS) [4] (see sidebar discussion for member companies). This map-out of the future industry milestones and their timing, as well as the key technology metrics and anticipated best-guess required innovations needed to sustain that growth, govern much of what happens in the R&D side of the microelectronics industry – it is a big deal. A distillation of the 2001 SIA roadmap (literally hundreds of pages of charts and verbiage) is shown in Table 1.4, and project outward in time until 2016 where the memory and microprocessor sectors are headed. The picture is sobering. A leading-edge 2-Gbit DRAM (dynamic random-access memory) today will be 24 times larger by 2014 (48 Gbit). A best-of-breed 5-GHz microprocessor today will be 19 GHz by 2014. All running off of a 0.5-V battery! You can easily imagine that there is much involved in sustaining a Moore's law growth pattern while marching to the ITRS drumbeat, and these nuances will be addressed in subsequent chapters.

Historical Anecdote #3: SIA

Currently, the SIA has 15 board member companies: Advanced Micro Devices, Agere Systems, Altera, Analog Devices, Conexant Systems, Cypress Semiconductor, IBM, Intel, Intersil, LSI Logic, Micron Technology, Motorola, National Semiconductor, Texas Instruments, and Xilinx. There are, however, over 90 SIA member companies, representing virtually all of the global microelectronics industry players.

1.6 Silicon: The master enabler

Simply put, the transistor, and our Communications Revolution, would not exist without the truly unique virtues of a class of solid materials called "semiconductors." No semiconductors, no transistors; no transistors, no electronics; no electronics, no Communications Revolution. This field of human endeavor is historically termed "microelectronics," "micro" being short for "micrometer" or

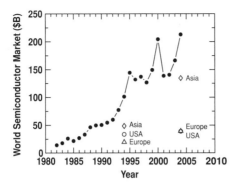

Figure 1.11 Evolution of global semiconductor market (in billions of $US). Regional breakdowns are shown for 1994 and 2004 (after ITRS).

"micron," one millionth of a meter. At present, microelectronics is rapidly evolving into "nanoelectronics," "nano" being short for "nanometer," one billionth of a meter. I term this merger "micro/nanoelectronics," a field that can be loosely defined as the science of producing electronics from semiconductors at a highly miniaturized scale.

Consider: Global semiconductor sales (including transistors, and all microelectronics–nanoelectronics derivatives: memory, microprocessors, etc.) in 2000 were $204,000,000,000 ($204B), a serious chunk of change (Fig. 1.11) [4]. Observe that this production is increasingly dominated by the Far East, especially Japan, Taiwan, Korea, and mainland China. Global sales of the downstream consumer electronic "stuff" (e.g., cell phones, laptop computers, CD/DVD players) built from these semiconductor microchips totaled over $1,128,000,000,000 ($1,128B) in 2000 [5]. For reference, the Gross Domestic Product (GDP), roughly speaking the economic output of a given country, was $9,817,000,000,000 for the United States in 2000; $1,870,000,000,000 for Germany [6]. Semiconductors, and the microelectronics and nanoelectronics derived from them, are clearly a major driver, arguably, THE major driver, of the world's economic engine. This trend is rapidly accelerating.

Let's briefly consider these magical materials – semiconductors.

Semiconductors are a class of materials (either elements or combinations of elements) that conduct electricity (electrons) with only modest efficiency, and to a degree falling between that of "conductors" (think metals) and "insulators" (think glass or plastic or rubber). Don't worry, we will dig deeper into this in subsequent chapters – let's keep it simple for now. Thus, one can consider a semiconductor to be a "sort-of" conductor (a "semi"-conductor – hence the name). Good. Now, a particular semiconductor, the element silicon (Si), captured well over 95% of this $204B global semiconductor market. This is no coincidence. Silicon is unique among all of nature's building materials, and even unique among semiconductors, making it ideal for producing 10,000,000,000,000,000,000 transistors in 60+ years!

Fortuitously, we live in a silicon world (hence the title of this book). This statement is both literally and figuratively true. Silicates, the large family

Figure 1.12   **The ubiquitous sand castle (courtesy of agefotostock.com, © Mario Bonotto).**

of silicon–oxygen-bearing compounds such as feldspar [$NaAlSi_3O_6$], beryl [$BeAl_2(Si_6O_{18})$], and olivine [$(MgFe)_2SiO_4$], make up 74% of the Earth's crust. Silicon is the third most abundant element on Earth (15% by weight), after iron (35%) and oxygen (30%). One need go no further than the beach and scoop up some sand to hold some silicon (Fig. 1.12).

Some of the remarkable virtues of the semiconductor silicon include the following (others will be delineated once we develop a bit more understanding):

- Silicon is wonderfully abundant (there are a lot of beaches in the world!), and can be easily purified to profoundly low background impurity concentrations [easily better than 0.000001 parts per million (ppm)], making it one of the purest materials on Earth when used to produce transistors.
- Silicon crystals (read: a regular arrangement of atoms inside the material, and the preferred form for building electronics, as we will see) can be grown amazingly large and virtually defect free – at present, to a gargantuan size of 12 in (300 mm) in diameter by roughly 6 ft long, weighing thousands of pounds! These perfect silicon crystals (called a silicon "boule" – rhymes with jewel) are literally the largest, perfect crystals on the face of the Earth. It is (perhaps) a shame that crystallized silicon, despite the fact that it shares an identical structure with that of crystallized carbon (our beautiful diamond gemstones), ends up with an opaque, uninspiring, grayish-silver hue (Fig. 1.13).
- Silicon has excellent thermal properties, allowing for the efficient removal of dissipated heat. This is key, because even a tiny amount of heat (say 0.001 W)

Figure 1.13   Picture of the end of a 200-mm Si boule.

provided by each of 10,000,000 transistors in a microprocessor, say, adds up to a lot of heat (10,000 W = meltdown!). If that heat cannot be efficiently removed, then the microchip temperature rises uncontrollably, degrading both reliability and performance.

- Like most of the technologically important semiconductors, silicon atoms crystallize in a "diamond" structure (Fig. 1.14), just like crystallized carbon (the true diamond!). This is a very stable and mechanically robust crystal configuration, and many of the desirable properties of silicon are directly linked to this underlying crystal structure.
- Silicon is nontoxic and highly stable, making it in many ways the ultimate "green" material, although in truth the gases needed to complete the fabrication process (di-borane, phosphine, and arsine) fall decidedly into the "nasty"

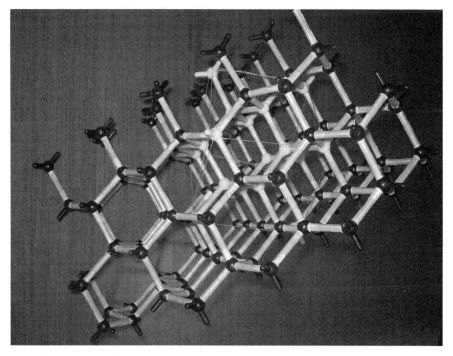

Figure 1.14   Schematic view of the atomic structure of a Si crystal.

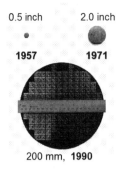
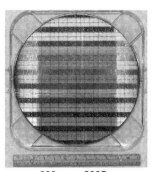

Evolution of silicon wafer size (courtesy of Intel Corporation).

category. We'll revisit the "greenness" of micro/nanoelectronics in the last chapter.

- Silicon has excellent mechanical properties, facilitating ease of handling during the micro/nanoelectronics fabrication process. For a 300-mm (12-in) diameter silicon boule, for instance, this allows the silicon wafers to be cut to roughly 0.80-mm thickness (producing ultraflat "potato chips," known as "wafers" in microelectronics jargon), maximizing the number of wafers per silicon boule. Think cost. This mechanical stability also minimizes wafer warpage with fabrication. Absolute flatness of the silicon wafer during exposure to extremely high temperatures (think $1,000^\circ$C!) is vital – no Pringle's potato chips allowed! The evolution of silicon wafer size from 1957 (the earliest days of high-quality silicon boules) to the present is depicted in Fig. 1.15.

- Perhaps most important, an extremely high-quality insulator (more formally called a "dielectric") can be trivially grown on silicon simply by flowing oxygen across the wafer surface at an elevated temperature (or even sitting it on the shelf for a few short minutes). This dielectric, silicon dioxide (SiO_2, "quartz" to geologists) is one of nature's most perfect insulators (1 cm of SiO_2 can support 10,000,000 V across it without being destroyed!). SiO_2 can be used in many creative ways in both transistor design and the fabrication of transistors, as will be seen.

Simply put, it is a remarkable fact that nature blessed us with a single material embodying all of the features one might naively close one's eyes and wish for when considering what is actually needed to populate the Earth with 10,000,000,000,000,000,000 transistors in only 60+ years. From a semiconductor manufacturing standpoint, silicon is literally a dream come true.

1.7 Micro/nanoelectronics at the state-of-the-art: 90-nm complementary metal-oxide-semiconductor technology

As a final topic in this first background, "big picture" chapter, it is helpful to go ahead and garner a good feel for the "state-of-the-art" in micro/nanoelectronics.

Figure 1.16   A 300-mm, 90-nm CMOS wafer (courtesy of Intel Corporation). (See color plate 3.)

Pardon me, state-of-the-art? Geek lesson #1: "State-of-the-art" is a common term used by technologists and speaks to the highest level of technical achievement representative of the world's best players in a given field, at that particular instant of time. Clearly it is a highly dynamic definition. I smile when I think that technology is considered by its practitioners to be an "art form" – I concur; you soon will too! At present, the state-of-the-art in micro/nanoelectronics is 90-nanometer (nm) complementary metal-oxide-semiconductor (CMOS) technology, fabricated (manufactured) on 300-mm-diameter silicon wafers (Fig. 1.16). There are upwards of two dozen or more companies practicing 90-nm CMOS in high-volume manufacturing on 300-mm wafers around the world today. CMOS represents a type of transistor, a "field-effect" transistor (FET), and is the workhorse for microprocessors (the brains) and memory (the storage) used in constructing computers, among other things. We'll talk about how transistors actually do their business in Chap. 8. Why 90 nm? Well, the component transistor is roughly 90 nm across. Said another way, 90 nm = 0.000,000,090 m. 90 billionths of a meter! Yep, pretty small (new concepts of "small" and "fast" will be discussed at length in the next chapter). Given that the atom-to-atom spacing in a silicon crystal is a little over 0.5 nm, this means that a 90-nm transistor at the state-of-the-art is only about 180 atoms across! In a Pentium 4 Intel microprocessor (Fig. 1.17), there are over 42,000,000 90-nm CMOS transistors. All in a centimeter-sized piece of silicon crystal. Several hundred such microprocessors will be built together on a single 300-mm silicon wafer. A 52-Mbit (52,000,000) static random-access memory (SRAM) chip for a computer is shown in Fig. 1.18, and contains over 300,000,000 90-nm CMOS transistors. Remember: In 1947, one transistor the size of your thumbnail; today, only 60+ years later, 300,000,000 transistors in

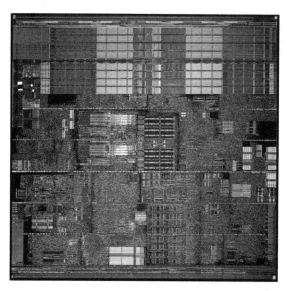

Figure 1.17   A bird's-eye view of a Pentium 4 microprocessor fabricated in 90-nm CMOS technology. This integrated circuit contains over 42,000,000 transistors and operates at a greater than 2.0-GHz frequency. The integrated circuit is roughly 1.0 cm on a side in size (courtesy of Intel Corporation). (See color plate 4.)

roughly the same amount of space. A modern miracle you definitely need to be aware of.

You might be reasonably wondering, what does the actual transistor in 90-nm CMOS technology look like? Good question! Figure 1.19 shows the beast. The active dimension (where the transistor "action" occurs) is only about 50 nm wide. And how would one actually connect together 300,000,000 transistors in one square centimeter? Good question. Simple answer: VERY carefully. In practice,

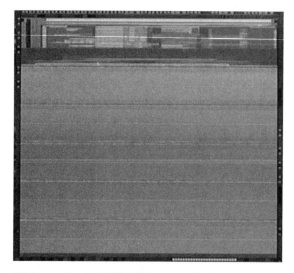

Figure 1.18   A bird's-eye view of a 52-Mbit SRAM memory chip fabricated in 90-nm CMOS technology. This integrated circuit contains over 300,000,000 transistors and is roughly 1.0 cm on a side in size (courtesy of Intel Corporation).

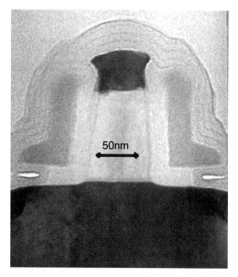

Figure 1.19 A cross-sectional view of a FET from a 90-nm CMOS technology. The active transistor region is 50 nm, 50 billionths of a meter (courtesy of Intel Corporation).

one uses multiple layers of metal "wires" as "interconnects" (Fig. 1.20). In this 90-nm CMOS technology, there are actually eight different metal interconnect levels for wiring of transistors, each of which is less than a millionth of a meter (a micron = μm) wide. Those eight metal layers are contained in a vertical height above the silicon wafer surface of only 20 millionths of a meter (20 μm). How do we make such thing? Stay tuned, I'll soon tell you. It is estimated, though, that on a single centimeter-sized 90-nm CMOS microprocessor, if we took all of the metal interconnect wiring needed to have each transistor "talk" to one another and stretched those wires end to end, it would be greater than 3 miles long [10]!

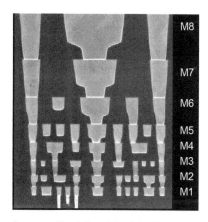

Figure 1.20 A cross-sectional view of the eight different layers of metal interconnects on a 90-nm CMOS technology. From the bottom of metal layer 1 (M1) to the top of metal layer 8 (M8) is about 20 μm, 20 millionths of a meter (courtesy of Intel Corporation).

Lest you think this is the end of the road, refer to the SIA Roadmap in Table 1.4. Yep, just the beginning! Hold on tight.

1.8 A parting shot

I began this book with a rather bold (outrageous?!) claim – the transistor, happened on and discovered by clever human minds in 1947, is the single most pivotal technological invention the Earth has yet known. The transistor's indispensable role in modern life was easily demonstrated by a simple thought experiment: Imagine subtracting them from the world and then asking how life would change. Simple answer: Life as we know it would grind a halt, instantaneously! The remarkable advances in semiconductor technology, the transistors built from them, and the downstream electronics and computer infrastructure we now take for granted, are proceeding along well-orchestrated exponential growth paths, fueling the rapid onrush of the Communications Revolution, and *homo sapiens'* march into a new Age of Civilization – the Information Age. These are exciting times. For some, bewildering times, given the inescapable pace of these technological changes. These are times of great promise for humankind, a feasible path, perhaps, for finally leveling the playing field of the world's enormous disparities; of bringing humanity closer to a single human family. Inevitably, these are also times of potential peril should we fail to embrace a responsible use of our newly minted technology. Never before have techno-geeks (yep, us engineers and scientists) played such a key role in the evolution of our species. I like that thought.

I would argue that everyone, not simply the technically literate among us driving this mad dash to the sea, should be (must be!) aware of the "hows" and "whys" of this micro/nanoelectronics revolution. We owe it to ourselves to be well versed in its beautiful creations, the seemingly magical nuances of form and function, and the marvelous array of ever-emerging applications and unintended uses that are now happening on a daily basis. I hope you find this book a useful contribution and guide to your budding knowledge of these world-changing events. Enjoy the ride!

References and notes

1. J. T. Last, "Two communications revolutions," *Proceedings of the IEEE*, Vol. 86, pp. 170–175, 1998.
2. See, for instance, *http://www.geohive.com*.
3. "How much information? 2003," *http://www.sims.berkeley.edu/research/projects/how-much-info-2003/execsum.htm#method*.
4. Internal Roadmap for Semiconductors (ITRS), *http://public.itrs.net/*.
5. Cahners Instat Group, *http://www.instat.com*.
6. See, for instance, *http://www.eia.doe.gov/emeu/international/other.html*.
7. G. E. Moore, "Cramming more components onto integrated circuits," *Electronics Magazine*, April 19, 1965.
8. G. E. Moore, "Our revolution," Semiconductor Industry Association presentation, 2002.
9. Semiconductor Industry Association (SIA), *http://www.sia-online.org/home.cfm*.
10. J. D. Meindl, J. A. Davis, P. Zarkesh-Ha, C. S. Patel, K. P. Martin, and P. A. Kohl, "Interconnect opportunities for gigascale integration," *IBM Journal of Research and Development*, Vol. 46, pp. 245–263, 2002.

2 A Matter of Scale

The "Deep Field" image of the universe from the Hubble Telescope (courtesy of NASA). (See color plate 5.)

To See The World in a Grain of Sand
And a Heaven in a Wild Flower,
Hold Infinity in the Palm of Your Hand,
And Eternity in an Hour.

William Blake

2.1 The tyranny of numbers

How big is the universe? Good question! If we take the Hubble Telescope that orbits the Earth and point it into deep space (a dark patch in the night sky) and just let it "look" for a long time, we obtain an image of the farthest reaches of the universe, to days of old just after the primordial "Big Bang" (see the opening figure). The "smudges" you see in this Hubble image are actually galaxies, not stars (e.g., our own Milky Way galaxy), each galaxy containing billions of individual stars. The known universe is about 1×10^{27} m across and is estimated to contain roughly 1×10^{21} stars (1,000,000,000,000,000,000,000). The "really big" has a face. But how big is it really?

Frankly, we humans are jaded by numbers. This is an unfortunate fact because it muddies our ability to appreciate the truly "grand scales" of the universe. How big is the universe? You say, "really big." How big is an atom? You say, "really small." But how big is big, and how small is small? To fully comprehend the micro/nanoelectronics revolution, we must wrestle with these important questions. We humans, despite our impressive smarts, have great conceptual difficulty grappling with distance and time scales, the "grand theme" if you will, of the universe. Very large and very small numbers can quickly tie our minds in knots (see sidebar discussion).

Deep Thoughts #1: Counting in the Animal Kingdom

We are not alone in the fact that very large and very small numbers can rapidly confuse us. I vividly recall an episode I saw as a kid of the television show *Wild Kingdom* that addressed this very fact. The host, Marlin Perkins, and sidekick were observing wildlife from an observation blind in the Serengeti Plain in Africa. The animals being observed could not differentiate between two humans arriving at the blind and only one leaving (the observer remaining) and two humans coming and going, and thus were lulled into believing that no one remained in the blind to watch their every move. The animal brain is not especially hardwired for counting! They have been known to have trouble with calculus too!

Partly this is because we are victims of mathematical convenience. Consider. I can use scientific notation to trivially write (very compactly, hence the utility for scientists and engineers) 1×10^{12}. Said another way, this is 12 "orders of magnitude" (12 powers of 10). But it really is 1,000,000,000,000 (1 followed by 12 zeros $= 10 \times 10 \times 10 \times 10 \times 10 \times 10 \times 10 \times 10 \times 10 \times 10 \times 10 \times 10$). One trillion. Okay, so what is 1,000,000,000,000? Well, 1,000,000,000,000 pennies is $10 billion dollars (a serious chunk of change!). Or, 1,000,000,000,000 people is 158 times the number of men, women, and children on planet Earth (about 6.5 billion folks). Or, 1,000,000,000,000 is about 100 times the distance (in meters) from the Earth to the Sun.

Now think small. How about 1×10^{-12}? That is, $1/1 \times 10^{12} = 1/1,000,000,000,000 = 0.000,000,000,001 = 0.1 \times 0.1 \times 0.1 \times 0.1 \times 0.1 \times 0.1 \times 0.1 \times 0.1 \times 0.1 \times 0.1 \times 0.1 \times 0.1$. Well, 0.000,000,000,001 meters (m) is about 1/10,000,000 the diameter of a human blood cell. Or, 0.000,000,000,001 is about 1/100,000 of the wavelength of blue light (in meters). Or, 0.000,000,000,001 m is about the length of 100 carbon atoms lined end on end. It may surprise you to know that today we can rather easily view individual atoms. Like the deep reaches of the universe, the really small atomic world also has a distinct face (Fig. 2.1). Although in the strictest sense we cannot, even in principle using the

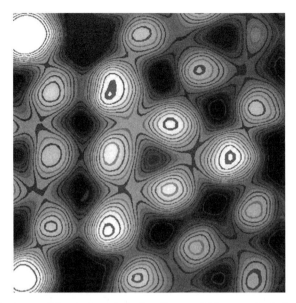

Figure 2.1   Image of the atomic surface of a silicon crystal from an atomic force microscope (courtesy of Professor Mannhart, University of Augsburg). (See color plate 6.)

most power optical microscope conceivable, "see" an atom, we can today easily "image" an atom if we are clever (in essence, we infer its physical presence and translate that inference into something our brain can visually interpret, i.e., an image). In this context, there is a fundamental difference between "seeing" and "imaging" in the micro/nano world that should be clarified.

2.2 "Seeing" versus "imaging" the infinitesimal

When we humans "see" an object with our eyes, visible electromagnetic (EM) waves ("light" in common parlance) of a very particular range of wavelengths [0.4–0.7 μm in common usage = 0.000,000,4 (violet) to 0.000,000,7 (red) meters – refer to sidebar discussion] "scatters" (bounces off) the object we are viewing, travels through space, and is focused through our eyes' cornea and lens onto our photosensitive retinal cells ("rods" for black and white, "cones" for color). Those retinal stimulations (transformed from optical inputs into chemical processes and then into electrical impulses) are sent along the optic nerve to the visual cortex of the brain and interpreted (in amazingly complex and subtle ways). Physically, however, I can "see" objects only when the size of the object I am viewing with my eyes is much larger than the wavelength of the light that I am using to see it with. Thus I can easily "see" a tree, because its 45 ft of height is large compared with the 0.4–0.7-μm wavelength of visible light my eye is seeing it with. But I cannot, for instance, "see" a state-of-the-art transistor directly, and certainly not an atom, even in theory. Why is this?

> ### Geek Trivia: ROYGBIV
>
> Recall from high school days the memory aid "ROYGBIV" – Red–Orange–Yellow–Green–Blue–Indigo–Violet – for the colors of the visible EM spectrum. Do appreciate that the human eye is actually capable of distinguishing over 10,000,000 different color permutations of these basic 7. Pretty remarkable!

One of the core principles of optics, "diffraction" (Christiaan Huygens, 1629–1695), involves the bending of EM waves around objects, and fundamentally limits the resolving power of all optical magnifying devices (including, obviously, microscopes). Significant diffraction effects occur when the size of the object being viewed is of the same order as the wavelength of the light being used to view it. That is, the object is big enough to effectively bend (hence distort) the light illuminating it. For instance, if a transistor is 0.25 μm wide (250 nm = 2.5×10^{-7} m), and I am trying to "see" it with blue light (with a 0.50-μm wavelength), I am facing serious fundamental problems. The only choice is to find some sort of wave with a wavelength smaller than the size of the object we wish to observe. Clearly, x-rays, because their wavelengths can be far shorter than those of visible light (e.g., 0.1 nm), would do the job, but in practice building effective x-ray "lenses" is problematic. The other logical choice is to use electrons!

From basic quantum theory (feel free to consult Appendix 2 for a reminder if needed) the electron is a "particle wave" and has an associated wavelength, in a manner analogous to light. For example, for an electron moving at 1/1,000 of the speed-of-light ($c = 3 \times 10^8 = 300{,}000{,}000$ m/s in vacuum), its "wavelength" is 2.43 nm ($2.43 \times 10^{-9} = 0.000{,}000{,}002{,}43$ m), much smaller than the transistor that we wish to "see" (say 0.25 μm). This is an easily attainable velocity for an electron. A scanning electron microscope (SEM) "shines" highly accelerated electrons through high-voltage steering grids (in effect, electron "lenses") onto the object being imaged, and those scattered electrons are detected in much the same way that your television tube works. An electron-sensitive detection layer (the "retina" of the SEM) lights up when scattered electrons strike it, and the intensity is proportional to the number of electrons striking the screen, and can thus form a "ghost image" of the object on the screen. The SEM is the standard instrument today for viewing micro/nanoscale objects such as transistors. Obviously, in resorting to the use of electrons to view the micro/nano world, we cannot in any meaningful way claim that we actually "see" the object, as our human retinas are not sensitive to those requisite wavelengths. Instead, we "image" those objects by using electrons. Adding color to such "images" is purely artistic enhancement (transistors don't look blue or green!), and thus SEM images are typically shown in black and white (some examples follow).

If we wish to probe below the micro/nano scale down into the atomic realm, we again confront a fundamental problem, because the wavelengths of even highly accelerated electrons are no longer small compared with the size of the atoms we wish to observe (an atom is roughly 0.1–0.2 nm in diameter, and thus smaller

Figure 2.2 STM image of the (100) atomic surface of nickel with the IBM logo spelled in xenon atoms. (Courtesy of International Business Machines Corporation. Unauthorized use is not permitted.)

than the electron wavelength at 2.4 nm from the preceding example). It should be noted for clarity that special high-resolution transmission electron microscopy (TEM) techniques can (just barely) image atomic arrangements along crystalline lattice planes, but not with sufficient detail to study individual atoms.

So how do we see an atom? By being very clever (we humans are uniquely gifted with ingenuity). There are at present two major techniques utilized in imaging actual atoms: The scanning tunneling microscope (STM), which uses "tunneling" electrons between the atom and an atomically sharp metal "tip" placed exceedingly close to the atomic surface and swept (scanned) across the surface to obtain an atomic image. Gerd Binnig and Heinrich Rohrer of IBM (see cover art of Chapter 12) won the Nobel Prize in Physics in 1986 for its invention [1]. In 1990 the STM was used in humankind's first successful attempt to directly and controllably manipulate (move) single atoms (as anticipated by Richard Feynman 30 years earlier [3]), yielding the famous image of the IBM Corporation logo written in xenon atoms (Fig. 2.2 [2]). The atomic force microscope (AFM) is a close cousin of the STM and utilizes a highly sensitive cantilevered "spring" to sense the atomic forces of each atom as the metal tip is scanned across the surface of the sample. The AFM was invented by Binnig and co-workers in 1986 [4]. Both atomic imaging techniques are in wide use today, opening a path toward visualizing the atomic world. Figure 2.1 is a AFM image of the surface of a silicon crystal, of obvious relevance to micro/anoelectronics. The range of colors added in Fig. 2.1 (color plate 6) is purely artificial and arbitrary and is used only to help visualize the atom – it has no color *per se*. The distance between each silicon atom (the bright circles in Fig. 2.1) is roughly 0.5 nm (0.000,000,000,5 m).

2.3   The distance scale of the universe

Back now to the master distance scale (in meters) of the universe (Fig. 2.3). Some highlights: According to our best astronomical data, the universe is roughly 1×10^{27} m (15 billion light years) across; 12 orders of magnitude smaller (1×10^{12}) is 1.0 light year, the preferred astronomical distance unit (the distance that light moving at 3×10^8 m/s travels in 1 year), at about 1×10^{15} m;

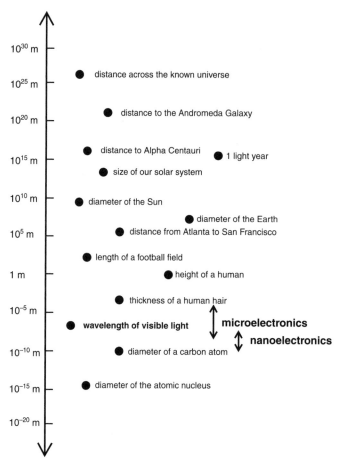

Figure 2.3 The distance scale of the universe.

12 orders of magnitude smaller is the length of a football field; 12 more orders of magnitude smaller is the diameter of a carbon atom, the basic building block of life; and finally, 5 orders of magnitude even smaller than the carbon atom is the diameter of the atomic nucleus (atoms are mostly empty space). Think about this for a moment. The approximate ratio of the distance (in meters) across the expanse of the universe (the largest practical distance scale in the universe) to the diameter of the atomic nucleus (the smallest practical distance scale in the universe) is 1,000,000,000,000,000,000,000,000,000,000,000,000,000,000 (1×10^{42}). Let's call this the "universal distance-scale parameter." Forty-two orders of magnitude! That is a lot of zeros! And it feels like an awfully big number. But how big? Consider a grain of salt from your salt shaker. If you look closely at one grain of salt, you see that it is shaped like a cube and is roughly 0.1 mm on a side. We can easily show that there are about 1×10^{22} atoms in a 1.0-cm^3 block of salt, or 1×10^{16} atoms in one grain of salt. Thus, 1×10^{42}, our universal meter scale parameter, would be equivalent to the number of atoms in 1×10^{26} grains of salt (1×10^{42} divided by 1×10^{16}). These 1×10^{26} grains of salt could be formed

into a giant cubic block of salt roughly 28.8 *miles* on a side! And that massive cube of salt would weigh 126,584,850,100,000 tons. That's a lot of salt! Clearly, 1×10^{42} is a big number by any stretch.

Observe in Fig. 2.3 that the day-to-day realm of micro/nanoelectronics is literally micron sized: $1 \times 10^{-6} = 0.000,001$ m (10,000 silicon atoms stacked end to end) – hence the name "microelectronics" (micrometer-sized electronics). Nanoelectronics is nanometer sized: $1 \times 10^{-9} = 0.000,000,001$ m (10 silicon atoms stacked end-to-end) – hence the name "nanoelectronics" (nanometer-sized electronics). The many miracles described in this book you are holding inhabit this micro/nanoscale world. On this zoomed-out grand distance scale of the universe, they are clustered about a more familiar and useful measure of small – the wavelength of visible light (0.4–0.7 μm) – and decidedly on the "tiny" end of the distances found in the universe.

2.4 The micro/nanoelectronics distance scale

Zooming in explicitly to the micro/nanoelectronics world, we plot a scale ranging from only the size of atoms (0.1 nm) to 1 cm, a size I could obviously hold between my fingers (Fig. 2.4). Now the world of micro/nanoelectronics comes more solidly into focus. A state-of-the-art transistor (e.g., 90-nm CMOS) is about the same size as a virus; 100 times smaller than a human blood cell; 1,000 times smaller than the thickness of a sheet of paper. In the 60+ years since the demonstration of the first transistor in 1947, the size of the transistor has decreased from about 100 μm (thickness of a human hair) to 65 nm, the size of a small virus, a decrease in size of over 1500×. As discussed in Chap. 1, the fundamental driving force behind this dramatic size reduction over time in the functional building block of the micro/nanoelectronics revolution is intimately tied to Moore's law.

It is instructive to "see" an actual microelectronic device, and we now zoom in from a macroscopic view (I can see it with the naked eye) to a microscopic view (it is invisible to the naked eye). This staged zoom-in starts from a 200-mm (8-in) silicon wafer (Fig. 2.5) and goes down to the smallest component found on the silicon wafer, inevitably the transistor itself. Figure 2.6 shows an optical micrograph (microphotograph) of a 6.0 mm × 4.5 mm silicon "test die" used for research on next-generation wireless transceiver integrated circuits. The tiny white squares are aluminum "contact pads" that allow one to (very carefully!) connect an external instrument to the circuit and measure its characteristics. There are roughly 70 identical dies repeated on this single 200-mm wafer. There are a variety of individual circuits on this silicon die, one of which is circled. This integrated circuit performs a RF "mixing" function (combining two radio waves of differing frequencies) and is about 1.0 mm × 1.0 mm in size (i.e., still macroscopic). An optical micrograph of the RF integrated circuit is shown in Fig. 2.7. The eight contact

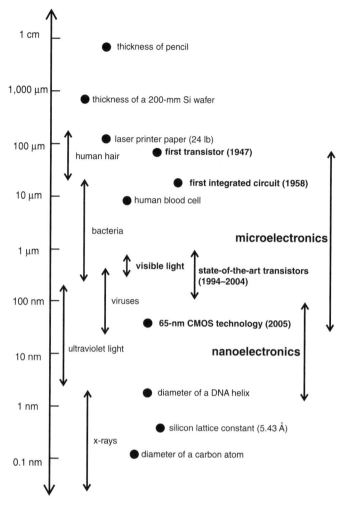

Figure 2.4 The distance scale of micro/nanoelectronics.

pads are clearly visible, and the complicated "guts" of the circuit can be now seen (the three circular spirals are mini-inductors). At this level of optical magnification, the transistors cannot be easily seen. We now switch over to an SEM image (using electrons), and can finally "see" individual transistors, metal wiring, and resistors (Fig. 2.8). The transistor (circled) is roughly 5.0 μm in diameter. A further magnification is shown in Fig. 2.9. If we now look at the transistor in cross section (literally break the wafer at a point running through the transistor – you can imagine this is not trivial to do!), we finally see the "guts" of the transistor itself (Fig. 2.10).[1] Observe that there is significant substructure and superstructure below and above the surface of the silicon wafer of this transistor (indicated by the blue line in color plate 7). Oxide-filled deep and shallow trenches electrically

---

[1] This image is of the other major type of transistor: A bipolar junction transistor (BJT). The image of the 90-nm CMOS transistor is shown in Chap. 1.

200 mm

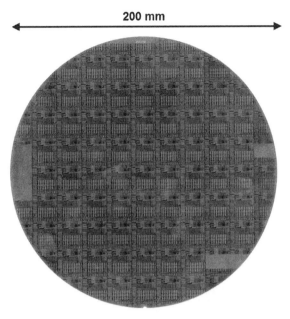

Figure 2.5 Plan (top) view optical micrograph of a 200-mm silicon wafer.

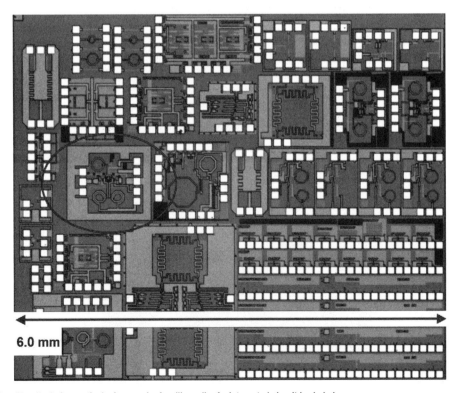

6.0 mm

Figure 2.6 Plan (top) view optical micrograph of a silicon die. An integrated circuit is circled.

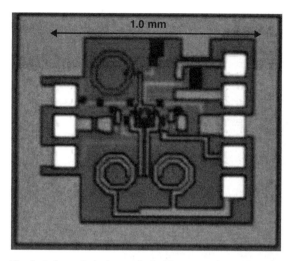

Plan (top) view optical micrograph of an integrated circuit. The "blurry" aspects of this micrograph are due to refraction from various composite layers in the circuit.

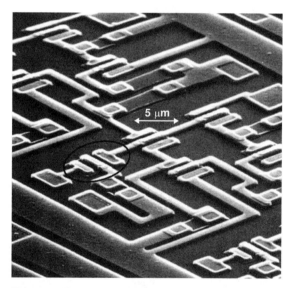

Higher-magnification plan (top) view of a SEM image of a integrated circuit. The transistor is circled. (Courtesy of International Business Machines Corporation. Unauthorized use is not permitted.)

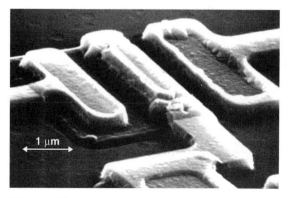

Higher-magnification plan (top) view of a SEM image of a transistor. (Courtesy of International Business Machines Corporation. Unauthorized use is not permitted.)

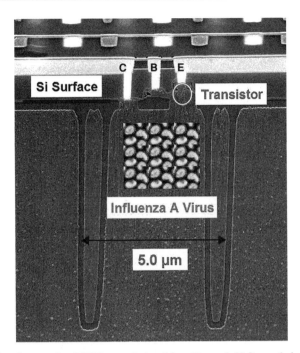

Figure 2.10 Cross-sectional SEM image of a transistor with a scaled Influenza A virus shown for comparison. (Courtesy of International Business Machines Corporation. Unauthorized use is not permitted.) (See color plate 7.)

isolate one transistor from another, and various levels of metallization layers are used to electrically connect the three terminals (collector, base, and emitter – C, B, E) of the transistor to other components of the circuit. Interestingly, 95% of the physics of the operation of the transistor occurs inside the circle under the "E" label, roughly 0.15 µm wide. For comparison purposes, a "family" of Influenza A virus (sorry I couldn't find an example a bit more cuddly), each with a diameter of about 0.10 µm, is superimposed on the transistor at the correct scale. State-of-the-art transistors are literally microbial sized.

2.5 The time and frequency scales of micro/nanoelectronics

The remarkable advances associated with the micro/nanoelectronics revolution rest squarely on the fact that transistors can be simultaneously made both very, very small, and very, very fast. Hence we must necessarily grapple with the speed scale that the universe presents to us if we want to fully appreciate "How fast is fast?" Consider the master time scale (in seconds) of the universe (Fig. 2.11). Some highlights: According to our best astronomical estimates, the universe is about 4.3×10^{17} s old (13.7 billion years); 8 orders of magnitude younger is the average human lifespan (2.4×10^{9} s); 9 orders of magnitude younger still is the time it takes to blink your eyes (a fraction of a second); 9 orders of magnitude younger is the time it takes the microprocessor in your laptop computer to execute a computational step ($1.0 \text{ ns} = 1 \times 10^{-9}$ s); and finally,

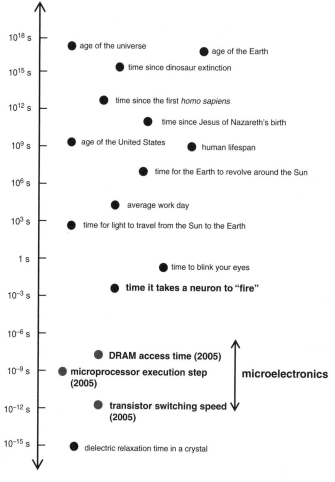

Figure 2.11   The time scale of the universe.

6 orders of magnitude even younger is the dielectric relaxation time for an electron in a crystal [how fast the electron can "feel" and respond to an imposed electric field – 1.0 femtosecond (fs) $= 1 \times 10^{-15}$ s].

The approximate ratio of the age (in seconds) of the universe (the longest practical time scale in the universe) to the dielectric relaxation time of electrons in a crystal (the shortest practical time scale in the universe) is 100,000,000,000,000,000,000,000,000,000,000 (1×10^{32}). Let's call this the "universal time-scale parameter." Thirty-two orders of magnitude! As with our universal distance-scale parameter, that is a lot of zeros, and it feels like an awfully large number. But how large? Well, if 1×10^{32} s, for instance, were compressed into the length of 1 year (365 days), we would complete our human lifespan and die on January 1, only 0.747 fs (0.000,000,000,000,000,747 seconds) after the stroke of midnight! Not much time to celebrate New Year's!

As can be seen in Fig. 2.11, the intrinsic switching speed (i.e., how fast I can open and close the transistor when it operates as a "switch" – think of toggling

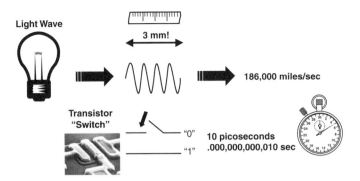

Transistor switch "racing" a beam of light.

your light switch) of a state-of-the-art transistor is on the picosecond time scale $(1 \times 10^{-12} = 0.000,000,000,001 \text{ s})$.

At present, the world record for switching speed in a silicon-based transistor is about 3 picoseconds (ps). This obviously sounds really fast, but (you get the picture!) how fast is fast? Consider the following thought experiment to give us an intuitive feel for "fast" in the context of transistors (Fig. 2.12). Let's imagine taking a transistor switch that is capable of toggling from "on" (a logical "1") to "off" (a logical "0") in 10 ps. I want to compare this switching speed with the fastest speed in the universe: the speed of light in vacuum $(3 \times 10^8 \text{ m/s} = 186,000 \text{ miles/s})$. Reminder: According to Einstein's theory of Special Relativity, no object can travel faster than the speed of light because the mass of that object will tend to infinity as its velocity approaches the speed of light. Hence the speed of light is the "hard" upper bound for fast in the universe. In effect, let's imagine the transistor switch "racing" a beam of light. If I take a stop watch and click it on and off as the transistor switch closes and opens, I can then ask the following question. How far does the beam of light travel (moving at 186,000 miles/s) during the 10-ps transistor switching event when my stop watch clicked from on to off? Answer? – only 3.0 mm! In essence, the transistor switch has stopped the light beam, the fastest thing in the universe, dead in its tracks! That's fast! As we will see, achieving such speeds in micro/nanoelectronics is fairly routine today, a modern miracle by any reckoning.

Closely related to the time scale of the universe, and very important in the context of micro/nanoelectronics applications, is the frequency scale of the universe (Fig. 2.13). For traveling (moving) periodic waves (e.g., electromagnetic waves or sound waves or water waves), it is very convenient to speak of the "frequency" of the wave; that is, the number of wave cycles per second – 1/s dimensionally, or reciprocal time = hertz (Hz). The frequency (f) of the wave is simply its velocity (the propagation speed of the wave, v) divided by its wavelength (the distance it takes for a wave to repeat itself, λ), or $f = v/\lambda$ (Fig. 2.14). For instance, we know that visible light has a wavelength of 0.4–0.7 μm. Because all electromagnetic radiation moves (in vacuum) at the speed of light $(3 \times 10^8 \text{ m/s})$, this means that visible light has a frequency range of 4.3×10^{14}–7.5×10^{14} Hz

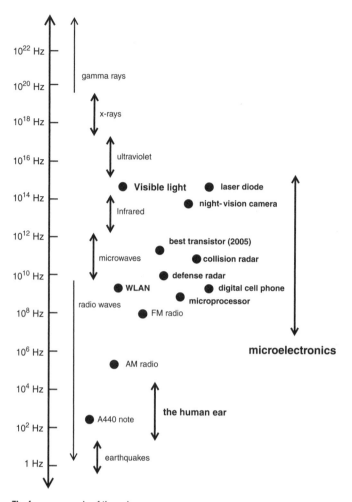

Figure 2.13   The frequency scale of the universe.

[or 0.4–0.7 terahertz (THz) (1×10^{15})]. Similarly, the human ear has a frequency range (at least when we're young!) of roughly 20–20,000 Hz. On a piano, the note A above middle C is perceived by the human ear as a pure tone at 440 Hz, often labeled A440. Refer to sidebar discussion on the frequencies used for seeing and hearing.

Deep Thoughts #2: Seeing and Hearing

Interestingly, there are over 11-orders-of-magnitude difference in the frequency response of our two human sensory organs that can beautifully detect periodic waves: our ears and our eyes. One might logically wonder what occurred in the course of human evolution to "push" the human proto-eye and proto-ear to these particular (fairly narrow) frequency bands. In the auditory case this was likely due to the constraints imposed by the cochlear transducer (middle ear bone), which depends on mechanical resonances in bones to detect "sounds," but is not so obvious for our eyes. Actually, the frequency range of the human sense organs are fairly limited compared with those of the rest of the animal kingdom. Some insects, for instance, can see well beyond the sensitive visible spectrum of the human eye into the ultraviolet (e.g., honey bees) and infrared (e.g., beetles), and many animals can hear to considerably lower (e.g., most insects) and higher frequencies (e.g., dogs and bats) than humans. I have often joked with my students that if we humans could "see" in the radio wave band of a few 100 MHz to a few GHz, the entire world would be literally aflame because of the presence of cells phones and other portable wireless devices doing their magic at those frequencies.

Unlike for the distance and time scales, the universe offers no hard practical boundaries in the frequency domain for EM waves, and EM waves can run the gamut from radio waves ($< 1 \times 10^{10}$ Hz), to microwaves ($1 \times 10^{10} < f < 1 \times 10^{12}$ Hz), to infrared ($1 \times 10^{12} < f < 4 \times 10^{14}$ Hz), to the narrow visible spectrum ($4 \times 10^{14} < f < 7 \times 10^{14}$ Hz), to ultraviolet ($7 \times 10^{14} < f < 1 \times 10^{17}$ Hz), to x-rays ($1 \times 10^{17} < f < 1 \times 10^{19}$ Hz), and finally to gamma rays ($> 1 \times 10^{19}$ Hz). Refer to Fig. 2.13.

There are, however, practical bounds on the intrinsic frequency capabilities of modern microelectronic (and photonic) devices that range roughly from 1.0 GHz (1×10^9 Hz–GHz) to 1.0 THz, although clearly electronic products based on microelectronic devices can also operate at lower frequencies (e.g., AM radio or FM radio), down in the 100s of kilohertz (kHz) to a few hundred megahertz (MHz), respectively. Useful frequency references to remember in the context of our discussion of micro/nanoelectronics include the carrier frequency used in your digital cell phone, which at present is 900 MHz to 2.4 GHz, the frequency of the microprocessor clock in your PC (perhaps 2.0–3.0 GHz if you have a good machine), and the wireless local-area network (WLAN) carrier frequency for your wireless router at home (either 2.4 or 5.8 GHz). All such technologically important frequencies are in the "few" GHz range. This choice of a "few" GHz for transmission of data or voice communications is not coincidental, but rather optimized for "clean" transmission bands for minimal interference over a given practical distance (from rain, atmospheric interference, absorption by atmospheric gases, etc.). The transistors used to build the integrated circuits in such systems are capable of functioning up to around 100 GHz at the state-of-the-art (W-band). Obviously, optical-based (photonic) microelectronic devices that produce (or detect) EM radiation are closer in frequency to that of the visible end of the spectrum (THz), and include important applications such as laser diodes (e.g., red light, 0.7 THz), light-emitting diodes (e.g., green light, 0.5 THz), or night-vision systems (infrared, 0.01 THz).

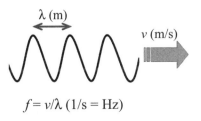

Figure 2.14   Definition of frequency for a traveling wave.

2.6 The temperature and energy scales of micro/nanoelectronics

The final grand scales that are needed to ground our discussion of the micro/nanoelectronics world are the temperature and energy scales of the universe. Temperature and energy are one and the same thing and can be used interchangably, depending on convenience. Energy, simply stated, is the capacity for doing "work," where, formally, work is the application of a force to move an object a given distance. Energy can come in one of two forms, either kinetic energy (the energy associated with the movement of an object) or potential energy (an object's latent energy, as associated with its position (e.g., because of gravity or an electric field) or chemical configuration (e.g., we might burn an object, say wood or gasoline, to release its latent potential energy). One of our most cherished laws of the universe is the "conservation of energy." That is, energy can neither be created nor destroyed as the universe evolves, only rearranged – it is conserved.

One of the common difficulties in properly appreciating energy scales rests with its dimensionality. In the MKS system, the preferred unit of energy is the joule ($1.0 \text{ J} = 1.0 \text{ kgm}^2/\text{s}^2$). Perhaps a more familiar energy unit is the calorie ($1.0 \text{ J} = 0.239 \text{ cal}$), although human "dietary calories" (the labels on our food packages) are actually expressed in units of kilocalories (kcal) (1,000 cal) – this clearly gives new meaning to a 500-cal (dietary) Twinkie (you are really eating 500,000 cal)! For American/British audiences, $1.0 \text{ J} = 0.949 \times 10^{-3}$ BTU (British thermal units), 42.78×10^{-7} kW hours (kwh), and 0.769×10^{-8} gal of gasoline (burned). Interestingly, though, for reasons soon to be seen, energy in the micro/nanoelectronics world is best expressed in "electron volts" (eV). One eV is the amount of kinetic energy a single (only 1!) electron gains when accelerated through a potential of only 1 V. That is, $1.0 \text{ eV} = 1.602 \times 10^{-19}$ J! And $1.0 \text{ eV} = 3.829 \times 10^{-20}$ cal. Said another way, you burned 13,060,000,000,000,000,000,000,000 eV when you ate that Twinkie! An electron volt feels like a very tiny number, and so it is. Consider this hands-on comparison. If I take a incandescent 100-W lightbulb and turn it on for only 1 h, I will consume a total of 22,500,000,000,000,000,000,000,000 (2.25×10^{25}) eV of energy! An electron volt is indeed a tiny amount of energy. And yet, amazingly, the operation of our microelectronic devices will center on eV-sized energies (e.g., the energy bandgap of silicon = 1.12 eV at room temperature).

Fundamentally, energy and the absolute temperature, measured in Kelvins (K), are related only by a scale factor, known as Boltzmann's constant (k) after its discoverer, $k = 8.617 \times 10^{-5}$ eV/K, such that energy (eV) $= kT$. See the sidebar discussion. Hence temperature is just a different measure of energy – they are the same thing. The "thermal energy" (kT), as it's known, figures prominently in the physics of micro/nanoscale systems, as we shall see. At "room temperature" (300 K = 26.85°C = 80.33°F), for instance, the thermal energy is a miniscule 0.0259 eV.

> **Geek Trivia: Kelvins**
>
> A pet peeve. Absolute temperature is measured in Kelvins (K), *not* degrees Kelvin (°K). Note that 0 K, "absolute zero," is $-273.15°C$. To convert between Kelvins, degrees Celcius, and degrees Fahrenheit, use $T(°C) = T(K) - 273.15$ and $T(°F) = 9/5 \times T(°C) + 32$.

> **Deep Thoughts #3: kT**
>
> By any reckoning, kT is the most fundamental parameter of the universe, because it governs the behavior of both the Fermi–Dirac and the Bose–Einstein distribution functions, which control the behavior and structure of all particles and hence matter in the universe, be it on Earth, or 5,000,000 light years away. Pretty fundamental!

Temperature is clearly a more intuitive human concept than energy. Our sensory organs can far more easily tell the difference between 0°C (freezing) and 40°C (frying) than it can discern the differences between 2 J and 2,000 J. Still, common misconceptions regarding temperature persist and should be dispelled at this point. Absolute zero (0 K $= -273.15°C = -459.67°F$) is the lowest possible temperature, but (a) 0 K cannot be attained, even in principle, according to the well-established laws of thermodynamics (0 K is the mythical mathematical infinity of the temperature world), and (b) all atomic motion does not cease at 0 K, from the laws of quantum mechanics.

Now to the master temperature–energy scale of the universe (Fig. 2.15). Some highlights: The core of our Sun is a little over 15,000,000 K (blistering), the center of our galaxy even hotter (but tougher to measure given that a gigantic black hole sits there). Nearly 6 orders of magnitude smaller than this (5×10^5) is room temperature (300 K), our convenient reference point for life on Earth, and micro/nanoelectronics; 100 times smaller still is the cosmic background temperature of the universe, at 2.73 K, the 13.7-billion-year-old afterglow remnants of the Big Bang. Now 2.73 K feels mighty close to absolute zero, but the temperatures of the universe actually decrease wildly from there. At present, the lowest recorded laboratory temperature is 500 pK, a whopping 10 orders of magnitude below the cosmic background! A pause for reflection is in order. Following the themes of our other universal scales, the approximate ratio of the highest to lowest attainable temperatures (in K) in the universe is about 1,000,000,000,000,000,000 (1×10^{18}). Let's call this the "universal temperature–energy-scale parameter." Like the other universal-scale parameters, it's an awfully big number.

Okay, so what do temperature and energy have to do with the micro/ nanoelectronics world? Lots! Figure 2.16 shows a zoom-in of the master temperature–energy scale, with highlights shown for relevance to the micro/

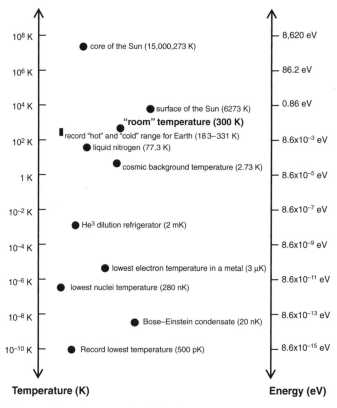

Figure 2.15 The temperature and energy scales of the universe.

nanoelectronics world. Some key points: (1) The usable temperature realm of conventional electronics ranges from −55°C to + 125°C, the so-called "mil-spec" (military specification) temperature range. Commercial electronics (things you and I would buy) is more in the temperature range of 0°C to + 85°C. (2) There is currently a push to design and operate electronic devices and circuits at increasingly lower temperatures (as low as 77.3 K, liquid nitrogen, or even 4.2 K, liquid helium), and to temperatures as high as +500°C. There are many unique applications that require such "extreme environment electronics," including, for instance, exploration of the Moon, where temperatures in the sunshine are at about +120°C, whereas temperatures in the permanent dark spots (craters, for instance), can be as low as −230°C, far outside the realm at which conventional electronics will reliably operate.

From an energy perspective, two additional important points can be gleaned for understanding micro/nanoelectronics: (1) the energy bandgap (E_G), *the* most fundamental parameter (after kT!) governing the propeties of semiconductors and the electronic and photonic devices built from them, spans the range of only a "few" eV (0.66 eV for germanium, 1.12 eV for silicon, 3.0 eV for SiC, and 5.5 eV for diamond). Given that an eV, as just discussed, is a woefully tiny amount of energy compared with the energy levels associated with everyday life, and that these tiny energies are in fact *required* for building the micro/nanoelectronic

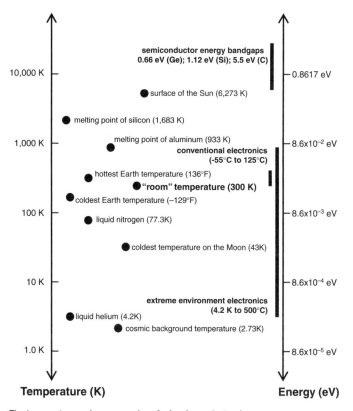

Figure 2.16 The temperature and energy scales of micro/nanoelectronics.

devices fueling the Communications Revolution, this is indeed a remarkable fact. (2) The other relevant energy metric of the micro/nanoelectronics world is the so-called "power-delay product" (PDP). Given that power is measured in watts (J/s), and switching delay is measured in seconds, this means that the PDP is a quantifiable measure of the energy expended in switching a transistor from an "off" state to an "on" state and vice versa. Minimization of the PDP of digital logic has been the root driving force behind the evolution and rampant successes of the past 40+ years of the micro/nanoelectronics industry. To get a feel for PDP numbers, the record speed for a single digital logic gate is currently at about 3 ps while consuming about 0.001 W (1 milliwatt, mW) of power, giving a PDP of about 3 femtojoules (3×10^{-15} J). Even though this is a tiny amount of energy, note that it is still the equivalent of 18,727 eV, far above the eV-sized energies of the underlying materials themselves.

2.7 Seeing is believing?

"Unless I see . . . I will not believe." Recall the words of perhaps history's most celebrated skeptic, Thomas, called Didymus (John 24:25). We humans are born skeptics, no doubt about it. I have argued that transistors, those smallest of building blocks fueling the advance of the micro/nanoelectronics revolution, and thereby touching, and in many cases rapidly defining, virtually all human pursuits on planet Earth in the 21st century, cannot be seen, at least in a traditional sense, even in principle. Can you believe without seeing? You must. Our day-to-day human existence brushes up against only the very narrowest of ranges of numbers that define small and fast. And yet the universe knows better – and offers so much more for your appreciation. We humans, endowed with our marvelous creativity, have engineered, cooked up in magical ways, objects that necessarily stretch our intuitive understanding and day-to-day sense of scale, to the breaking point. The micro/nanoelectronics world, infinitesmally small and blindingly fast, is rapidly redefining the human sense of scale in many fascinating ways.

Should you care, you nonspecialists not directly involved in the creating and machining of micro/nanoelectronics? You had better care, else the world will drag you along, kicking and screaming, unaware and unappreciative, still blinking in your bewilderment, and you will be the lesser for it. In the end humankind will have little choice but to embrace the creations and advances of the micro/nanoelectronics juggernaut. Don't miss the wonderful opportunity to celebrate the miracles of scale that surround you.

Now that we have a better grasp of the nuances of "small" and "fast," and some appreciation of the grand scale and themes of the universe, we are ready to plunge ahead into the micro/nanoelectronics world. Hold on tight!

References and notes

1. See, for instance, *http://nobelprize.org/physics/educational/microscopes/scanning/*.
2. D. M. Eigler and E. K. Schweizer, "Positioning single atoms with a scanning tunneling microscope," *Nature (London)*, Vol. 344, pp. 524–526, 1990.
3. R. P. Feynman, "There's plenty of room at the bottom," keynote speech at the annual meeting of the American Physical Society, December 29, 1959; *http://www.zyvex.com/nanotech/feynman.html*.
4. See, for instance, *http://spm.phy.bris.ac.uk/techniques/AFM/*.

3 Widget Deconstruction #1: Cell Phone

The new Bat phone (courtesy of Nokia).

I Want To Know God's Thoughts.
The Rest Are Details.

Albert Einstein

3.1 With a broad brush

The telephone, in its many forms, is firmly embedded in the American lexicon: Dick Tracy had a wrist phone; Batman, a Bat phone (for the NEW Bat phone, see the chapter-opening figure!); and the comedic spy Maxwell Smart, a shoe phone. The telephone. So ubiquitous, so seemingly innocuous as to almost escape our attention, telephone traffic (voice and data) embodies by far the largest fraction on global electronic information flow (96.6% of the 17,905,349 TB produced in 2002). The telephone is as integral to modern life as it gets, and richly deserving of a careful look by all. Not surprisingly given its electronic nature, micro/nanoelectronics technology evolution has (and still is) reshaping the face

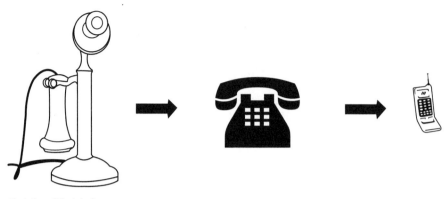

Figure 3.1 Evolution of the telephone.

of telephony. Figure 3.1 shows the migratory path from the old-timey phone (recall the great scene from *It's a Wonderful Life* involving a phone like this), to the click-click-click "rotary" phone (I am old enough to have used one in my home as a kid, but they are now relegated to antique stores or retrotechnology boutique shops), to the now-ubiquitous cellular telephone (aka the "cell phone").

Geek Trivia: First Cell Phone Call

The first public telephone call placed on a portable cellular phone was made by Martin Cooper (then general manager of Motorola's Communications Systems Division) on April 3, 1973. Following this public demonstration using a "brick-like" 30-oz phone, Cooper led a 10-year quest to bring the portable cell phone to market. Motorola introduced its 16-oz, $3,500 "DynaTA" phone into commercial service in 1983 (only 23 years ago!). It took seven additional years before there were a million subscribers in the United States [1].

Cell phones are everywhere (Fig. 3.2), as are their co-conspirators, those unsightly cell towers. Don't get me started on the cell towers masquerading as trees – abominations! An example of a more aesthetically pleasing cell tower "sculpture" is shown in Fig. 3.3. It is estimated that in 2006 there were over 2 billion cell phones in use worldwide (2,168,433,600 to be precise!), and there now exists such a thing as cell phone "etiquette" (sadly, all too frequently ignored, to our collective chagrin). Consider: "The *Zagat* restaurant guide reports that cell phone rudeness is now the number one complaint of diners, and *USA Today* notes that 'fifty-nine percent of people would rather visit the dentist than sit next to someone using a cell phone'" [2]. Although such notions may make you smile, smirk, or growl, depending on your disposition, do appreciate that therein lies a case-in-point of technology (micro/nanoelectronics technology, as we will soon see) actively shaping social interaction, societal mores, and just maybe the evolutionary path of civilization. Isn't it intrinsically worth knowing something about how these little buggers work? You bet!

Figure 3.2 The ubiquitous cell phone and cell tower (courtesy of agefotostock.com © R. Definsky (left) and © M. Robinson (right)).

Figure 3.3 An example of cell tower "sculpture" forming the outside chapel at Saint Ignatius House Retreat Center, in Atlanta, GA. (See color plate 8.)

Historical Aside: Who Invented the Telephone?

Most technological landmarks have interesting and nuanced histories, and the telephone is certainly no exception [3]. The telegraph, invented in 1831 by Samuel Morse, demonstrated that communication at a distance by electrical means was possible – a pivotal event in the history of human interaction. Still, the required transmission means of rapidly tapped "dots" and "dashes" (i.e., Morse Code) was cumbersome, to say the least. Wouldn't it be much better to simply electrically transmit and receive the human voice? Indeed it would. In the 1870s, two young inventors, Elisha Gray and Alexander Graham Bell, independently designed devices that could transmit speech electrically (i.e., the telephone). Both men rushed their respective designs to the patent office within hours of each other, but Bell won the footrace and patented his telephone design first (U.S. Patent 174,465). Gray and Bell entered into a protracted legal battle over the invention of the telephone, which Bell eventually won [sadly, we all know who Bell was (and the subsequent "Ma Bell" monopoly), but most do not know the name of Gray]. On June 2, 1875, Bell, while experimenting with his new technique called "harmonic telegraph," discovered he could actually hear sound over the wire by modulating the electrical current in the wire (like that of a "twanging clock spring"). Once he figured out how to convert a voice into electrical pulses (and back again) he was home free. Bell's notebook entry on March 10, 1876, describes his first successful experiment with the telephone (he was only 29). Speaking through the instrument to his assistant, Thomas A. Watson, in the next room, Bell uttered the now-famous words, "Mr. Watson – come here – I want to see you." The rest, as they say, is history.

From the initial invention of the telephone (see sidebar discussion), which you would not physically recognize today as a phone (it was an unwieldy beast – see a picture in Chap. 4), to the present, its form and function have rapidly evolved. What began as a simple voice transmission–reception device can now support (depending on your phone) an exceptionally diverse array of functionality, including an appointment book; a calculator; e-mail; an Internet connection; games (grrr . . .); a TV; a camera; an MP3 music player; and a global positioning system (GPS) receiver (to name a few). A state-of-the-art cell phone is thus an amazingly sophisticated object, especially because we want it also to be portable (i.e., run off batteries for a sufficiently long period to be useful) and to possess a very small "form factor" (i.e., be tiny enough to hold in your palm).

3.2 Nuts and bolts

So what exactly is a cell phone? At its heart, a cell phone is a sophisticated two-way radio. Such a two-way radio has both a "transmitter," which sends voice signals, and a "receiver," which receives voice signals. The sum of these is two radio functions is called a transmitter–receiver or simply "transceiver." Many other electronic circuits are wrapped around this radio to support the various other functions of the cell phone beyond telephony (e.g., e-mail or text messaging); but yep, it's just a radio. And no, it would not work without transistors. Radio has its own fascinating roots (see Chap. 4), but cellular telephony is at its core a beautiful and highly successful merger, albeit a nontrivial merger, of the telephone and the radio.

Before talking about how cellular technology works, we need to define a few basic radio concepts. First, a reminder of what radio actually is. Radio waves are the part of the EM "spectrum" with wavelengths ranging from a few 10s of centimeters to 100s of meters or longer ($< 1 \times 10^{10}$ Hz–10 GHz in frequency). For example, an FM radio broadcast might be at 94.1 MHz (94,100,000 Hz) or an AM radio broadcast at 790 kHz (790,000 Hz). Cellular radio frequencies are in the 900-MHz (900,000,000 Hz) range for first-generation (1G) analog cellular phones or in the 1.9-GHz (1,900,000,000 Hz) range for second-generation (2G) digital cellular phones, higher than the radio frequencies you listen to on your car stereo.

Consider the simplest types of two-way radios, say a "walkie-talkie" or a Citizen's Band (CB) radio. These are "half-duplex" radios, meaning that the two radios (one sending voice at point A; one receiving voice at point B) share the same radio frequency, a "channel" (a predefined frequency) for moving information. A walkie-talkie may have only 1 channel, and a CB radio perhaps 40 channels that one manually selects ("meet me on channel 6, good buddy"), but a modern cell phone may use thousands of distinct channels, and the switching among these many channels is transparent to the cell phone user. Cell phones, by contrast, use a "full-duplex" radio, meaning that the cell phone uses one channel for sending and a different channel for receiving, a decided advantage given that two or more people can then talk or listen at the same time.

Radios transmit and receive your voice via a "modulated (altered) carrier" frequency. The voice information is "encoded" into the high-frequency EM signal traveling at the speed of light (186,000 miles per second) by radio 1 and sent, and is then "decoded" by radio 2 on arrival so you can hear it. This voice modulation might be accomplished by changing the frequency of the EM carrier wave (i.e., frequency modulation – hence, FM) or by changing the magnitude (amplitude) of the EM carrier wave (i.e., the older, and often static-filled, amplitude modulation – hence, AM). In the case of cell phones, a more complicated carrier modulation scheme is used. The "range" of the radio, how far it can communicate effectively, depends on the signal output power of the transmitter and sensitivity of the receiver. The range might be 1 mile for a walkie-talkie using a 0.25-W transmitter or 5 miles for a 5-W CB radio.

In the context of cell phones, channel "bandwidth" is the range of frequencies that the transmitted–received signal uses (refer to sidebar discussion on bandwidth). Thus bandwidth is expressed in terms of the difference between the highest-frequency signal component of the channel and the lowest-frequency signal component of the channel. For instance, a cell phone channel might have a 30-kHz bandwidth at a carrier frequency of 900 MHz.

Technical Aside #1: Bandwidth

Just to confuse things, bandwidth has several common connotations. For instance, bandwidth is also often used as a synonym for "data transfer rate" – the amount of data that can be carried from point A to point B during a given time period (usually 1 s). This kind of bandwidth is usually expressed in bits (of data) per second (bps). For example, a modem that works at 57,600 bps (57.6 kbps) has twice the bandwidth of a modem that works at 28,800 bps. In general, a data link with a high bandwidth is one that can carry a lot of information (bits) very rapidly, and is a good thing for, say, rapid file downloads and file exchange.

Okay, back to the concept of cellular telephony. If I would like to talk to a friend by mobile phone, and I would like to do so over long distances, I have a serious fundamental problem because the transmitted power of the modulated radio signal falls off as 1 over the distance squared ($1/r^2$) between the radios (from basic EM theory; see Appendix 2 if you need a reminder). Hence, transmitting over long distances takes very high-power transmitters to send the EM signal and very big antennas to receive the EM signal (e.g., an FM radio station might use a 50,000-W transmitter on a 100-m-high radio tower to broadcast over a metropolitan area of 10s of miles). This is clearly not something you can carry around in your pocket! Instead, for cell phones, one breaks the area in question into overlapping modular "cells" (think bee honeycombs) of perhaps 10 square miles, each with a "base station" of small transmitter power (10s of watts) on a centrally placed cell tower (Fig. 3.4). A single mobile telephone switching office (MTSO) connects all of the

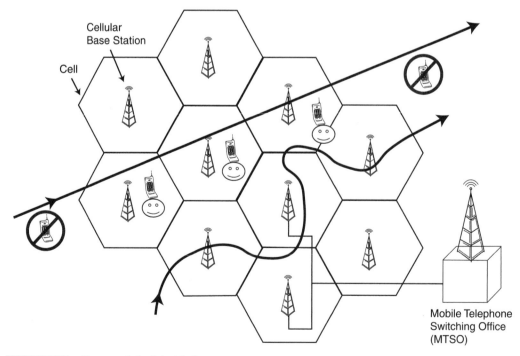

Figure 3.4   The concept of cellular telephony.

cell base stations, typically over a landline (the MTSO routes both cell calls and landline calls), to give adequate cell "coverage." As you and your phone move through the cell and away from the base station, the base station recognizes this movement (by a weakening signal), and "hands off" to the next cell before it loses contact with you, hopefully in a seamless fashion that is transparent to the user. This process is repeated as you move from cell to cell to cell while driving through the city. Welcome to mobile communication.

A cellular service provider is given a certain number of channels to distribute among its available cells (say 100). Because each cell phone uses two channels for a phone call (one to talk and one to listen), this means that 50 people can be on the phone at the same time in the same cell. Because the transmitted power of the base station is small, the signal does not travel very far, and the cellular company can reuse those 100 channels in another (nonadjacent) cell. All cell phones have special codes associated with them, which the base station uses to identify the phone, the user, and the service provider. You can't hide from Big Brother if you are using a cell phone! This simple fact has produced significant societal implications regarding privacy issues and is particularly relevant in combating terrorism, for instance. More on this in a later chapter.

Let's explore, then, how a cell phone call actually occurs, a process that is FAR more complex than you likely imagined! When you power on the phone, its receiver "listens" for a system identification (SID) code being broadcast on the "control channel" from the cellular base station you happen to be in. If your phone does not find a SID, it will declare "out of range" – no cell coverage – and you are out of luck. Assuming it does receive a valid SID, the phone compares the information in the signal received with that programmed into the phone. If the SIDs match, for instance, the cell phone knows it is communicating with its own "home" system. The phone also transmits, along with the SID, a "registration request," and the MTSO then keeps track of your phone's location in a database in real time, so that it knows what cell you are in if it wants to, say, ring your phone (or report you to the police!).

Now, let's say you have your phone on and are receiving service, and your BFFL (best-friend-for-life) dials your number. The MTSO receives the incoming call and then locates you (via its database). Knowing that your cell phone is listening, the MTSO then picks a frequency pair (two channels) that it, via the local cell tower, and your phone will use to handle the call, and sends that information to the phone over a suitable control channel. Once the phones and the cell tower switch to those defined frequencies, then the call is "connected," your phone rings, and your BFFL tells you to pick some milk on the way home; all while you are driving your car (hopefully using a hands-free headset and concentrating on the road!).

As you move to the edge of your current cell while talking, the cell's base station is constantly noting your transmitter's "signal strength" and will observe that it is weakening ("Can you hear me now?"). Meanwhile, the base station in the cell you are moving toward is also monitoring your phone and its signal strength and sees your signal strength increasing. The two base stations contact each other via the MTSO, and at the appropriate optimal point the MTSO will send another message to your phone on a control channel telling it to switch frequencies (channels) again (to one not being used in the new cell). The old cell now drops you ("click!"), and the new cell picks you up, thus effectively "handing off" your phone to the next cell. This hand-off sequence happens as you move from cell to cell and could in principle happen many times during a long call while you are driving.

How about the now-infamous "roaming" charges on your cell phone bill? Let's say the cell you are in belongs to your own (home) service provider, but the next cell you are moving into belongs to a different (competing) service provider. Clearly you don't want the MTSO to drop your call during hand off just because you need competitive companies to temporarily get along! If the SID of the new cell does not match your phone, then the MTSO knows that you are in roaming mode. The MTSO handling the cell you are roaming in now contacts the MTSO of your home system, validates you, and then tracks you as you move through its cell, treating you as "one of its own," albeit for a major cost addition! Roaming internationally where cellular standards may vary significantly can be problematic, but is increasingly being addressed by service providers by actually including multiple radios in each cell phone supporting each needed standard; say, one for home and one for abroad!

As you consider the many gyrations involved in actually making a mobile cell phone call, it is in fact pretty remarkable that it generally occurs in an error-free manner. You should consider that hidden complexity carefully the next time you are tempted to gripe about a dropped call!

So, do you have a "1G" phone or a "2G" phone, and is it AMPS or GSM or CDMA? Confusing, huh? In 1983, the 1G cell phone standard called AMPS (advanced mobile phone system) was introduced, approved by the FCC, and first deployed in Chicago (nice piece of Geek Trivia). AMPS uses 832 channels (790 voice channels and 42 data channels – for control channels, paging, etc.) ranging from 824 MHz to 894 MHz, each with a 30-kHz bandwidth. Interestingly, the choice of a 30-kHz channel bandwidth was not arbitrary, but was instead chosen to emulate the voice quality achieved on landline phones (back then, the competition). The transmit and receive channels are separated by 45 kHz to minimize channel-to-channel interference. AMPS was the first cellular standard to be widely offered commercially and, even with its more modern variants, is known today as an "analog" cellular standard. "Analog" means that the cell phone's radio transmits and receives your voice as an analog signal (a modulated EM carrier) and is not digitized (changed to "1s" and "0s") when sent or received. Although still very common in the United States, analog phones cannot offer the more advanced (digital) cellular services such as e-mail or Web browsing that many of us have come to expect.

"Digital" cell phones represent the 2G of cellular technology. Why go digital? Fundamentally, analog signals cannot be nearly as easily compressed (made smaller) and manipulated as digital signals. The theory and practice of

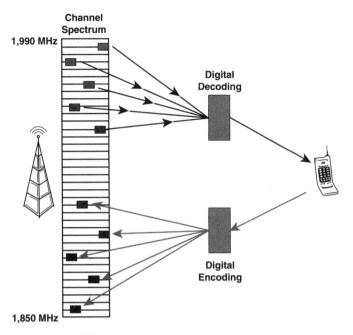

Figure 3.5   The concept of CMDA cell phones.

compression and decompression of digital information for communication are known as digital signal processing (DSP) and is a relatively new and vibrant subdiscipline of electrical engineering that has its roots in cellular telephony. Effectively, when DSP techniques are used, more channels can be squeezed into the available bandwidth, and higher data transmission rates can be achieved, enabling a 2G phone to enjoy significantly more features than a 1G phone. Clearly, the data rates needed to support Web browsing on a cell phone, for instance, are significantly higher than those for simply talking on the phone. Digital 2G phones enable this.

Many current 2G phones rely on frequency-shift keying (FSK) to send (digital) data back and forth using conventional AMPS (an obvious cost savings). FSK uses two frequencies (one for "1s" and one for "0s"), alternating rapidly between the two (hence the name). Sophisticated DSP techniques are used to encode the analog voice signal into a highly compressed digital stream, send it, and decode it back to an analog voice signal, all while maintaining an acceptable signal quality.

There are three common techniques used by more advanced 2G cell phone networks to transmit their digital information: frequency-division multiple access (FDMA), time-division multiple access (TDMA), and code-division multiple access (CDMA) [4]. As the names suggest, FDMA puts each call on a separate frequency, TDMA assigns each call a certain time interval on a designated frequency, and CDMA assigns a unique code to each call and "spreads" it out over the available frequencies (CDMA is illustrated in Fig. 3.5, and is an example of "spread-spectrum" communications). The "multiple access" in FDMA, TDMA, and CDMA simply refers to the fact that many users can utilize the cell simultaneously (obviously a good thing in a high-population-density area). The resulting advantages of 2G phones over 1G phones are compelling. TDMA, for instance, has triple the channel capacity of a 1G analog network and operates either at 800 MHz (called the Interim Standard IS-54 band) or at 1.9 GHz (the IS-136 band), or both. The global system for mobile (GSM) communications is the prevalent 2G cellular standard in Europe, Australia, and much of Africa and the Far East. GSM communications use TDMA, and interestingly, incorporate data "encryption" (intelligent scrambling of bits) of its transmitted digital information for added security.

Third-generation (3G) cell phones are the latest rage and are intended to be true multimedia platforms (aka "smartphones"), enabling voice, e-mail, Web browsing, video, games, GPS, etc., in a mobile environment. Transmission data rates to and from a 3G phone must be very high to support such functionality, and in current 3G networks they can be in the range of 3 Mbps (this bandwidth is equivalent to downloading a 3-min MP3 song to your phone in about 15 s). Even good 2G phones, by contrast, have data rates in only the 144-kbps range (an 8-min download of the same song – clearly impractical for busy people). To accomplish this, 3G networks use even more sophistication than 2G networks: CDMA-2000, WCDMA (wideband CDMA), and/or TD-SCDMA (time-division synchronous CDMA), all sophisticated, DSP-intensive, spread-spectrum techniques [4]. Solutions ranging between the capabilities of 2G and 3G phones are often labeled 2.5G.

Mind boggling, huh? Think about this next time you are surfing the Web on your cell phone while stuck in traffic! The evolutionary end of the cell phone is nowhere in sight. It represents perhaps the original "killer-app" in the consumer electronics market. Implicit in this discussion of the evolution of cellular telephony from 1G to 2G to 3G (and beyond) is that the sophistication of micro/nanoelectronics needed to build each new generation of cell phone has grown exponentially. Said another way, Moore's law growth trends in the integrated circuit industry have effectively defined the evolution of cellular telephony. Let's look now a little closer at what is actually inside that cell phone.

3.3 Where are the integrated circuits and what do they do?

Based on a "complexity per unit volume" metric, cell phones are arguably the most intricate and sophisticated objects you and I encounter on a daily basis. In addition to their complex two-way radios, modern cell phones must process literally millions of calculations per second to compress and send and decompress and receive a "voice (or data) stream" for the user, not to mention all of the electronic overhead needed to support the other high-level applications of say a 3G phone. Let's pull off the covers and take a tour of what actually lurks inside that cell phone hiding in your pocket. We start from a general perspective and then zoom in on a particular cell phone example for some concrete visual images.

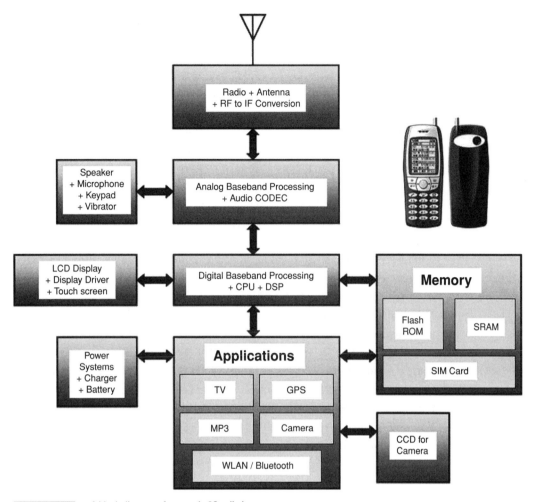

Figure 3.6 A block diagram of a generic 3G cell phone.

At a high level, a 3G cell phone can be broken into the following major subsystems (refer to Fig. 3.6 [5]):

- *Antenna:* The antenna is a carefully designed piece of metal that efficiently sends and receives the modulated EM waves to and from the radio at the appropriate carrier frequency (e.g., 1.9 GHz). A 3G cell phone might have several antennae, in fact; one for the radio, one for the GPS receiver, and one for the WLAN connection.
- *Display:* The display is the main visual interface between you and the phone. It consists of a color liquid-crystal display (LCD) and may be touch-screen activated, as in a PDA (personal digital assistant). A piece of electronics called a "display driver" is used to interface between the phone and the LCD.
- *Microphone, Speaker, Keypad, Vibrator:* These elements can be viewed as the main sensory "transducers" of your cell phone [i.e., they change the form of (transduce) sound (microphone), for instance, or pressure (keypad), into electricity and back again]. The microphone and speaker manipulate incoming

and outgoing speech; the keypad enables dialing, text input, or Web surfing; and the "vibrator" announces calls when you are operating in silent mode (often, woefully, not so silent!).

- *Radio:* The usefulness of a cell phone begins and ends with the radio, because the radio transceiver must seamlessly interface between the phone and the EM world for signal reception and transmission. The radio contains "frequency translation" electronic circuitry to convert the (high) radio frequency (RF) of the EM carrier (e.g., 1.9 GHz) to an "intermediate frequency" (IF) that is more easily manipulated by the subsequent electronics of the analog baseband processing unit (e.g., 20 MHz). This RF-to-IF conversion is obviously a two-way process, and is called either "down-conversion" (RF-to-IF) on the "receive" (incoming call) path or "up-conversion" (IF-to-RF) on the "transmit" (outgoing call) path.

- *Memory:* The sophistication of a cell phone requires multiple forms of memory where information can be stored (more details on the various forms of electronic memory can be found in Chap. 6). Memory in a 3G phone might include "Flash ROM," a software-controlled, electrically alterable, low-power read-only memory (ROM); "SRAM" (static random-access memory), a high-speed, high-density memory that does not need to be electrically refreshed (hence static); capability to accept a plug-in flash "memory stick"; and a subscriber identity module (SIM) card, which contains information on your identity that the phone (and the cell tower) can recognize when needed.

- *Analog Baseband Processing:* The analog baseband processing unit is responsible for converting incoming analog (continuous in time) signals into digital signals ("1s" and "0s") via an analog-to-digital converter (ADC), which allows the information content to be processed much more efficiently by the digital signal processor and microprocessor. The opposite process of converting digital signals to analog signals occurs by means of a digital-to-analog converter (DAC) and is needed to transmit data. The analog baseband processor also engages various filtering functions to improve signal fidelity, and includes the "Audio CODEC" (for COmpress–DECompress), which manipulates the incoming and outgoing voice information and interfaces with the speaker or microphone for speech decoding and synthesis.

- *Digital Baseband Processing:* The digital baseband processing unit is essentially a microprocessor (likely more than one) that serves as the brains of the cell phone, functioning as a "command and control center," if you will. The digital baseband processor(s) takes care of general housekeeping tasks, provides seamless interfacing and communication with all of the various subsystems, and implements all DSP functions needed for encoding and decoding of calls. The digital baseband processor likely has its own SRAM and DRAM (dynamic RAM) memory on-chip for its immediate needs.

- *Power System:* There is much more to the power system of a cell phone than just a battery. Given that battery life between charging events is the primary constraint on phone portability, a great deal of effort is expended in minimizing battery drain. For instance, a 3G phone has built-in electronic intelligence to infer when a given component or subsystem in the phone can be powered down and put into "sleep" mode, thus saving battery life. Similarly, a cell phone that is simply receiving ("listening") within a cell draws orders-of-magnitude less power than one that is actively transmitting ("talking"), and the digital baseband processors will know this and act accordingly to maximize battery life.

- *Applications:* The applications that run on a 3G cell phone are many and highly varied, including: a GPS receiver for location awareness; an MP3 digital music player (Moving Picture Expert Group, Layer 3 – MPEG3, or just MP3 for short); a digital camera that uses a charge-coupled device (CCD); a wireless local-area network (WLAN) for connection to the Internet (e.g., via a "Bluetooth" WLAN radio); a host of games; and believe it or not, even television. It cannot, as yet, drive the car (relax, give it time!).

Okay, so where are the transistors in this beast? Each of the various sub-systems just outlined will physically consist of one or more integrated circuits, each of which is built from transistors. There are actually 100s of millions of transistors in a 3G cell phone. For instance, your phone might have 80 M of SRAM (> 480,000,000 transistors). The microprocessor for the digital baseband unit might have 10,000,000 transistors, exclusive of its on-chip SRAM and DRAM. The DAC of the analog baseband processing unit might have 10,000 transistors. The two-way radio has even fewer transistors, perhaps only in the 100s. As shown in Fig. 3.7, if we zoom in on the radio of the cell phone, one of the critical elements in the receiver is the low-noise amplifier (LNA). The LNA is responsible for amplifying (making larger) the incoming EM carrier wave from the antenna (which might be only microvolts (μV's, one millionth of a volt) in magnitude, making the signal large enough to be manipulated by the radio circuitry. It must perform this signal amplification at high (RF) frequencies without adding excessive noise to the EM signal, or else the information it contains (voice, data, or both) will be lost. The LNA may look deceptively simple (refer to Fig. 3.7), because it requires in this case only two transistors, two inductors (the two circular shapes in the die photo), one capacitor, and one resistor; but if only one of those transistors is removed (or is damaged), the entire cell phone is worthless.

We turn, finally, to visually exploring the integrated circuits in a real 3G cell phone [6]. WARNING: DO NOT ATTEMPT THIS AT HOME! Figures 3.8–3.12 show a "naked" 3G cell phone, covers conveniently pried off (yes, this violates your warranty!), highlighting the "mother-board" and the various integrated circuits of the major subsystems. Yep, it's pretty complicated. Next time you drop your phone on the concrete and can't for the life of you imagine why it will no longer speak to you, recall this inherent complexity. And to think your cellular service provider will give it to you for pennies . . . yeah, right!

There you have it – the cell phone. A remarkable level of complexity in a very tiny space. The important take-away? Appreciate that the transistor and the various types of integrated circuits constructed from them (radio, ADC, microprocessor,

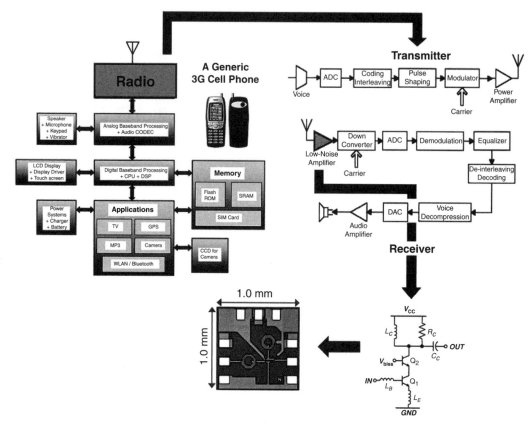

Figure 3.7 Breakdown of a generic 3G cell phone from the block level to the radio, focusing in on the low-noise amplifer (LNA) implementation as a silicon integrated circuit. The requisite transistors of the LNA are Q_1 and Q_2 in the circuit schematic.

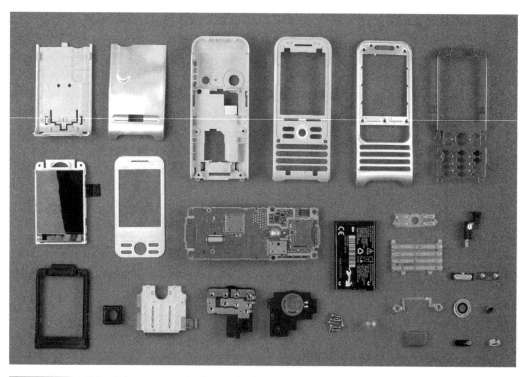

Figure 3.8 Stripped-down components of a generic 3G cell phone (courtesy of Portelligent, Inc.). (See color plate 9.)

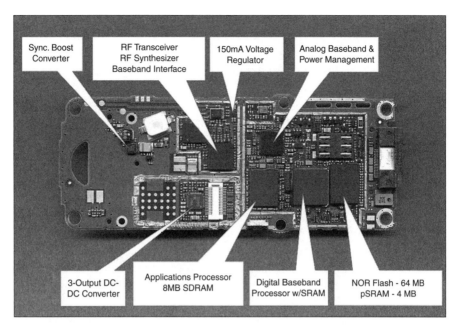

Figure 3.9 Front view of a generic 3G cell phone with the covers removed, showing the various integrated circuit components (courtesy of Portelligent, Inc.).

SRAM, CODEC, etc.) are the key enablers of this technological marvel. As transistor technology has matured by means of its Moore's law growth path, so has the cell phone's functionality and sophistication – these two evolutionary progressions are essentially joined at the hip. One can argue about the intrinsic merits of cell phones (I will in a subsequent chapter), and whether ultimately they represent a step forward or a step backward for civilization, but one cannot

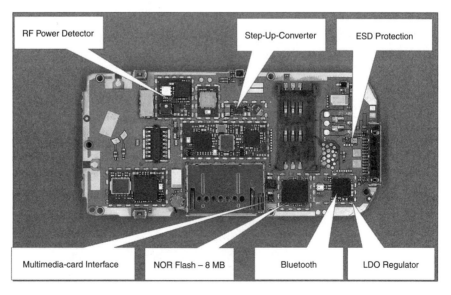

Figure 3.10 Rear view of a generic 3G cell phone with the covers removed, showing the various integrated circuit components (courtesy of Portelligent, Inc.). (See color plate 10.)

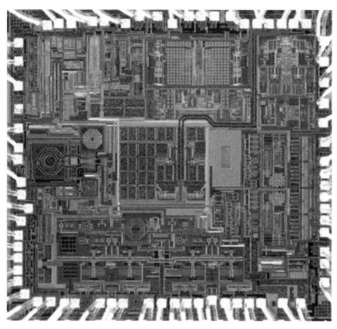

Figure 3.11   A zoom-in, showing a "delidded" integrated circuit, in this case one of the RF transceivers (courtesy of Portelligent, Inc.). (See color plate 11.)

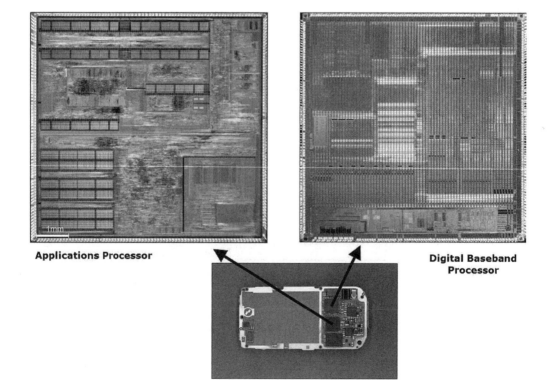

Figure 3.12   A zoom-in, showing "delidded" integrated circuits, in this case the application processor with embedded memory and the digital baseband processor (courtesy of Portelligent, Inc.). (See color plate 12.)

argue whether or not they are here to stay. The original killer-app for consumer electronics. 1G, 2G, 3G. What will a 6G cell phone look like in 2020? Your guess is as good as mine; but I am hoping that it will make my coffee AND drive me to work! Cutting the grass wouldn't be a bad idea either!

References and notes

1. See, for instance, *http://inventors.about.com/cs/inventorsalphabet/a/martin_cooper.htm*.
2. C. Rosen, "Our Cell Phones, Ourselves," *The New Atlantis*, Number 6, pp. 26–45, 2004. See *http://www.thenewatlantis.com/archive/6/rosen.htm*.
3. See, for instance, *http://inventors.about.com/library/inventors/bltelephone.htm*.
4. See, for instance, *http://electronics.howstuffworks.com/cell-phone.htm*.
5. See, for instance, *http://www.necel.com/en/solutions/applications/mobile/3g/01.html*.
6. I am deeply indebted to Portelligent, Inc., for allowing me to use their images of a 3G cell phone. You can visit them at *http://www.teardown.com/*.

4 Innumerable Biographies: A Brief History of the Field

Photograph of the first transistor (courtesy of Science and Society Picture Library).

History Is The Essence
Of Innumerable Biographies.
Thomas Carlyle

In its November 2004 "Special 40th Anniversary Issue," *IEEE Spectrum* published a fascinating article titled "The View From the Top" [1]. In it were identified the 40 most influential across-all-disciplines technologists (obviously a subjective choice, but spanning Nobel Laureates, leading academics, CEOs of major high-tech companies, etc.; you get the idea: movers and shakers), who were asked to respond to the following question: "What has been the most important technology [invented or discovered] in the last 40 years?" A full 38 out of 40 answered "microelectronics technology." Forgive me if you are one of those

40 – I have taken some liberty with your words. If you answered, "integrated circuits" or "transistors," or "microchips," or "the computer," or "semiconductors," or something ultimately derived from these, such as "the Internet," or "the Web," or "wireless communications," I have lumped your answer into "microelectronics technology," because clearly none of these "higher-order" technological inventions could have occurred without the prior invention of the transistor and the subsequent maturation of microelectronics as a pervasive technology. Said another way, all of these secondary technologies ultimately trace their roots to microelectronics. Since the transistor itself was invented in 1947 [2, 3] (i.e., not within the past 40 years), to be more precise I've called the defining technological discovery of the past 40 years "microelectronics technology," but the seminal invention that has driven the entire microelectronics field was clearly the transistor itself.[1]

So how did the transistor and microelectronics technology come to be? Excellent question. Surprising answer. Read on.

4.1 What history can teach us

To my mind, there are four compelling reasons to examine the historical foundations of any given technological field.

- *Know the roots of your chosen profession.* This is true whether you are a concert pianist, a professional baseball player, a computer programmer, a materials scientist, an architect, a chef, or an integrated circuit engineer. The better you know the roots of your field, the better able you will be to place yourself in its context and find the best path forward. If you don't aspire to accomplish something meaningful (and hopefully new) with your career . . . you are in the wrong profession! Besides, for something as influential as micro/nanoelectronics on the evolution of civilization, I would argue that this particular history should be known by all, not just technologists. Yep, it is worthy of classroom book learning. I strongly suspect that 1,000 years hence the 50 years of intense technological innovation from roughly 1950 to 2000 surrounding this field will burn very brightly in the historical record.
- *Learn how your predecessors innovated.* It is always instructive for the next generation to glimpse how the pioneers of their chosen profession accomplished all that they did. Seeing how others innovate will make innovation easier for you. Some take-aways from this particular history: (1) Serendipity often plays a major role in discovery. This is an important lesson for all budding innovators to glean early on and should give all of us mere mortals extra incentive. (2) Multidisciplinary training matters a lot. (3) Proper handing

---

[1] Curious to know what the other two folks thought? Think biology: "the discovery of DNA" and "gene sequencing." Not bad answers, but keep in mind that no useful technology can be derived from DNA or gene sequencing without the (extensive) use of computers; hence (to my mind anyway), such biologically inspired technologies rest squarely on the back of the transistor and microelectronics technology.

of your inventions is key to defining their future. (4) Business acumen and selling-savvy can be awfully important to success in technological fields.

• *History is defined by people, not events.* This is the point of Thomas Carlyle's quotation opening this chapter and runs counter to many people's naive notion of what history is. There are clearly exceptions to this history-shaping rule of people-define-events (not vice versa), but they are rare. You and I can change history. We can revector the many evolutionary threads that span human history. This is truly an empowering idea. To my mind, history is best approached not as a dry collection of facts and dates, but rather as a fascinating melange of unusual personalities.

• *Kids, history can be fun!* Okay, believe it or not, the history of technology can make for some pretty darn interesting reading. This may surprise the masses, but, yep, it's true – history can indeed be fun! Given the wacky personalities of the best and brightest minds walking the planet on any given day of the week, this shouldn't surprise you really. The history of micro/nanoelectronics is indeed often soap-opera-ish, and chocked full of eccentric personalities, massive egos, and backbiting. Fact is often stranger than fiction!

4.2 The uniqueness of microelectronics

From a broad-brushed historical perspective, many aspects of the development of microelectronics technology are strikingly unusual. For instance, most technologies have traditionally proceeded from "art" to "science," rather than the other way around. In the steel industry, for instance, literally centuries of empirical craftsmanship (the "art") transpired before even the most rudimentary scientific underpinnings were established (the "science"). People knew how to produce high-quality steel and fashion it into all sorts of sophisticated objects, from tools to guns, long before they understood why steel worked so well. This knowledge base was largely empirically driven – as they say, practice makes perfect. Part of this inherent difficulty in applying science to steel production lies in the exceeding complexity of the materials problems involved (ditto for the production of glass, for instance – glass is perhaps *the* quintessential millennia-old technology). Steel is simply a very complicated material system, chock full of nuances and unknown variables (this is no longer true, obviously). Microelectronics, by contrast, represents one of the rare instances in the history of technology in which "science preceeded the art" [4].[2]

Part of this difference between microelectronics and all other technologies lies in the fundamental materials problems involved. For microelectronics we are dealing with a near-perfect single crystal of regularly arranged atoms of a single

---

[2] I gratefully acknowledge my significant debt to Warner and Grung's ideas and historical perspective on the development of the transistor and the evolution of microelectronics technology. I consider their treatment of the various nuances, the many funky personalities, and the fascinating twists and turns surrounding the early days of microelectronics to be one of the very best I've seen.

element – silicon – about as simple and ideal a materials problem as one can imagine to apply the scientific method to. By most accounts, and for obvious reasons, today silicon is far and away the best understood material on Earth. In such "simple" problems, trivial (literally, back of the envelope) theory can often work astoundingly well in capturing reality. For instance, imagine introducing a tiny number of phosphorus atoms into silicon (e.g., add 1 phosphorus atom for every 100 million silicon atoms). *Question:* How much energy does it take to release the extra electron bound to the incorporated phosphorus atom in the crystal (this is called the "dopant ionization energy" and is very important information for building transistors and circuits)? Hint: A scratch of the head and a puzzled look would be appropriate here for most readers! *Answer:* This problem can be well described by a simple "toy model" of a hydrogen atom (the Bohr model, 1913), something every first-year college science and engineering student learns, and in two seconds of simple calculations we have an answer accurate to within a factor of three! Correct answer? 45 meV. Toy-model answer? 113 meV. This accuracy is rather amazing if you stop and think about it, given the complex quantum-mechanical nature of the actual problem involved. Science works quite well in such idealized venues, and in the end this turns out to be pivotal to the ultimate success of microelectronics technology, because, as a circuit design engineer, for instance, I do not have to stop and worry about performing a detailed first-principles solution to Schrödinger's equation (clueless? – join the rest of us) to know how to bias my transistor in the electronic circuit I am constructing. If I did, there clearly wouldn't be many cell phones or computers around!

It is interesting that microelectronics technology in fact has followed the popular (but alas commonly misperceived) direction of technological innovation – first discover the science, and when the science is mature, then hand it off to the engineers so they can refine and develop it into a viable technology for society's happy use. A glance at the historical record indicates that this common view of how technological innovation *should* ideally proceed, however appealing, is rarely borne out in practice. Microelectronics is one of the very few exceptions [5].

> The transistor was born with a rich legacy of science, but virtually nonexistent art. By art in this context is meant the accretion of skills, techniques, and engineering approaches that would permit the fabrication of microscopic structures called for by transistor inventors for feasibility demonstrations and design engineers for subsequent refinements.

Just who exactly were the purveyors of the requisite art of this fledgling microelectronics empire? Interestingly, and unique at least to the time, these individuals were trained scientists (physicists primarily), but working principally as engineers, all the while searching for (profitable) inventions. They were a new breed, modern Renaissance people of sorts, a fascinating amalgam of scientist – engineer – inventor. As we shall see, this proved to be a rather formidable combination of traits for the pioneers in this field, and a remarkably few individuals ushered in the many high-technology fruits we all enjoy today. More on their names (and personality quirks!) in a moment.

Before launching into the specific people and events that led to the invention of the transistor, setting the stage for the subsequent micro/nanoelectronics revolution, it will be helpful to examine the broader context and underlying motives that helped spawn this fascinating new field. What drove them inexorably to the transistor, and why?

If I Have Seen Further Than Others,
It Is By Standing On The Shoulders Of Giants.

Isaac Newton

4.3   The shoulders we stand on

Some Giants of Electrical Engineering

Ironically, the foundations of electrical engineering (EE), as a now-entrenched discipline, rest squarely on the shoulders of physicists, not electrical engineers! Skeptical? Start naming the units we commonly employ in EE: hertz (Heinrich Hertz, 1857–1894, physicist), volts (Alessandro Volta, 1745–1827, physicist), ohms (Georg Ohm, 1789–1854, physicist), amps (Andre-Marie Ampere, 1775–1836, physicist), farads (Michael Faraday, 1791–1867, physicist), Kelvins (William Thomson, Lord Kelvin, 1824–1907, physicist), gauss (Karl Friedrich Gauss, 1777–1855, physicist and mathematician), newtons (Isaac Newton, 1642–1727, physicist extraordinaire), and joules (James Joule, 1818–1889, physicist). This list could go on. There are exceptions, of course; watts (James Watt, 1736–1819, inventor and engineer) and teslas (Nikola Tesla, 1856–1943, inventor and engineer) come to mind. The fact that the many fruits of engineering rest on the triumphs of scientific theory cannot be overstated. Science and engineering are highly complementary, and to my mind equally noble, pursuits. Remember this the next time you hear EE undergraduates grousing about their woes in a physics or chemistry course!

Although one might reasonably expect the history of transistors to be firmly tied to the evolution of computers – that is, electrical computing – it is in fact much more closely linked to the development of telephones – electrical communication [5]. The evolution of electrical communication (moving information from point A to point B by manipulating electrons), the foundations of which predate the 19th century, began with the pioneering work on battery power sources by Alessandro Volta (1774) and Hans Christian Oersted (1819), the development of a primitive electrical signaling system by Baron Pavel Schilling (1823), the discovery of electromagnetic induction by Michael Faraday (1831), and the combination of Faraday's principle with Schilling's signaling system by Wilhelm Weber and Carl Friedrich Gauss (1833). This set the stage for Charles Wheatstone's British patent on a primitive telegraph (1837) and Samuel Morse's (Fig. 4.1) more practical version of a telegraph system (also 1837 – Fig. 4.2), culminating in the

Figure 4.1   Samuel Finley Breese Morse (1791–1872) was born in Charlestown, MA, and was a trained portrait painter. He was also a keen inventor, and in 1832 conceived the idea of a magnetic telegraph using a single-wire system, which he exhibited to Congress in 1837. He created a dot-and-dash code for sending messages that used different numbers representing the letters in the English alphabet and the 10 digits, which became known as Morse Code. In 1844 an experimental telegraph line was completed between Washington and Baltimore, and Morse sent the first historical telegraph message: "What hath God wrought?" (courtesy of Science and Society Picture Library).

Figure 4.2   This early Morse telegraph key (ca. 1850–1870) was used by the U.S. Post Office Telegraph Service. This key, which functions as a simple switch for generating "dots" and "dashes," is quite cumbersome in shape and size. As Morse keys evolved, they became increasingly sophisticated, developing into carefully engineered instruments. The Morse system required a skilled operator using a Morse key to interrupt the current from a battery into a wire circuit. At the other end, a buzzer converted the electrical signals into sounds, which were manually converted back into letters by a trained operator (courtesy of Science and Society Picture Library).

Figure 4.3 Alexander Graham Bell (1847–1922) was born in Edinburgh, Scotland, but emigrated to the United States, settling in Boston where began his career as an inventor. After experimenting with various acoustic devices, Bell (at age 29!) produced the first intelligible telephone transmission with a message to his assistant on June 5, 1875. He patented this device as the telephone in 1876, a patent he had to defend against Elisha Gray, who had filed a similar patent only hours later. Bell, who became an American citizen in 1874, also invented the phonograph (1880) and the gramophone (1887). Among Bell's other achievements was the "photophone," which transmitted sound on a beam of light – a precursor of today's optical-fiber systems – and the invention of techniques for teaching speech to the deaf (courtesy of Science and Society Picture Library).

first successful electrical transmission of information across a wire, with the phrase "What hath God wrought?" from Washington, D.C., to Baltimore, MD, in 1844 (in Morse Code of course!). The Age of Electrical Communication was launched. The path from telegraphy to telephony traces its roots to William Sturgeon's development of the horseshoe magnet (1824), followed by Joseph Henry's conception of a magnetic armature (1830), allowing him to ring a bell remotely by using an electromagnet (1831) – think telephone "ringer." Philipp Reis, working in Germany in the 1850s, devised a primitive system to transmit music tones and coined the term "telephone." Elisha Gray developed practical switches and relays for a rudimentary routing system of such electrical signals, producing a "musical telegraph" (1874). Finally, along came Alexander Graham Bell (Fig. 4.3), who beat Gray to the punch, filing a patent for a rudimentary voice transmitting device (aka the telephone – Fig. 4.4), literally hours before Gray (1876).

Geek Trivia

Reis arrived at "telephone" from "tele," meaning "at a distance," and "phone" for "phonics" (the study of sounds).

Meanwhile, the race to electrical communication through the air (i.e., wireless telephony) was on, resting squarely on James Clerk Maxwell's (Fig. 4.5) pivotal mathematical description and unification of Faraday's observations on electrical phenomena (1865) and Heinrich Hertz's (Figure 4.6) use of spark gaps (think spark plug) for generating electrical oscillations (voltage sine waves) and transmitting and receiving them through the air (1887).[3] The search for improvements by Hugh Thomson, Reginald Fessenden, Nikola Tesla, and Charles Steinmetz

---

[3] Hertz was also the one who put Maxwell's equations into their modern form, using concepts from vector calculus, and popularized their importance.

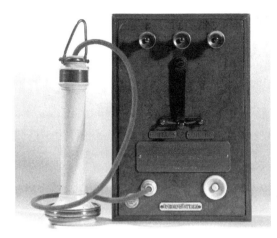

Figure 4.4 On March 7, 1876, Bell secured a patent for an electrical speech machine, later called the telephone. After showing the instrument at the American Centennial Exhibition in Philadelphia in June, news of Bell's invention spread throughout Europe. He was asked to demonstrate the telephone to Queen Victoria (1819–1901) at her residence, Osborne House, on the Isle of Wight in the south of England. On the evening of January 14, 1878, contact was made with nearby Osborne Cottage and terminals in Southampton and London. This pictured telephone and terminal panel were used at Osborne Cottage (courtesy of Science and Society Picture Library).

Figure 4.5 James Clerk Maxwell (1831–1879), one of the world's greatest theoretical physicists, was born in Edinburgh, Scotland, and studied mathematics at Trinity College, Cambridge. After graduating in 1854, he became a professor first at Aberdeen (1856) and then at King's College, London (1860). He was also the first Cavendish Professor of Physics at Cambridge University (1871). Maxwell's most important work was on the theory of electromagnetic radiation, and in 1873 he published "A Treatise on Electricity and Magnetism." He also formulated the kinetic theory of gases (courtesy of Science and Society Picture Library).

to Hertz's spark-gap apparatus culminated in Ernst Alexanderson's rotating machine (literally, an "alternator"), capable of generating robust EM waves (think of a modulated radio carrier wave). Meanwhile, Edouard Branly, in England, devised a so-called "coherer" (1892) that was capable of generating a variable resistance (then a novel concept). Guglielmo Marconi (Fig. 4.7) improved on the coherer, devising the all-important and now-ubiquitous antennae and subsequently succeeding in demonstrating wireless telegraphy over a distance of 2 km in Italy (1896 – Fig. 4.8), clearly a milestone in modern communications history.[4] By 1901 Marconi had refined his wireless system sufficiently to perform the first transatlantic telegraph communication. Radio days were fast approaching!

Ultimately it was the development of the vacuum tube that propelled the fledgling wireless communications field. Not surprisingly, vacuum tubes grew out of lightbulb technology. Incandescent lightbulbs were

[4] It is little recognized that this feat was independently accomplished in Russia by Popov, also in 1896, who sent the rather appropriate message of "Heinrich Hertz" [5].

Figure 4.6 Heinrich Hertz (1857–1894) discovered "Hertzian waves" (EM waves) and demonstrated that they behaved like light waves except for their wavelength. He was reputedly the first person to transmit and receive radio waves, and the unit of frequency of radio waves is named after him (courtesy of Science and Society Picture Library).

invented by Thomas Edison (Fig. 4.9), after his famously exhaustive search for a practical light source powered by electricity. Edison eventually succeeded in his quest after developing an efficient vacuum pump and then placing a carbonized bamboo fiber (yep, he tried a gazillion materials as filaments before the great eureka moment!) inside an evacuated glass enclosure (1876 – Fig. 4.10).[5] During Edison's subsequent refinement of his lightbulb, he serendipitously happened on the "Edison Effect," in which a second "cold" electrode or metal plate sealed inside the same evacuated glass bulb with a "hot" filament received a current when a positive voltage was placed on the cold electrode (1883). The "thermionic diode" (the first "vacuum tube") was born (and patented – Edison was the most prolific inventor in history, with over 1,000 patents to his credit, a remarkable feat by any measure, and a tribute to his creativity). Ironically, Edison went no further with his discovery, but simply noted the effect, and then quickly moved on to what he thought at the time to be bigger and better things. Edison's truncated foray into vacuum tubes represents an excellent case-in-point of a highly capable inventor–engineer who, while on the very brink of making a world-changing innovation, simply failed to grasp the potential impact and properly exploit his idea. The discovery of the Edison effect is today widely considered to be his principal scientific discovery, and can be properly regarded as the birth point of the field of electronics, no small claim to fame [5].[6]

Geek Trivia

The term "electronics" dates to the late 1920s, appearing in periodicals in both the United States and in England. Interestingly, not until 1963 did electrical and computer engineering, as a profession, change its name from the *Institute of*

---

[5] The Smithsonian Museum of American History in Washington, D.C. has a wonderful Edison exhibit, well worth visiting.

[6] It has been pointed out that Edison was at the time only one year away from opening his Pearl Street electrical generating station and was thus likely deluged with managerial responsibilities (aka the weight of the world) associated with that venture. Although certainly plausible, it should nevertheless be mentioned that Edison likely lacked the requisite formal education in mathematics and science to properly understand and develop this complex phenomenon [6]. *C'est la vie!* Remember this next time you complain to a friend about your woes in your calculus class. Trudge on, you may well need some math one day!

Figure 4.7   Gugliemo Marconi (1874–1937) was an Italian inventor and physicist. He discovered a method in which EM waves could be used to send messages from one place to another without wires or cables (aka wireless). Having read about Hertz's work with EM waves, he began to carry out his own experiments in wireless telegraphy, and in 1894 he successfully sounded a buzzer 9 m away from where he stood. Marconi shared the 1909 Nobel prize for physics with Karl Ferdinand Braun (courtesy of Science and Society Picture Library).

Radio Engineers (IRE) to its present form of *Institute of Electrical and Electronics Engineers* (IEEE). The IEEE is the largest professional society in the world, and currently embodies 39 distinct societies, contains over 365,000 members, sponsors over 125 technical journals and 300 annual conferences, and has defined and maintains over 900 active electronic standards (e.g., 802.11a for a WLAN).

Fortunately, others were willing and able to pick up the vacuum tube banner and run with it. British engineer John Fleming (Fig. 4.11) improved the thermionic

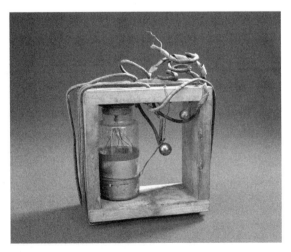

Figure 4.8   Marconi conducted his first experiments in wireless telegraphy in Bologna in 1895, and in 1898 transmitted a radio signal across the English Channel. On December 12, 1901, Marconi successfully transmitted a series of signals across the Atlantic, from Poldhu in Cornwall to Signal Hill, Newfoundland, using this experimental tuned transmitter. The test proved that long-distance radio communication was a practical possibility (courtesy of Science and Society Picture Library).

Thomas A. Edison (1847–1931) was a prolific American inventor who registered over 1,000 patents, many of which were related to the development of electricity. His inventions, in collaboration with his research staff, include the automatic telegraph, the electric lamp, the phonograph (gramophone), and the kineto-scope, an early form of cinematography (courtesy of Science and Society Picture Library).

diode vacuum tube ("vacuum tube" for short) and importantly used it to detect an EM radio signal (1904), creating the "Fleming valve" [so named because of its operational analogy to a heart valve, conducting current in only one direction (in modern parlance, a "rectifier")]. Fleming valves (Fig. 4.12) were cleverly used to demonstrate integral concepts to wireless communications, including transmission (sending) and detection (receiving) of modulated periodic EM "carrier" waves. The rest is history. On Christmas Eve, 1906, Reginald Fessenden (Fig. 4.13) used a Fleming valve-based apparatus to perform the first voice radio broadcast at Brant Rock, MA, using a modulated 1-kW! output from a 50-kHz Alexanderson alternator, with a microphone placed in series with the alternator and Marconi's antenna. Dumbfounded wireless telegraphers in the northeastern United States were treated to music and voice magically coming through their dot–dash–dot Morse Code receptions [5]! Also in 1906, Lee De Forest invented the "audion," a vacuum tube triode ("tri" for three, meaning three electrodes or "terminals"), the first vacuum tube amplifier (think small current in, large current out, clearly of singular importance in overcoming inherent attenuation of electrical signals).

Made by the American inventor Thomas Alva Edison (1847–1931), this lamp had a single loop of carbon that glowed when an electric current flowed through it. The glass bulb contained a partial vacuum, and thus there was so little oxygen in the bulb that the filament could get very hot without catching fire and burning up (courtesy of Science and Society Picture Library).

Vacuum tube amplifiers (Fig. 4.14) were refined through the development of the concept of both positive and negative "feedback" [Edwin Armstrong (1914) and Harold Black (1927), respectively], design techniques that are so fundamental to electronics that they are ubiquitous today. Two notorious problems associated with vacuum tubes were at this stage readily apparent – not only were they pretty challenging to make, but they had a pretty darn short lifespan. Said another way, tube electronics cost a lot, and their reliability was poor. Once a functioning tube was up and running, inevitably, either the vacuum leaked away and the filament burned out, or even a

Figure 4.11 Sir John Ambrose Fleming (1849–1945) was an English physicist and electrical engineer and professor of Electrical Engineering at University College, London. Between 1885 and 1926, Fleming invented the "thermionic valve," the first vacuum rectifier; he was also a pioneer in the application of electricity to lighting and heating and instrumental in the development of radio telegraphy and telephony. In 1899 Fleming became a technical adviser to the Marconi Company, helping design the Poldhu transmitter in Cornwall used in Marconi's 1901 transatlantic wireless transmissions (courtesy of Science and Society Picture Library).

gentle knock resulted in a shower of glass, all clearly serious concerns for building practical radios.

Meanwhile, as the budding vacuum tube electronics industry waxed, driven largely by the ultimate promise of wireless communications, Ferdinand Braun observed that a metal wire tip pushed up against a crystal of lead sulfide (aka galena) to form a "point-contact" similarly functioned as a electronic rectifier (1874). During the next three decades of refinement, this fundamentally different type of electronic rectifier came to be known as a "cat's-whisker diode," for obvious visual reasons (zany engineers!). Literally thousands of "solid-state" materials were evaluated as cat's-whisker diode candidates, but only a very select few worked, including silicon, copper oxide, germanium, selenium, and silicon carbide (the first one should sound familiar!). Interestingly, only what we now know as semiconductors made good cat's-whisker diodes, for reasons that proved very elusive.

Geek Trivia

The term "semiconductor" dates to 1911 in a paper in a scientific journal published in German [7].

Although cat's-whisker diodes were far smaller and much easier to make than vacuum tubes, they similarly suffered from abysmal reliability. The underlying operation principles of these solid-state diodes remained essentially opaque, and even as late as 1909 the rectification properties were erroneously considered to be a purely thermal phenomenon. A scientific pause now occurs to let the catastrophe of the Great War pass.

Post–World War I, most scientific research focused squarely on the atom and the nucleus and on the development of quantum theory as an explanatory paradigm. Although clearly of great practical importance, the advancement of

Figure 4.12 Fleming's original thermionic valve was made for a set of experiments on an effect seen in early electric lamps. The source of current is a heated negative electrode, the "cathode." Current flows through the vacuum to the positive electrode, the "anode." Fleming's valve exhibited the unique property of "rectification" (allowing current flow in only one direction when the voltage polarity was reversed). Fleming later developed his vacuum tube "valves" as sensitive detectors of radio signals (courtesy of Science and Society Picture Library).

electronics languished, primarily because of poor understanding of the underlying mechanisms. Electronics tinkerers abounded to support the emerging radio industry (think crystal radio sets), but the science of the subject lagged. This changed abruptly with the emergence of "radar" systems immediately prior to World War II (the Germans had radar, and England and the United States, among others, wanted desperately to equalize the situation, for obvious reasons). Radar, as we shall soon see, becomes the great catalysis of transistor development.

Figure 4.13 In 1907, Canadian Reginald Aubrey Fessenden (1866–1932) transmitted a human voice by radio for the first time. Fessenden was responsible for more than 500 patents regarding radio communication (courtesy of Science and Society Picture Library).

Geek Trivia

RADAR is an acronym for – RAdio Detection And Ranging.

In radar, one generates and bounces an EM radio wave off the object in question (e.g., an airplane), and detects the back-reflected EM "echo," observing the total travel time and direction of the echo source (Figs. 4.15 and 4.16). This can then be used for precisely locating moving objects (day or night!), clearly a real advantage in knowing when aircraft with black swastikas are coming to firebomb London during the Blitz, for instance, and highly useful in shooting something down should the need arise. Radar systems are pervasive today, for both commercial (think air traffic control) and military purposes. To get a significant radar echo, and hence precise target-location information, the wavelength of the EM wave should be very small compared with that of the object being imaged. For flying aircraft, this optimal EM frequency lies in the microwave regime (e.g., 5–30 GHz). Interestingly enough, cat's-whisker diodes respond at these microwave frequencies and thus proved particularly useful in building radar systems. Given that the war effort pushed countless scientists into engineering roles in a rather frantic manner,

Figure 4.14   Lee De Forest (1873–1961) was an American electrical engineer and inventor. In 1907, while working for the Western Electric Company, he patented the "thermionic triode," which he called the "audion" valve. The audion was essentially a refinement of Fleming's thermionic valve, but with an important additional (third – hence triode) electrode between the cathode and the anode, enabling amplification. The triode was able to amplify weak electrical signals, making it a crucial invention for the development of radio. For 50 years, triode valves were a fundamental component in all radio equipment, until being replaced with the transistor. In the United States, De Forest is known as "the father of radio" (courtesy of Science and Society Picture Library).

a veritable laser beam of capable theoretical attention was finally focused on developing solid-state electronics technology and, ultimately more important, understanding the science and physical mechanisms involved. Solid-state science was suddenly in vogue. By the late 1930s, valid theoretical descriptions of rectifying diodes was largely in place because of the pioneering efforts of physicists Nevill Mott (1939), Walter Schottky (1939), and Boris Davydov (1938), not coincidentally, located in England, Germany, and the Soviet Union, respectively!

4.4 The invention–discovery of the transistor

It was in this context, then, that in the mid-1940s, the famous "Transistor Three" assembled at Bell Telephone Laboratories (aka "Bell Labs"), in Murray Hill, NJ, and began their furious search for a solid-state amplifying device, a semiconductor triode (our soon-to-be transistor).[7] Hired by Director of Research

---

[7] For fascinating historical discussions on the invention of the transistor, the interested reader is referred to [5, 9–11].

Figure 4.15 Radar station control center at "Chain Home," England, 1940–1945. This was one of the first types of radar stations in the British air defense system. Radar revolutionized British offensive and defensive capabilities during World War II, playing a vital role in locating incoming German bomber formations during the Battle of Britain (1940). It worked by bouncing powerful, ultra-short radio waves (microwaves) off approaching aircraft. These men and women are working in the Chain Home receiver room with console (right) and receivers (left). There were two main types of radar stations in service in England during World War II, the East Coast and West Coast types. The former was characterized by 360-ft-high self-supporting transmitting towers, and the latter used 325-ft-high steel transmitting masts (courtesy of Science and Society Picture Library).

Figure 4.16 British Army demonstration of early mobile radar equipment, Hyde Park, London, August 17, 1945. Here the soldiers demonstrate the use of one of the earliest mobile radar platforms, in which "anti-aircraft searchlights are used in conjunction with the radar – this allows the beams to be switched straight on to an incoming target" (courtesy of Science and Society Picture Library).

Mervin J. Kelly (later president of Bell Labs) in 1936, William Shockley, a theoretical physicist, in 1945 became group leader of a new "semiconductor" group containing Walter Brattain, an experimental physicist extraordinaire (who joined Bell Labs in 1929) and John Bardeen, a theoretical physicist (who joined Bell Labs in 1945).[8] Kelly's dream, put to Shockley as a provocative challenge of sorts for his new team, was ". . . to take the relays out of telephone exchanges and make the connections electronically" [8]. Kelly was after a replacement of the notoriously unreliable mechanical telephone relay switch, but could have equivalently asked for a replacement of the equally unreliable vacuum tube amplifier of radio fame, the electrical equivalence of a solid-state amplifier and electronic switch being implicit [5]. Importantly (and sadly, quite unusual by today's standards) Kelly assembled this "dream team" of creative talent, and then promptly gave them *carte blanche* on resources and creative freedom, effectively saying "I don't care how you do it, but find me an answer, and find it fast!"

During the spring of 1945, Bell Labs issued an important *Authorization for Work* explicitly targeting fundamental investigations into a half-dozen classes of new materials for potential electronic applications, one of which, fortuitously, was semiconductors [principally silicon (Si) and germanium (Ge)]. The "Transistor Three" all had some experience with semiconductors, and thus logically focused their initial efforts on amplification using semiconductors, based on Shockley's pet concept of a capacitor-like metal–semiconductor structure not unlike that of a modern MOSFET (metal-oxide semiconductor field-effect transistor). This search direction culminated in more than a few of what Shockley termed "creative failures." The reason such devices failed was soon understood (by Bardeen) to be related to surface defects in the semiconductor crystal itself.

Armed with Bardeen's theory, Bardeen and Brattain began to look for ways to effectively "passivate" (remove the defects from) the surface of the silicon crystal, into which he had stuck two metal cat's whiskers for electrodes. Unfortunately, condensation kept forming on the electrodes, effectively shorting them out. In a creative leap, on November 17, 1947, Brattain dumped his whole experiment into nonconducting (distilled) water (he knew a vacuum would have worked better but later said that it would have taken too long to build). To his amazement, he thought he observed amplification. When Bardeen was told what had happened, he thought a bit, and then on November 21, suggested pushing a metal point into the semiconductor surrounded by a drop of distilled water. The tough part was that the metal contact couldn't touch the water, but must instead touch only the semiconductor. Ever the clever experimentalist, Brattain later recalled in 1964 [9], "I think I suggested, 'Why, John, we'll wax the point.' One of the problems was how do we do this, so we'd just coat the point with paraffin all over, and then we'd push it down on the crystal. The metal will penetrate the paraffin and make contact with the semiconductor, but still we'd have it perfectly insulated from

---

[8] Others in the team included chemist Stanley Morgan (co-leader with Shockley), physicist Gerald Pearson, and physicist Robert Gibney.

the liquid, and we'll put a drop of tap water around it. That day, we in principle, created an amplifier."

Once they'd gotten slight amplification with that tiny drop of water, Bardeen and Brattain figured they were onto something significant. They experimented with many different electrolytes in place of the water and consistently achieved varying degrees of amplification.

Geek Trivia

One favorite electrolyte of the transistor team was affectionately named "Gu" (goo), glycol borate, which they "extracted" from electrolytic capacitors by using a hammer, a nail, and a vise (use your imagination!) [5].

On December 8, Bardeen suggested they replace the silicon with (in retrospect) less defective germanium. They suddenly achieved an amplification of $330\times$, but unfortunately only at very low frequencies (rendering it unsuitable for the envisioned applications). Bardeen and Brattain thought the liquid might be the problem, and they then replaced it with germanium dioxide (GeO_2). On December 12, Brattain began to insert the point-contacts. To his chagrin, nothing happened. In fact, the device worked as if there were no oxide layer at all. As Brattain poked the gold contact repeatedly, he realized that no oxide layer was present – he had washed it off by accident. Furious with himself, he serendipitously decided to go ahead and play a little with the point-contacts anyway. To his intense surprise, he again achieved a small amplification, but importantly across all frequencies. Eureka! The gold contact was effectively puncturing the germanium and passivating the crystal surface, much as the water had.

During that month, Bardeen and Brattain had managed to achieve a large amplification at low frequencies and a small amplification for all frequencies, and now they cleverly combined the two. They realized that the key components to success were using a slab of germanium and two gold point-contacts located just fractions of a millimeter apart. Now suitably armed, on Tuesday afternoon, December 16, Brattain put a ribbon of gold foil around a plastic triangle and cut it to make the point-contacts. By placing the vertex of the triangle gently down on the germanium block to engage the point-contacts, he and Bardeen saw a fantastic effect – a small signal came in through one gold contact (the electron "emitter") and was dramatically amplified as it raced through the Ge crystal (physically the "base" of the contraption) and out the other gold contact (the electron "collector").

Geek Trivia

The three electrodes (terminals) of the transistor, the "emitter," the "base," and the "collector," were actually named by Bardeen, and those names have stuck.

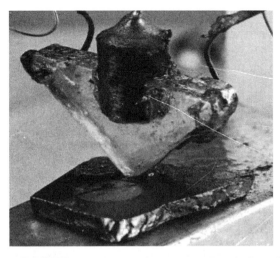 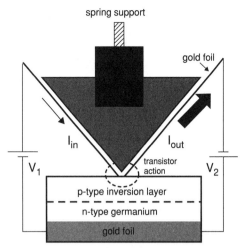

Figure 4.17 The famous point-contact transistor, the first solid-state amplifying device, invented in 1947 by John Bardeen and Walter Brattain at Bell Laboratories in the United States. Bardeen and Brattain discovered that, by placing two gold contacts close together on the surface of a crystal of germanium through that an electric current was flowing, a device that acted as an electrical amplifier was produced. (Photograph courtesy of Alcatel-Lucent). A schematic diagram of the device is shown on the right. (See color plate 13.)

The first point-contact semiconductor amplifier had been invented, or discovered, depending on your point of view (Fig. 4.17)! Bardeen's and Brattain's seminal paper on their transistor (remarkably only one and one-half pages long! – see Fig. 4.18) was published in *Physical Review* on July 15, 1948 [3].

The Transistor, A Semiconductor Triode

J. BARDEEN AND W. H. BRATTAIN

A three-element electronic device which utilizes a newly discovered principle involving a semiconductor as the basic element is described. It may be employed as an amplifier, oscillator, and for other purposes for which vacuum tubes are ordinarily used. The device consists of three electrodes placed on a block of germanium[1] as shown schematically in Fig. 1. Two, called the emitter and collector, are of the point-contact rectifier type and are placed in close proximity (separation ~0.005 to 0.025 cm) on the upper surface. The third is a large area low resistance contact on the base.

The germanium is prepared in the same way as that used for high back-voltage rectifiers.[2] In this form it is an N-type or excess semiconductor with a resistivity of the order of 10 Ω-cm. In the original studies, the upper surface was subjected to an additional anodic oxidation in a glycol borate solution[3] after it had been ground and etched in the usual way. The oxide is washed off and plays no direct role. It has since been found that other surface treatments are equally effective. Both tungsten and phosphor bronze points have been used. The collector point may be electrically formed by passing large currents in the reverse direction.

Each point, when connected separately with the base electrode, has characteristics similar to those of the high back-voltage rectifier. Of critical importance for the operation of the device is the nature of the current in the forward direction. We believe, for reasons discussed in detail in the accompanying letter,[4] that there is a thin layer next to the surface of P-type (defect) conductivity. As a result, the current in the forward direction with respect to the block is composed in large part of holes, i.e., of carriers of sign opposite to those normally in excess in the body of the block.

Reprinted with permission from J. Bardeen and W. H. Brattain, "The Transistor, A Semiconductor Triode," *Physical Review*, vol. 74, no. 2, pp. 230–231, July 15, 1948. Copyright 1948. The American Physical Society. Publisher Item Identifier S 0018-9219(98)00748-8.

[1] While the effort has been found with both silicon and germanium, we describe only the use of the latter.

[2] The germanium was furnished by J. H. Scaff and H. C. Theuerer. For methods of preparation and information on the rectifier, see H. C. Torrey and C. A. Whitmer, *Crystal Rectifiers*. New York: McGraw-Hill, 1948, ch. 12.

[3] This surface treatment is due to R. B. Gibney, formerly of Bell Telephone Laboratories, now at Los Alamos Scientific Laboratory.

[4] W. H. Brattain and J. Bardeen, *Phys. Rev.*, vol. 74, July 15, 1948.

Fig. 1. Schematic of semiconductor triode.

When the two point contacts are placed close together on the surface and dc bias potentials are applied, there is a mutual influence which makes it possible to use the device to amplify ac signals. A circuit by which this may be accomplished is shown in Fig. 1. There is a small forward (positive) bias on the emitter, which causes a current of a few milliamperes to flow into the surface. A reverse (negative) bias is applied to the collector, large enough to make the collector current of the same order or greater than the emitter current. The sign of the collector bias is such as to attract the holes which flow from the emitter so that a large part of the emitter current flows to and enters the collector. While the collector has a high impedance for flow of electrons into the semiconductor, there is little impediment to the flow of holes into the point. If now the emitter current is varied by a signal voltage, there will be a corresponding variation in collector current. It has been found that the flow of holes from the emitter into the collector may alter the normal current flow from the base to the collector in such a way that the change in collector current is larger than the change in emitter current. Furthermore, the collector, being operated in the reverse direction as a rectifier, has a high impedance (10^4 to 10^5 Ω) and may be matched to a high impedance load. A large ratio of output to input voltage, of the same order as the ratio of the reverse to the forward impedance of the point, is obtained. There is a corresponding power amplification of the input signal.

The dc characteristics of a typical experimental unit are shown in Fig. 2. There are four variables, two currents and two voltages, with a functional relation between them. If two are specified the other two are determined. In the plot of Fig. 2 the emitter and collector currents I_e and I_c are taken as the independent variables and the corresponding voltages, V_e and V_c, measured relative to the base electrode, as the dependent variables. The conventional directions for

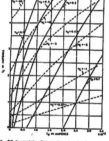

Fig. 2. DC characteristics of an experimental semiconductor triode. The currents and voltages are as indicated in Fig. 1.

the currents are as shown in Fig. 1. In normal operation, I_e, I_c, and V_e are positive, and V_c is negative.

The emitter current, I_e, is simply related to V_e and I_c. To a close approximation

$$I_e = f(V_e + R_F I_c) \qquad (1)$$

where R_F is a constant independent of bias. The interpretation is that the collector current lowers the potential of the surface in the vicinity of the emitter by $R_F I_c$, and thus increases the effective bias voltage on the emitter by an equivalent amount. The term $R_F I_c$ represents a positive feedback, which under some operating conditions is sufficient to cause instability.

The current amplification factor α is defined as

$$\alpha = (\partial I_c / \partial I_e)_{V_c} = \text{const.}$$

This factor depends on the operating biases. For the unit shown in Fig. 2, α lies between one and two if $V_c < -2$. Using the circuit of Fig. 1, power gains of over 20 dB have been obtained. Units have been operated as amplifiers at frequencies up to 10 megacycles.

We wish to acknowledge our debt to W. Shockley for initiating and directing the research program that led to the discovery on which this development is based. We are also indebted to many other of our colleagues at these laboratories for material assistance and valuable suggestions.

Figure 4.18 The paper published by Bardeen and Brattain debuting their invention of the point-contact transistor to the world. They acknowledge Shockley's contribution in the last paragraph. [Reprinted with permission from J. Bardeen and W. H. Brattain, "The transistor, a semiconductor triode," *Physical Review*, Vol. 74, no. 2, pp. 230–231, July 15, 1948. Copyright (1948) by the American Physical Society.]

Geek Trivia

The name "transistor" was coined by J. R. Pierce, following an office ballot (betting pool?!) – all cool widgets MUST have cool widget names! He started with a literal description of what the device actually does electrically, a "trans-resistance amplifier," which he first shortened to "trans-resistor," and then finally "transistor" [5]. *Finito.*

After an additional week of experimental refinement, the formal transistor demonstration was repeated for Shockley's semiconductor group (Gibney, Shockley, Moore, Bardeen, and Pearson, with Brattain at the wheel) and importantly, two of the Bell Lab's "Brass" (a research director, R. Bown, and Shockley's boss, H. Fletcher – ironically, Kelly was not informed for several more weeks), on Tuesday afternoon, December 23, 1947, just before everyone left for the Christmas holidays (it was already snowing), which is now regarded as the official date stamp for the invention of the transistor and the birth of the Information Age [5,9].

In a telling comment, Shockley later said, "My elation with the group's success was balanced by the frustration of not being one of its inventors" [5,8]. It is ironic, then, that the most famous picture of the event contains all three men, with Shockley front-and-center hovering over the transistor as if it were his own baby (Fig. 4.19). On the famous portrait of the Transistor Three, "[John Bardeen]

Figure 4.19 The Transistor Three: William Shockley (seated), Walter Brattain (right), and John Bardeen (left) (photograph courtesy of Alcatel-Lucent).

said to me, 'Boy, Walter really hates this picture.' I said to him at the time, 'Why? Isn't it flattering?' That's when he made this face at me, and shook his head . . . He said to me, 'No. That's Walter's apparatus, that's our experiment, and there's Bill sitting there, and Bill didn't have anything to do with it.' " – Nick Holonyak, in an interview with Frederick Nebeker, June 22, 1993.

Meanwhile . . . Shockley spent New Year's Eve alone in a hotel in Chicago, where he was attending the American Physical Society Meeting. He spent that night and the next two days essentially locked in his room, working feverishly on his idea for a new transistor type that would improve on Bardeen's and Brattain's point-contact device. Why? Primarily because he felt slighted.[9] Shockley believed he should have received credit for the invention of the transistor, given that his original idea of manipulating the surface of the crystal set the stage for Bardeen's and Brattain's eventual success . . . and of course because he was the boss! Not surprisingly, the Bell Labs' lawyers (and management) didn't agree. And so, over New Year's holiday, Shockley fumed, scratching page after page into his lab notebook, primarily focusing on his new idea for constructing a transistor from a "sandwich" of different semiconductor layers all stuck together. After some 30 pages of scribbling (e.g., see Fig. 4.20), the concept hadn't quite gelled, and so Shockley set it aside.

As the rest of the semiconductor group worked feverishly on improving Brattain's and Bardeen's point-contact transistor, Shockley remained aloof, concentrating on his own ideas, secretive to the extreme. On January 23, unable to sleep, Shockley sat at his kitchen table early in the morning when the lightbulb suddenly switched on. Building on the semiconductor "sandwich" idea he'd come up with on New Year's Eve, he became quickly convinced that he had a solid concept for an improved transistor. It would be a three-layered device. The outermost pieces would be semiconductors with too many electrons (n-type), and the piece in the middle would have too few electrons (p-type). The middle semiconductor layer would act like a faucet – as the voltage on that part was adjusted up and down, he believed it could turn current in the device on and off at will, acting as both a switch and an amplifier.

Predictably, Shockley didn't breathe a word. The basic physics behind this semiconductor amplifier was very different from Bardeen's and Brattain's device, because it involved current flowing directly through the volume of the semiconductors, not along the surface, and Shockley proceeded to derive its operational theory. On February 18, Shockley learned that his idea should in fact work. Two members of the group, Joseph Becker and John Shive, were conducting a separate experiment whose results could be explained only if the electrons did in fact

---

[9] It has been pointed out that perhaps the most important consequence of the invention of the point-contact transistor was its direct influence on Shockley as an incentive to discovery. "This experience and the resulting emotion wound Shockley's spring so tightly that in the following 10 years, he was the author of the most remarkable outpouring of inventions, insights, analyses, and publications that [any] technology has ever seen. It is only a slight exaggeration to say that he was responsible for half of the worthwhile ideas in [all of] solid-state electronics" [4].

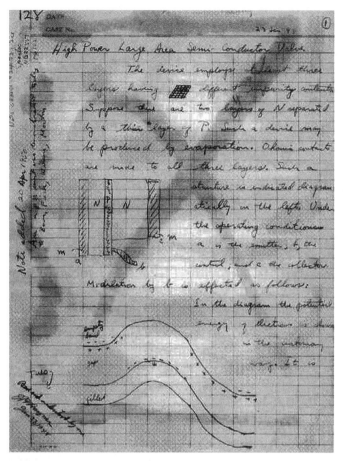

Figure 4.20   Page 128 of Shockley's technical journal, showing his idea of the bipolar junction transistor (photograph courtesy of Alcatel-Lucent).

travel right through the bulk of a semiconductor, the linchpin to the practicality of Shockley's idea. When they presented their findings to the rest of the group in their normal weekly team meeting, Shockley literally leaped from his seat and for the first time shared his idea for a "sandwich" transistor. It became painfully obvious to all that Shockley had been hoarding his secret for weeks and would likely have continued to do so for some time. In Shockley's own words, "Shive forced my hand" [4]. Bardeen and Brattain sat stunned as they realized they had been deliberately kept in the dark.

Needless to say, Shockley was quick to file the now-immortal patent for his "bipolar junction transistor" (BJT) (Fig. 4.21), U.S. Patent 2,502,488, filed in June of 1948, and issued April 4, 1950 ... yep, solo. (For some general discussion on patents and the process of filing a patent, see the addendum sidebar discussion at the end of the chapter). Shockley had the last laugh.[10] With both idea and a basic

---

[10] Interestingly, the BJT patent also almost casually mentions the basic idea of the "heterojunction" bipolar transistor using a wide-bandgap emitter (Si), with a narrow-bandgap base and collector (Ge), anticipating the now enormously important field of bandgap-engineered devices.

Patented Apr. 4, 1950   **2,502,488**

UNITED STATES PATENT OFFICE

2,502,488

SEMICONDUCTOR AMPLIFIER

William Shockley, Madison, N. J., assignor to Bell
Telephone Laboratories, Incorporated, New
York, N. Y., a corporation of New York

Application September 24, 1948, Serial No. 50,895

9 Claims. (Cl. 179—171)

1

This invention relates to means for and methods of translating or controlling electrical signals and more particularly to circuit elements utilizing semiconductors and to systems including such elements.

One general object of this invention is to provide new and improved means for and methods of translating and controlling, for example amplifying, generating, modulating, etc., electric signals.

Another general object of this invention is to enable the efficient, expeditious and economic translation or control of electrical energy.

In accordance with one broad feature of this invention, translation and control of electric signals is effected by alteration or regulation of the conduction characteristics of a semiconductive body. More specifically, in accordance with one broad feature of this invention, such translation and control is effected by control of the characteristics, for example the impedance, of a layer or barrier intermediate the two portions of a semiconductive body in such a manner as to alter advantageously the flow of current between the two portions.

One feature of this invention relates to the control of current flow through a semiconductive body by means of carriers of charge of opposite sign to the carriers which convey the current through the body.

Another feature of this invention relates to a body of semiconductive material, means for making electrical connection respectively to two portions of said body, means for making a third electrical connection to another portion of the body intermediate the first two portions, and circuit means including power sources whereby the influence of the third connection may be made to control the flow of current between the other connections.

A further feature of this invention resides in a body of semiconductive material comprising two zones of material of opposite conductivity type separated by a barrier, means for making external electrical connections respectively to each zone and means for making a third connection to one of the zones adjacent the barrier for controlling the flow of current between the other two connections.

Another feature of this invention involves a semiconductive body which may be used for voltage and power amplification when associated with means for introducing mobile carriers of charge to the body at relatively low voltage and extracting like carriers at a relatively high voltage.

2

Other objects and features of this invention will appear more fully and clearly from the following description of illustrative embodiments thereof taken in connection with the appended drawings in which:

The single figure shows in section a device suitable for practicing the invention, in connection with an appropriate circuit.

The device of this invention is in some respects like devices illustrated and described in the application of W. Shockley Serial No. 35,423 filed June 26, 1948. Reference is made to said application for a general description and discussion of pertinent principles, suitable materials and certain terms which may be employed in the description and claiming of this invention.

The terms N-type and P-type are applied to semiconductive materials which tend to pass current easily when the material is respectively negative or positive with respect to a conductive contact thereto and with difficulty when the reverse is true, and which also have consistent Hall and thermoelectric effects.

The conductivity type (either N- or P-type) of a semiconductor may be determined in one way by minute quantities of significant impurities as discussed in the previously noted application Serial No. 35,423. Energy relations within the semiconductor may also determine the conductivity type as more fully discussed in the application of J. Bardeen and W. H. Brattain Serial No. 33,466 filed June 17, 1948.

The term "barrier" or "electrical barrier" used in the description and discussion of the device of this invention is applied to a high resistance interfacial condition between contacting semiconductors of respectively opposite conductivity type or between a semiconductor and a metallic conductor whereby current passes with relative ease in one direction and with relative difficulty in the other.

The device shown in the figure comprises a body or block of semiconductive material, for example germanium, containing significant impurities. The block comprises two zones 10 and 11 respectively of N- and P-type materials separated by the barrier 12. The opposite ends of the block are provided with relatively large area connections 13 and 14 respectively, which may be metallic coatings such as solder, vapor deposited metal coatings, electroplated metal coatings or the like. Means for making connection to one of the zones closely adjacent to the barrier 12 may comprise a point contact 15 bearing on the surface of the block close to the barrier. The contact

Figure 4.21   Page 1 of Shockley's seminal bipolar junction transistor patent, U.S. Patent 2,502,488, issued April 4, 1950 (courtesy of the U.S. Patent Office).

theory of operation in place, it is ironic that the requisite art of making Shockley's BJT did not yet exist, and thus the BJT, ultimately a far more manufacturable and important transistor than the original point-contact device, was not actually demonstrated until April 12, 1950, and the article published in 1951 (Fig. 4.22) [13].

Brattain, Bardeen, and Shockley (justifiably) shared the Nobel Prize for physics in 1956 for their collective contributions in inventing their respective transistors (my guess is that they didn't chum around afterward in a three-way backslapping celebration!). Post transistor, Brattain went on to finish his career at Bells Labs, but Bardeen left Bell Labs in 1951, anxious to pursue different theoretical interests,

Figure 4.22   The first BJT (photograph courtesy of Alcatel-Lucent).

culminating in his *second!* Nobel Prize for his theory of superconductivity (a still-standing record).

4.5 Newsflash!

After Bardeen and Brattain's transistor patent was safely filed on June 17, 1948, Bell Labs finally prepared to announce their landmark discovery to the world. The announcement was to be two-pronged assault: (1) In the last week of June Shockley called the editor of *Physical Review*, and told him he was sending him three papers that (hint, hint) needed to appear in the July 15 issue (this less-than-three-week publication cycle must be a record for a technical journal). One was the fundamental transistor paper [3], one offered an explanation the observed transistor action (also by Brattain and Bardeen), and a third (now largely forgotten) was authored by Pearson and Shockley (go figure). (2) Bell Labs scheduled a major press conference in New York for June 30. Not surprisingly, a mandatory military briefing was held first on June 23, with all in attendance forced by Bell Labs President Oliver Buckley to raise their right hands and swear that they would say nothing until June 30 (I'm not joking). The plot thickens. On Friday, June 25, Admiral Paul Lee, Chief of Naval Research, called Buckley and told him that the Navy had in fact already made a similar discovery, and he wanted to make the June 30 spectacle a joint Bell Labs–U.S. Navy announcement. Major panic attack! The next day, Shockley and company, armed with a cadre of patent lawyers in tow, met with the Navy, and after an intense line of questioning of engineer Bernard Salisbury by Shockley (you can easily visualize this!), the Navy backed off, ultimately conceding that the Salisbury gadget was not in fact a transistor, and that the Bell Labs' team had been first. Whew!

The press conference went on as planned as a Bell Labs' solo act. Sadly, however, with Bown making the formal announcement and the ever-entertaining Shockley selected for fielding all of the reporter's questions, the seminal contributions of Bardeen and Brattain were not prominently featured in the limelight. Reactions from the popular press to the transistor were decidedly lukewarm, but Bell Labs was not done. The next step, in hindsight exceptionally shrewd, was to send out hundreds of letters, not to the press, but to scientists, engineers, and radio manufacturers, inviting them to Bell Labs on July 20 for a demonstration of the new widget. The reception to this event was anything but mixed. Wowed by the discovery, requests for sample transistors began pouring in from all over. Kelly once again acted wisely, quickly forming a brand-new Bell Labs group, headed by Jack Morton, aimed explicitly at developing the technological infrastructure for manufacturing transistors in volume – the requisite "art," if you will. By mid-1949, after significant sweat, over 2,700 "type-A" point-contact transistors had been produced and distributed to interested parties [9].

Fundamental problems with transistor reliability (at the time little better than their vacuum tube competition) and control of performance parameters, however, soon forced Morton's team to embrace Shockley's BJT as the preferred transistor architecture for mass production (surely Shockley was smiling by now!), and progress then came rapidly. Western Electric was the first company selected to manufacture transistors, paying a $25,000 patent licensing fee for the privilege. All companies agreeing to pay this (in hindsight ridiculously cheap) fee were invited to attend an exclusive "find-out-how-we-make-em" meeting at Bell Labs in the spring of 1952, and representatives from 26 U.S. and 14 foreign companies were in attendance. 1953 was proclaimed "The Year of the Transistor" by *Fortune* magazine. To place things in perspective, the combined transistor production from U.S. companies in 1953 was about 50,000 a month, compared with nearly 35 million vacuum tubes, but *Fortune* already glimpsed a bright future for the transistor and the potential for producing "millions a month" [9]! The selling price for a single transistor in 1953? About $15.

End of story? Nope. Read on.

4.6 How the West was won

By any measure, the ever-flamboyant Shockley charged ahead full steam, pulling much of Morton's team and the rest of Bell Labs with him, near-single-handedly developing the requisite know-how needed to both understand and produce practical transistors (his BJT of course). Ultimately, he also influenced many key developmental aspects of the fledgling microelectronics industry. William Shockley: larger than life, extraordinarily brilliant, workaholic, the quintessential scientist–engineer–inventor amalgam [5]. Imagine: Shockley tooled around New Jersey in a two-seater, open-cockpit British MG sports car, dabbled in magic as a hobby (he once pulled a bouquet of flowers from his sleeve at a technical meeting), loved

Figure 4.23   Current photograph of the building where the Shockley Semiconductor Laboratory started operation in 1955 in Palo Alto, CA, effectively launching what came to be known as "Silicon Valley."

to coin aphorisms, and quite ably embodied the famous "3C" mentality many at Bell Labs espoused (capable, contentious, and condescending) [14]. Capable he definitely was. Worshipped by some, hated by many, discounted by none. Intense in personality style, but importantly also lucid in his technical explanations, he was always willing and able to champion a good idea (preferably his own). Two key Shockley contributions of this time frame to the microelectronics fabrication "art" include the now-pervasive ideas for using "photoresist" for stenciling patterns on semiconductors (modern-day photolithography) and using ion implantation (essentially a mini particle accelerator) as a means to controllably introduce impurities into semiconductors (both are discussed in Chap. 7) [5].

Never satisfied with his many technical successes, Shockley ultimately opted to leave Bell Labs in 1955, in pursuit of business entrepreneurship, presumably to "make his million." Just south of San Francisco, Shockley assembled a collection of bright, young, like-minded scientist–engineer–inventors (think Robert Noyce, Gordon Moore – only one of whom was older than 30!), and in February of 1956 started his own company, aptly named (surprise!) "Shockley Semiconductor Laboratory." Shockley's first digs were in a converted Quonset hut located at 391 South San Antonio Road (really!), leased for $325 per month, but he soon moved into Stanford University's Industrial Park, in Palo Alto, in what would soon become known as "Silicon Valley" (Fig. 4.23). Noyce later related what it was like getting a call from Shockley to join his new team: "It was like picking up the phone and talking to God."

Transistor God? Perhaps. Business mogul? . . . alas, no (pay attention, students, there is an important lesson to glean!). With his frighteningly talented young minds assembled, Shockley made his first (and fatal) miscalculation: He decided

Figure 4.24   Current photograph of the Fairchild Semiconductor building in Palo Alto, CA, which produced the first silicon integrated circuits.

that his first product would not be the logical choice – his own BJT – but instead, a two-terminal, four-layer *p-n-p-n* diode (the precursor of the modern thrysistor). His underlings (correctly) argued that circuit designers would much prefer a (three-terminal) BJT for both switching and amplification needs. Shockley disagreed, and, predictably, his word was final. After a frustrating two years, eight of the Shockley Lab employees obtained independent backing from Sherman Fairchild, and launched Fairchild Semiconductor, also in Palo Alto (Fig. 4.24). Shockley began lovingly referring to them as the "traitorous eight" [5]! Shockley's lack of "people skills" in handling his employees is now legend. Throwing in the towel, he left his company in 1959 to join the faculty at Stanford University, and in April of 1960, Clevite Transistor purchased his brainchild, and "losses from Shockley in the remaining nine months will amount to about $400,000" [9]. Read: Not pretty.[11]

Some morals to the story? Sure.

• Technical smarts and business acumen do not necessarily go hand in hand.
• People skills matter.
• Stop and seriously consider the opinions of those who may disagree with you. They may have a valid point.
• Ego can be a double-edged sword (Hint: An overindulged ego can most definitely do you in).

---

[11] It is a sad to note that, for all of his brilliance and the clearly seminal role he played in the development of the transistor and microelectronics, Shockley is perhaps best remembered for his sadly misplaced, but predictably strongly held, conviction of the causal link between race and intelligence. Ostracized, and increasingly lonely, Shockley died of prostate cancer in August of 1989.

4.7 The integrated circuit

Not surprisingly, once transistor technology fell into the hands of business-savvy scientist–engineer–inventors, the field took off like gangbusters. Fairchild Semiconductor soon prospered by hitching their wagon to the BJT. Meanwhile, Gordon Teal, an early Shockley collaborator, left Bell Labs in 1952 for family-health reasons and settled back in Texas, joining a then-obscure geophysical company named . . . Texas Instruments (aka TI). Teal went to work on transistors, and by 1954 TI had a functioning germanium BJT production line, which they used to produce the world's first commercial "transistorized" radio, the Regency. Also in 1954, Teal attended a technical meeting in which a series of speakers waxed poetic on the impossibility of moving from germanium to silicon as the preferred semiconductor for transistor manufacturing. With dramatic flare, when Teal came up to speak, he promptly tossed a handful of the first silicon transistors onto the table and introduced the method of making them, the Bell-Labs-invented Teal–Little process of growing junctions from a semiconductor "melt" [5]. TI was the first to step onto the now-pervasive silicon playing field. In hindsight, this proved to be a giant leap for microelectronics commercialization.

By the late 1950s, rumors abounded on the possibility of making a complete "circuit in silicon" – that is, fabricating multiple transistors on a piece of silicon and connecting them together with deposited metal "wires" to form a complete "monolithic" (Greek, meaning "single stone," i.e., in one piece of semiconductor) "integrated circuit" on a single piece of silicon. The now-famous IC. Independent and parallel efforts scaled the IC mountain. Jack Kilby at TI came up with the idea of making an IC in silicon and importantly realized that integration of a practical electronic circuit would require building the requisite passive elements (resistors, capacitors, and inductors) in the silicon itself. On September 12, 1958, he succeeded in building the first integrated digital "flip-flop" circuit.[12] Meanwhile, Robert Noyce at Fairchild focused on the best way to create metal interconnections for building an IC with many components, introducing the "planar process" for doing so on January 23, 1959 (Fig. 4.25). Predictably, an ugly patent dispute emerged (Noyce's more narrowly focused IC patent was actually issued first, on April 25, 1961), but after a protracted legal battle (go figure), both Kilby and Noyce are today (fairly) recognized as the co-inventors of the integrated circuit.

By March of 1961, Fairchild had introduced a series of six different ICs and sold them to NASA and various equipment makers for $120 each [9]. In October of 1961, TI introduced its Series 51 Solid Circuits, each of which had roughly two dozen elements. Later that month they introduced a "minicomputer" consisting of 587 such Series 51 ICs. This primitive "computer" (little more than a calculator) weighed in at only 280 g and was the size of a sardine can [9]. Contrasting that achievement with ENIAC, the world's first (tube-based) digital computer is instructive (see Fig. 4.26). The world most definitely took

---

[12] Kilby shared the Nobel Prize in physics for this work in 2000.

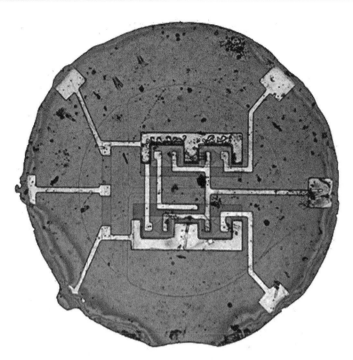

Figure 4.25   One of the first planar ICs produced at Fairchild, in this case a silicon bipolar digital "flip-flop" circuit (photograph courtesy of Fairchild Camera and Instrument Corporation).

Figure 4.26   The first digital computer, ENIAC. Shown are Glen Beck (background) and Betty Snyder (foreground), programming ENIAC in BRL building 328 at the University of Pennsylvania. Photo taken by K. Kempf (courtesy of U.S. Army).

note. *Business Week* observed that "The impending revolution in the electronics industry [due to the invention of the IC] could make even the remarkable history of the key transistor look like the report of a preliminary skirmish." Importantly (and often overlooked), in May of 1961, then-President John F. Kennedy inadvertently created an near-instantaneous IC market with the announcement of his vision to put a man on the Moon. Clearly, size and weight were *the* deciding factors for spaceflight, and the IC would thus necessarily play a pivotal role. The IC race was on!

Not Invented Here

It is with sad irony that I point out that Bells Labs never succeeded as a major player in IC technology. Why? They fell victim to their own 3C mentality. Jack Morton, the capable (and strong-willed) leader of the Bell Labs transistor effort, squashed all attempts at monolithic integration. Morton argued forcefully (and he was the boss – sound familiar?) that transistor integration was a fundamentally terrible idea. People were well aware by this time of the so-called interconnection "tyranny of numbers." That is, as more and more discrete transistors were connected together to build a useful electronics system, the wiring alone became a reliability nightmare. The IC was one approach to solve this; namely, move the bulky wires down onto the silicon itself. Morton considered this a flat-out dumb idea. He reasoned that if the yield (percentage of functioning devices out of the total we make) was, say, only 50%, then if we make 10 transistors on the same piece of silicon to form an IC, the net yield of the final circuit would only be 0.5^{10}%, about 1/10th of 1%, hardly worth the effort. Bell Labs laughed (often loudly) as the other players in the IC field pushed forward, supremely confident in their reasoning. They were profoundly wrong. As pointed out in [5], Bell Labs was ultimately convicted by an NIH mentality – Not Invented Here (read: must not be a good idea if we didn't think of it first). *C'est la vie!*

4.8 The rest of the story

To celebrate its 35th anniversary in 1965, *Electronics* magazine asked the CEO of the newly formed IC player, Intel, to write an article speculating on the future of IC technology. Noting that the complexity of ICs since their introduction in 1962 had doubled yearly to 50 components in 1965, he saw no fundamental reason why this trend should not continue indefinitely, projecting that this (exponential) growth rate should enable over 65,000 components per IC by 1975! Finding no physical reason why this level of complexity could not be achieved in practice, he further daringly proclaimed that this exponential growth in ICs would be *likely* to happen. He concluded that, "The future of integrated electronics is the future of electronics itself," ... ultimately making "electronic techniques more generally available throughout all society," and leading to "such wonders as home computers" ... and "portable communications equipment" [15]. A

Figure 4.27 The first microprocessor, the Intel 4004 (courtesy of Intel Corporation). (See color plate 14.)

remarkably prescient vision for only 3 years after the first IC demonstration. That man? Yep, Gordon Moore. You know, Moore's law. It was no real wonder then that Intel Corporation announced in 1971 the world's first "microprocessor" IC, the Intel 4004 (Fig. 4.27). The Intel 4004 represents the birth of the modern computer.

The rest is history. From transistor in 1947 . . . to integrated circuit in 1958 . . . to microprocessor in 1971. The game's afoot! A revolution is at hand! Let's push on.

Addendum: Filing a Patent

As I recall well, on the first hour of the first day of my first "real" job (at IBM Corporation in Yorktown Heights, NY) a form was put before me in which I essentially signed over any rights I had to any patentable invention I might conjure up in my working career with my new employer. My name would be on any patent I filed, yes, but they would own the rights to it. Refuse to sign? Only if you relished instant unemployment! The ubiquitous "patent," aka "intellectual property," aka "IP." High-tech industry very often lives (and dies) by its IP, and in many cases can be the bedrock upon which a new start-up company stands.

Edison has the record at 1,000 U.S. patents. I hold one U.S. patent[13] (and proud of it, thank you!). But what exactly is a patent? And what is involved in getting one? And why would one bother? Listen up, budding engineers and scientists, you need to know a few basics about patents!

In the words of the folks who control your IP future [12], "A 'patent' for an invention is the grant of a property right to the inventor, and is issued by the U.S. Patent and Trademark Office (USPTO). Generally, the term of a new patent is 20 years from the date on which the application for the patent was filed in the United States. The right conferred by the patent grant

(continued)

[13] J. D. Cressler and T. C. Chen, "Efficient method for improving the low-temperature current gain of high-performance silicon bipolar transistors," U.S. Patent 5,185,276, issued 1993.

Addendum: Filing a Patent *(continued)*

is, in the language of the statute and of the grant itself, 'the right to exclude others [herein lies its importance!] from making, using, offering for sale, or selling' the invention in the United States or 'importing' the invention into the United States. What is granted is *not* the right to make, use, offer for sale, sell or import, but, instead, is the right to exclude others doing so. Once a patent is issued, the patentee (you, or more commonly, your company) must enforce the patent without the aid of the USPTO." Enforcing a patent will inevitably mean a "patent infringement lawsuit," exceptionally costly, and no trifling matter for a cash-poor start-up company.

In the United States, there are actually three types of patents that can be granted: (1) "Utility" patents, which "may be issued to anyone who invents or discovers any new and useful process, machine, article of manufacture, or composition of matter, or any new and useful improvement thereof," (2) "design" patents, which "may be granted to anyone who invents a new, original, and ornamental design for an article of manufacture," and (3) "plant" patents, which "may be granted to anyone who invents or discovers and asexually reproduces any distinct and new variety of plant."

So how many patents are out there? As of 2005 the USPTO had granted 6,836,899 utility patents, 500,396 design patents, and 15,460 plant patents. These are tiny numbers compared with the actual numbers of patent applications filed. Read: It's not easy at all to get the USPTO to issue you a patent.

In the language of the USPTO statute, any person who "invents or discovers any new and useful process, machine, manufacture, or composition of matter, or any new and useful improvement thereof, may obtain a patent," subject to the conditions and requirements of the law. Interestingly, patent law specifies that the subject matter must be "useful," referring to the condition that the subject matter of the patent has a useful purpose (try defining that!). Hence, a patent cannot be obtained for a mere idea or suggestion. For an invention to be patentable it also must satisfy strict criteria for being "new," which provides that an invention cannot be patented if "(a) the invention was known or used by others in this country, or patented or described in a printed publication in this or a foreign country, before the invention thereof by the applicant for patent," or "(b) the invention was patented or described in a printed publication in this or a foreign country or in public use or on sale in this country more than one year prior to the application for patent in the United States. ...." In legalese, your idea must also be "non-obvious to one skilled in the art," meaning, if I am working in the same field as you (say, transistor design), and I am a *bona fide* expert, your idea cannot seem to me to merely be an "obvious" extension over the "prior art" (ideas that came before you). You also obviously have to be the first to come up with the idea. Remember Bell and Gray – hours matter! Moral: Most of us know Bell, very few of us know Gray. You are generally issued an "invention notebook" when you join a company, and you are encouraged to write down any ideas you may have, and have that entry "witnessed" by a colleague, who can then testify (in principle in a court of law) to the precise time and date stamp of your brainstorm.

Does determining what is and what is not patentable require a veritable army of technologically trained lawyers? You bet! And their hourly billing rate will likely make you blush. (FYI – "technologically trained lawyer" is not an oxymoron! A B.S. in engineering + a J.D. from law school is a hot ticket in the patent law field to seriously big bucks.) Although the USPTO charges only a few hundred dollars to actually file your patent, when all is said and done (assuming that rare success from start to finish), a patent can easily cost in the tens of thousands of dollars. A much cheaper alternative is therefore often used by companies. They will file a "Patent Disclosure" on your idea, which is the next best thing to owning the patent because, by disclosing your new idea to the public it becomes "prior art" and is hence no longer patentable by anyone else, effectively guaranteeing your right to use it (although you obviously cannot exclude anyone else from using it). One might imagine that large high-tech companies use their "patent portfolios" to beat up their competition by not allowing others to use their cool ideas. Typically this is not the case, because it is often prohibitively expensive (darned technologically trained lawyers!) to prove another company is "infringing" on your patent. Instead, Company A will often "license" their patent to Company B . . . yep, for a hefty fee, and this patent "royalty stream" becomes yet another (often large) source of revenue. Patents are definitely worth knowing about!

References and notes

1. "The View From The Top," *IEEE Spectrum,* pp. 36–51, November 2004.
2. See, for instance, *http://www.bellsystemmemorial.com/belllabs_transistor.html.*
3. J. Bardeen and W. H. Brattain, "The transistor, a semiconductor triode," *Physical Review,* Vol. 74, no. 2, pp. 230–231, 1948.
4. R. M. Warner, Jr., "Microelectronics: Its Unusual Origin and Personality," *IEEE Transactions on Electron Devices,* Vol. 48, pp. 2457–2467, 2001.
5. R. M. Warner, Jr. and B. L. Grung, *Transistors: Fundamentals for the Integrated-Circuit Engineer,* Wiley, New York, 1983. See especially Chap. 1, and the copious references within.
6. S. Weber (Editor), *Electronics,* Fiftieth Anniversary Issue, Vol. 53, 1980.
7. J. Konigsberger and J. Weiss, *Annalen der Physik,* Vol. 1, p. 1, 1911.
8. W. Shockley, "How we invented the transistor," *New Scientist,* p. 689, 1972.
9. M. Riordan and L. Hoddeson, *Crystal Fire,* Sloan Technology Series, 1998.
10. T. R. Reid, *The Chip,* Random House, New York, 2001.
11. P. K. Bondyopadhyay, P. K. Chaterjee, and U. Chakrabarti (Editors), "Special Issue on the Fiftieth Anniversary of the Transistor," *Proceedings of the IEEE,* Vol. 86, No. 1, 1998.
12. See, for instance, *http://www.uspto.gov/.*
13. W. Shockley, M. Sparks, and G. K. Teal, "p-n junction transistors," *Physical Review,* Vol. 83, p. 151, 1951.
14. J. L. Moll, "William Bradford Shockley, 1910–1989," in *Biographical Memoirs,* Vol. 68, National Academy, Washington, D.C., 1995.
15. G. E. Moore, "Cramming more components onto integrated circuits," *Electronics,* pp. 114–117, 1965.

5 Semiconductors – Lite!

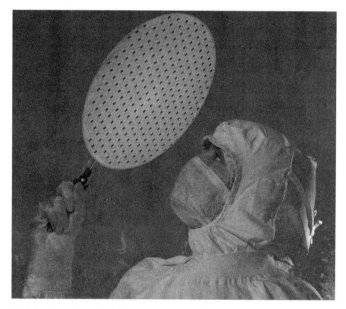

A fabrication engineer holding up a 300-mm silicon wafer. (Photograph courtesy of International Business Machines Corporation. Unauthorized use is not permitted.) (See color plate 15.)

> *Man's Mind Stretched To A New Idea*
> *Never Goes Back To Its Original Dimensions.*
> Oliver Wendell Holmes

The goal of this chapter can be simply stated: What is it about semiconductors that makes them so compelling for realizing the transistors that are driving our Communications Revolution? To answer this, we need to develop a language of sorts of semiconductors and a few important physical concepts. Can I promise minimal math? Yep. Can I assure you of conceptual simplicity? Well, yes and no, but I promise I will be gentle! Read on.

5.1 What are semiconductors?

From an electronics perspective, there are three broad classes of solid materials: (1) conductors, (2) insulators ("dielectrics" would be a slightly fancier, but

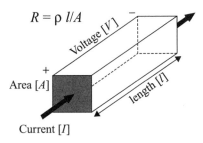

$$R = \rho \, l/A$$

Voltage [V]

+

Area [A]

length [l]

Current [I]

Figure 5.1

Illustration of a slab of semiconductor with the definition of resistivity.

equivalent, word), and (3) semiconductors. Conductors are aptly named – they are good conductors of electricity. Said another way, electrons can move freely through conductors in response to an applied voltage (V, in volts V), generating a current (I, in amps A), because current is simply a moving charge. Said yet another way, conductors do not provide much "resistance" (R, in ohms Ω) to electron motion in response to an applied voltage (V). This is conveniently captured by the pleasingly simple relation known as "Ohm's law," $V = IR$. Ohm's law states that the voltage across and the current through a block of material are linearly related through an innate property of the material known as resistance. Thus, if a very large current (I) flows in a piece of material in response to a very small applied voltage (V), the resistance must be very small. You have a conductor. Think metals: copper, gold, aluminum, silver. Not so fast, though. Resistance can be a tricky thing, because it depends on the dimensions of the material in question, and, as we have already seen, the dimensionality of our transistors is far removed from the scale of everyday life. I can, for instance, fashion a block of a decent conductor sufficiently small (yep, think microns) that it exhibits a large resistance, seemingly contradicting our notion of what a good conductor should be.

To be more precise in our classification of electronic materials, we instead define the "resistivity" (ρ) of the material, using the equation $R = \rho l / A$, where for a block of the material in question, l (in centimeters) is the length of the sample, and A (in square centimeters) is the cross-sectional area the electrons move through (Fig. 5.1). As can be easily verified, ρ has dimensions of Ω cm, clearly different from resistance. In this case, Ohm's law ($I = V/R$) can be nicely generalized to be $J = \mathcal{E}/\rho$, where J is the current density ($J = I/A$, i.e., current per unit area) and $\mathcal{E} = V/l$ is the electric field (i.e., voltage per unit length). You might legitimately ask, "Why bother?" Well, unlike for resistance, resistivity is an "intrinsic" parameter of the material (meaning independent of sample dimension). That is, if I take two pieces of the same material, they will both exhibit identical resistivity, even though their actual resistance can differ significantly (Fig. 5.2). Clearly, resistivity is a more general (hence better) parameter for defining electrical differences between various materials.

Given this definition, conductors always exhibit a "small" resistivity, regardless of their size, and in practice are typically less than about 1×10^{-3} Ω cm (1 mΩ cm). Insulators, on the other hand, do not conduct electricity well, and hence exhibit a large resistivity, typically greater than about 1×10^{5} Ω cm (100 kΩ cm). Semiconductors, then, are a unique class of materials exhibiting a range of resistivities between those of metals and insulators [1, 2]. That is, 1×10^{-3} Ω cm

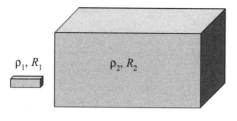

$$\rho_1 = \rho_2 \quad \text{BUT} \quad R_1 \neq R_2$$

Figure 5.2   Illustration of the independence of resistivity on sample dimensions.

(conductor) $< \rho_{semi} < 1 \times 10^5 \ \Omega$ cm (insulator). In short, semiconductors are defined to be those materials in nature that are semi-conductors; that is, "sort-of" conductors. Not a conductor, and not an insulator, but somewhere in between. So far so good.

5.2   What makes semiconductors so special?

Possessing an intermediate value of resistivity, in and of itself, suggests nothing with respect to why it is that semiconductors can be cleverly fashioned into useful devices. Why build our transistor from a semiconductor, and not a conductor, or even an insulator? Something must be missing. At the deepest level, the fundamental usefulness of semiconductors to realize a myriad of electronic and photonic devices is rooted in our ability to controllably manipulate its resistivity. With extreme precision. Across many orders of magnitude of resistivity. And importantly, we can do it in ways that naturally lend themselves to mass production in a manufacturing environment – read: cheaply. In point of fact, we can rather easily change a semiconductor from an insulator to a conductor . . . and vice versa! The same cannot be said for either insulators or conductors. Once a conductor, always a conductor. Not so for semiconductors.

In addition, we also have a veritable toolbox of means at our disposal to manipulate the resistivity of semiconductors. For example, we can (1) controllably introduce impurities into it ("dope" it), or (2) shine a light on it, or (3) apply a voltage to it, or (4) squeeze on it, or (5) change its temperature. Depending on what we are trying to accomplish, we may in fact use one or more of these means simultaneously to "morph" our semiconductor into an advantageous form to fit our immediate needs. One might logically wonder what it is about semiconductors that enables them to possess this key feature of a "tunable" resistivity. Ultimately it traces back to the existence of the so-called "energy bandgap" (E_g – "bandgap" for short), and its relative magnitude (not too big, not too small). The deepest physical underpinnings of the existence of the energy bandgap are . . . gulp . . . tied to the quantum-mechanical nature of electrons moving in a crystalline lattice. Sound scary? It's not too bad. Read on.

5.3 Types of semiconductors

Despite their uniqueness as electronic materials, semiconductors can come in an amazingly diverse array of flavors. There are both "elemental" and "compound" semiconductors, as well as semiconductor "alloys." Elemental semiconductors are easy. There is only one type of atom present in the crystal [e.g., silicon (Si) or germanium (Ge)]. In compound semiconductors there exists a definite ratio between the various elements present in the crystal. For example, in gallium arsenide (GaAs), there is a 1:1 ratio between gallium (Ga) and arsenic (As) atoms in the crystal. Compound semiconductors are further subdivided by the chemical "Group" they come from (think Periodic Table – see Fig. 5.3). There are Group III-V compound semiconductors (one element from Group III in the Periodic Table and one from Group V – e.g., GaAs), Group II-VI compound semiconductors (e.g., ZnO, zinc oxide), Group IV-VI compound semiconductors (e.g., PbS, lead sulfide), and even Group IV-IV compound semiconductors (e.g., SiC, silicon carbide).

In a semiconductor alloy, however, the ratios between the constituent elements can be varied continuously, forming a solid "solution" (one element is literally dissolved into another – slightly more energy is required than when dissolving salt into water, but the idea is the same). For example, in Si_xGe_{1-x}, x is the mole

Important Semiconductors

Group	II	III	IV	V	VI
		5 **B** boron	6 **C** carbon	7 **N** nitrogen	8 **O** oxygen
		13 **Al** aluminum	14 **Si** silicon	15 **P** phosphorus	16 **S** sulfur
	30 **Zn** zinc	31 **Ga** gallium	32 **Ge** germanium	33 **As** arsenic	34 **Se** selenium
	48 **Cd** cadmium	49 **In** indium	50 **Sn** tin	51 **Sb** antimony	52 **Te** tellurium
	80 **Hg** mercury		82 **Pb** lead		

Elemental

Si	silicon
Ge	germanium

Compound

IV-IV

SiC	silicon-carbide

III-V

GaAs	gallium arsenide
GaN	gallium nitride
GaP	gallium phosphide
GaSb	gallium antimonide
InP	indium phosphide
InAs	indium arsenide
InSb	indium antimonide

II-VI

ZnO	zinc oxide
ZnS	zinc sulfide
CdTe	cadmium telluride

IV-VI

PbS	lead sulfide
PbSe	lead selenide

Alloys

binary	Si_xGe_{1-x}	silicon-germanium
ternary	$Al_xGa_{1-x}As$	"al-gas"
	$In_xGa_{1-x}P$	"in-gap"
	$Hg_{1-x}Cd_xTe$	"mer-cat"
quarternary	$Al_xGa_{1-x}As_{1-y}P_y$	

Figure 5.3 Blowup of the Periodic Table of the elements, showing the domain of semiconductors. Also shown are the technologically important semiconductors, broken out by type.

fraction of the respective element (see sidebar) and can vary between 0 to 1; that is, from pure Ge ($x = 0$) to pure Si ($x = 1$) to, in principle, any composition in between. Thus in a $Si_{0.9}Ge_{0.1}$ alloy, there are nine Si atoms for each Ge atom (a SiGe alloy with 10% Ge), and this semiconductor is said to be a "binary alloy" (*bi* meaning "two" in Latin). Semiconductor alloys are especially important today because they enable us to create a virtually infinite number of "artificial" semiconductors (meaning, they cannot be found naturally occurring in nature – a rather sobering thought; more on this later), allowing us to carefully tailor the material to a given need or application. This process of conceiving and building artificial materials is called "bandgap engineering," for reasons that will soon be made clear, and is at the heart of both nanoelectronics and photonics. Practical semiconductor alloys can get even more complicated. For instance, $Al_x Ga_{1-x} As$ (aluminum gallium arsenide) is a "ternary alloy" (*ter* meaning "three times" in Latin). Those "in-the-know" would call this alloy "al-gas" (look again at the chemical composition and you will see why – zany material scientists!). Most of you own some "al-gas" – they are used to produce the laser diodes that might be found in your CD/DVD players and your laser printer. Try this one on for size: $Ga_x In_{1-x} As_{1-y} P_y$ (gallium indium arsenic phosphide) is a "quaternary alloy" (*quartum* meaning "fourth time" in Latin) and is used to make high-quality laser diodes used to drive long-haul fiber-optic transmission lines for Internet backbones.

Reminders From a Previous Life #1: The Mole

The ubiquitous mole of chemical fame. Formally, a "mole" (abbreviated mol) is the amount of substance of a system that contains as many elementary pieces as there are atoms in 0.012 kg of carbon-12 (the choice of which is an interesting story in itself – I'll save that for a rainy day!) The number of atoms in 0.012 kg of carbon-12 is known as "Avogadro's number" and is determined experimentally (6.022×10^{23} mol^{-1}). The relationship of the atomic mass unit (abbreviated amu) to Avogadro's number means that a mole can also be defined as that quantity of a substance whose mass in grams is the same as its atomic weight. For example, iron has an atomic weight of 55.845, so a mole of iron, 6.022×10^{23} iron atoms, weighs 55.845 g. Cool, huh? This notation is very commonly used by chemists and physicists. "Mole fraction" is the number of moles of a particular component of a given multicomponent material (e.g., a binary semiconductor alloy) divided by the total number of moles of all components.

Geek Trivia: The Mole

The name "mole" is attributed to Wilhelm Ostwald, who in 1902 introduced the concept of a mole as an abbreviation for molecule (derived from Latin *moles*, meaning "massive structure"). Ostwald used the concept of the mole to express the gram molecular weight of a substance.

5.4 Crystal structure

Interestingly, semiconductors such as silicon can be grown in three very different physical forms depending on the exact growth conditions used (e.g., by varying the growth temperature): crystalline, polycrystalline, and amorphous. As illustrated in two dimensions in Fig. 5.4, crystalline semiconductors have a well-defined periodic arrangement of atoms ("long-range" atomic ordering, typically over macroscopic dimensions); polycrystalline semiconductors ("poly" for short for those in-the-know) are composed of microcrystallite "grains" of crystalline material of typically submicron-sized dimensions, separated by "grain boundaries"; and amorphous semiconductors have a random arrangement of atoms (only "short-range" atomic ordering, typically over nanometer-sized dimensions). All three forms of semiconductors are in fact used in making modern electronic devices, but the cornerstone of the transistor world is ultimately the crystalline semiconductor, because only near-perfect crystals exhibit the requisite properties of a well-defined energy bandgap and a controllably tunable resistivity.

The periodic arrangement of atoms inside a crystal is called the "lattice" and can be one, two, or three dimensional (1D, 2D, or 3D, respectively), as illustrated in Fig. 5.5. From a mathematical perspective it is very useful to be able to generate the entire crystal lattice from a small subset of atoms, and this lattice "building block" is known as the lattice "unit cell," the edge length of which is known as the "lattice constant" a (careful: a is *not* the distance between atoms). The lattice constant of semiconductors is in the range of a fraction of a nanometer, or a "few" angstroms (abbreviated Å) (for instance, $a_{Si} = 0.543$ nm $= 5.43$ Å). One particularly relevant 3D unit cell is the so-called face-centered-cubic (FCC) unit cell, which can be used to construct a fcc lattice, hence fcc crystal, as shown in Fig. 5.5 (copper, for instance, crystallizes in a fcc lattice).

Technologically important semiconductor crystals such as silicon are more complicated in their atomic arrangements. Figure 5.6 shows the unit cell of the "diamond" lattice, the form in which silicon crystallizes (it is known as "diamond"

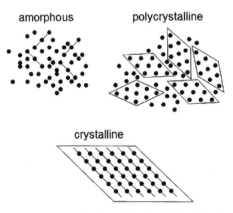

Figure 5.4   The different classifications of semiconductors based on their degree of atomic order: crystalline, polycrystalline, and amorphous.

1D lattice

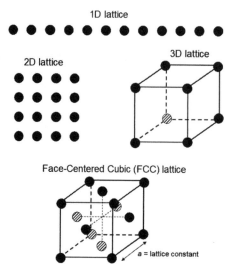

2D lattice

3D lattice

Face-Centered Cubic (FCC) lattice

a = lattice constant

Figure 5.5   The concept of 1D, 2D, and 3D lattices. Also shown is a 3D FCC lattice.

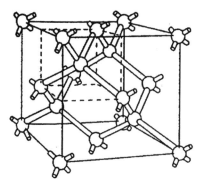

Figure 5.6   The unit cell of the diamond lattice, an FCC lattice with four interior atoms.

because it is of a form identical to that of crystallized carbon – our jewel diamonds). Figure 5.7 shows a famous end-on view through the diamond lattice, down one of the lattice "channels." The diamond unit cell is actually formed by two interpenetrating FCC unit cells, such that we end up with an outer FCC unit cell shell and four additional interior atoms (Fig. 5.6). Semiconductors such as GaAs crystallize in the "zincblende" lattice, which is identical to the diamond lattice, but with As as the exterior FCC atoms and Ga as the interior four atoms. Clearly the particular structure into which a given element crystallizes depends on its valence (outer-shell) electrons and the nature of those respective chemical bonds formed during that crystallization process. Silicon (atomic number $= 14$) has a $1s^2\, 2s^2\, 2p^6\, 3s^2\, 3p^2$ electronic orbital configuration (recall your chemistry!), with four valence electrons ($3s^2\, 3p^2$) covalently bonding with four neighboring silicon atoms (by means of a so-called "shared electron bond"), yielding a trigonal symmetry for the lattice that is exceptionally stable. (Recall: A diamond crystal is the hardest known substance – silicon is not far behind.)

5.5   Energy bands

Good. Semiconductors form crystals of a unique construction. It should not surprise you then that the physical structure of crystalline semiconductors

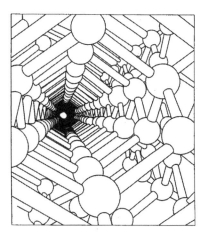

Figure 5.7   An end-on view through the silicon lattice. (After L. Pauling and R. Hayward, *The Architecture of Molecules*, Freeman, San Francisco, 1964).

(i.e., its diamond lattice configuration) is intimately tied to its resultant electronic properties, and particularly the existence of our requisite energy bandgap needed for building transistors. So, just how does the energy bandgap come to be? Hold on to something tightly, and read on!

To begin, we must first develop a simplified "model" that qualitatively captures the essential physics of a semiconductor crystal, but that also allows us to avoid the complex reality facing us; namely, at its core, a semiconductor crystal is an exceptionally complicated quantum-mechanical system with a number of components equal roughly to Avogadro's number (there are 5×10^{22} silicon atoms in a 1-cm^3 block of crystal). Please refer to the sidebar discussion on scientific "models." We call this particular model for semiconductors the "energy band" model. I will often refer to it as a "toy model" to emphasize its limitations. As we shall see, however, it also has great utility for us to intuitively understand many complicated phenomena in semiconductors.

As illustrated in Fig. 5.8, let's imagine a 2D view of a simplified silicon lattice. Each silicon atom covalently bonds to its four neighbors, and each atom is a lattice constant away from its neighbors. The basic question is this: What energy values (refer to the sidebar on energy) can an electron assume as it moves in this crystal? Wow! Tough problem! Let's see what our toy model says.

From basic quantum theory, we know that the potential energy of an electron orbiting an atomic core, as a function of radial distance from the nucleus, is funnel shaped about the atomic core (there is an attractive atomic potential

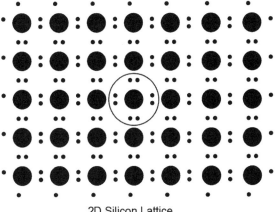

2D Silicon Lattice

Figure 5.8   Conceptual view of a hypothetical 2D silicon lattice, showing the four "covalent" (shared-electron) bonds.

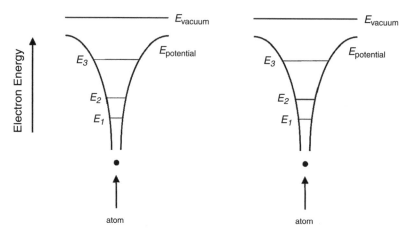

Conceptual view of the electron energy as a function of position for two isolated silicon atoms, each with discrete electron energy states.

between the electrons and the positively charged nucleus). We also know that the electron energies for each isolated atom are "quantized"; that is, they are discrete in value – an electron can have only certain precisely defined energies. Let's call those allowed energies E_1, E_2, E_3, etc. [we also define E_{vacuum} as a reference energy – for instance, releasing an electron from the atom (ionizing it) is equivalent to saying that we add enough energy to the electron to boost it from, say, E_3 to E_{vacuum}]. Quantum mechanics guarantees that Mr. Electron can be in "state" E_1 or state E_2, but nowhere in between. Both of these facts are conveniently represented in Fig. 5.9, which plots total electron energy as a function of position. If we now imagine moving two such isolated silicon atoms closer together, at some distance close enough (e.g., the lattice constant) for the atoms to "feel" each other's presence, the discrete energy states of the isolated atoms merge together to form electron states that are shared between both atoms (Fig. 5.10). The formerly discrete electron energy state tied to its host atom has morphed.

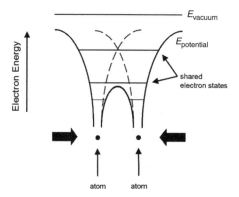

Conceptual view of the electron energy as a function of position for two silicon atoms brought into close proximity, now with shared discrete electron energy states.

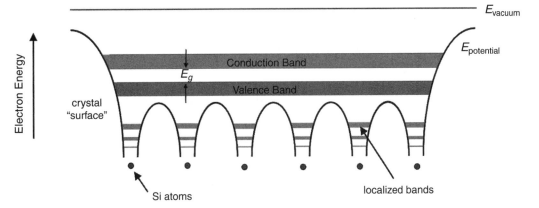

Figure 5.11 Conceptual view of the electron energy as a function of position for a semi-infinite 1D lattice of silicon atoms, now with both localized electron energy bands and, importantly, extended valence and conduction bands.

Fine. Now, if we imagine assembling an entire 3D lattice of such atoms (all 5×10^{22} of them), these shared electron states will broaden in energy to form extended "energy bands," consisting of a huge number (again, think something like 5×10^{22}) of allowed electron states in each band that stretch spatially across the crystal. Importantly, there is no barrier for an electron in the conduction band to move, say, from point A to point B – the electron states in the band are said to be "extended" states. In addition, a range of forbidden electron energies between the two bands remains.

This creation of energy bands is illustrated for a 1D cut in Fig. 5.11. The topmost extended band of electron states is called the "conduction band," and the band right below it is called the "valence band." The forbidden range of electron energies is called the "forbidden energy bandgap" or typically just "bandgap" for short. Our energy band model for semiconductors essentially begins here, and we now cut and paste the extended energy band view from the whole crystal, and instead focus on only the elements that are needed to understand basic transistor operation – this is the so-called "energy band diagram," as depicted in Fig. 5.12. The bottommost energy of the conduction band (known as the conduction band

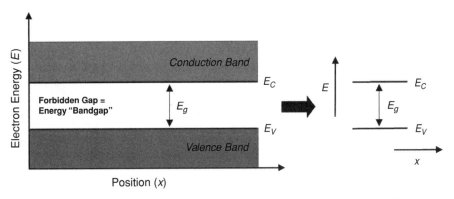

Figure 5.12 The energy band diagram: electron energy as a function of position as we move through the crystal.

"edge") is labeled E_C (E for energy, C for conduction band), and the uppermost energy of the valence band (the valence band "edge") is labeled E_V. The bandgap is labeled E_g and is clearly given by $E_g = E_C - E_V$. Bamm! We're in business.

Caffeine Alert #1 (Warning – Read at Your Own Risk!): Band Structure

As stated, the energy band model has great utility, but is nonetheless a toy model; an oversimplification. If we take a deeper look at this problem (i.e., use some real quantum mechanics on it!), we'll see that the *real* origin of energy bands and the companion bandgap we care about lie at the very deepest level with the movement of electrons in a system of periodic potentials (the atomic cores – our lattice). Provided the arrangement of forces (potentials) acting on an electron is periodic (regularly spaced), quantum mechanics will guarantee that electron energy bands will inevitably result. The classic solution to the band structure problem was first offered by Kronig and Penney in 1931 [4]. One can prove in fact by means of quantum mechanics that electrons moving in *any* periodic system will exhibit a band structure. Said another way, *all* crystals (semiconductor, metal, or insulator) will exhibit band structure. In this context, when I say "band structure," I mean more than simply the energy band diagram. Here, the resultant range of electron energies within the conduction band depends on the so-called "wave vector" of the electron. The electron wave vector ($\vec{k}$) is related to its quantum-mechanical wavelength (an electron is a particle wave), and that wavelength depends on the direction in the crystal in which the electron is actually moving. This directionality dependence shouldn't surprise you really, because the 3D silicon lattice is far from isotropic (that is, it differs in atomic arrangement depending on the direction you move within the lattice). Hence the electron's energy actually varies with direction. The energy band structure of silicon, for two technologically important directions, is shown in Fig. 5.13. It is the lowest energy in the conduction band (for all $\vec{k}$ directions) that defines E_C and the highest energy in the valence band (for all $\vec{k}$ directions) that defines E_V. As can be seen, the actual bandgap occurs for silicon in the Γ–X direction. If the maximum of the valence band and the minimum of the conduction band occur at the same value of $\vec{k}$, then the semiconductor is said to have a "direct" energy bandgap (GaAs is a good example); whereas, if they occur at different values of $\vec{k}$, the semiconductor has an "indirect" bandgap (Si is a good example). This direct vs. indirect bandgap difference is important and largely differentiates semiconductors that are better suited for electronic (electron-manipulating) than for photonic (light-manipulating) applications. In short, it is far easier to get light (photons) out of a direct bandgap material than out of an indirect bandgap material. When we move from the full energy band structure to the simplified energy band diagram, we essentially throw out the wave vector directionality, lock down E_C and E_V, and then plot how they move as a function of position. That is, band structure plots E as a function of $\vec{k}$, whereas the energy band diagram plots E as a function of x.

It is in fact the magnitude of the bandgap that defines whether we have a metal, or an insulator, or a semiconductor (Fig. 5.14). Remember, all crystals, because they have periodic arrangements of atoms, will exhibit a band structure, but it is

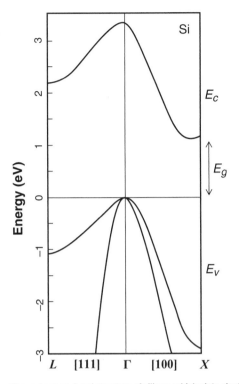

Figure 5.13   The real energy band structure of silicon, which plots electron energy as a function of the electron wave vector for two technologically important directions.

the resultant magnitude of E_g that differentiates the three. Metals have overlapping conduction and valence bands ($E_g = 0$), semiconductors have modest-sized bandgaps (typically in the vicinity of 1 eV – for silicon $E_g = 1.12$ eV at 300 K), and insulators have large bandgaps (typically > 7–8 eV).

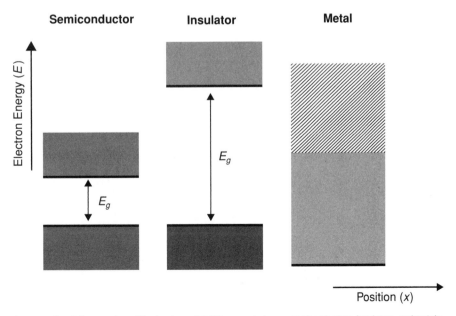

Figure 5.14   An energy band diagram view of the fundamental differences between semiconductors, insulators, and metals.

Models of Reality

When I was a kid, I loved to build model airplanes (Revell 1/32 scale World War II airplanes to be precise). The combination of youth, an active imagination, a well-constructed ME-109 and B-17, and access to firecrackers and smoke bombs (okay, and perhaps a little too much freedom!), made for some particularly engaging adolescent war games, but that is the subject for another day!

In science and engineering the formal notion of a "model" is important, and is alas often glossed over, so let's revisit that concept momentarily. A 1/32 scale ME-109 "model" is a scaled version of the real deal, but alas, built of plastic, glue, and paint. It has utility in the sense that it is not overly expensive, it can be fun to put together, I can hold the resultant masterpiece in my hand, play with it, etc. (okay – if it loses the dogfight with the Spitfire, it might even crash and burn!). In a similar manner, we invent models in science and engineering that provide some useful function. Generally, scientific models are simplified artificial constructs that are conceived for their utility in helping us understand very complex physical phenomena. Such models are often mathematical in form, because (a) math is an exceptionally powerful tool for reducing complex patterns and processes into fairly "simple" sets of descriptive equations, and (b) math is also inherently well equipped for prediction (e.g., by running the time variable in my equations forward on a computer).

How are models used in practice? Consider, for instance, the growth of a sunflower to maturity from its seed. This seemingly simple act clearly involves an exceptionally complex set of physical phenomena with hundreds of variables, and thus developing a complete understanding of the process that can capture all of the nuances going on inside a particular young sunflower swaying in a breeze in a field in Umbria, Italy, on a sunny day, is hopelessly beyond our reach. We might, instead, create a simplified mathematical model involving the chemical rate equations of a set of steps in the sunflower's photosynthesis process that is key to its metabolism and growth, which I might then apply to better understanding a careful set of growth data taken on a statistically meaningful sample of sunflowers grown under tightly controlled conditions in my laboratory.

There are clearly levels of sophistication, and hence validity, of such scientific models. An overly simplified model might, for instance, be good only for intuitive understanding (e.g., my sunflower grows taller with water, sunlight, and time, and dies if I remove them), yet lack the sophistication of quantitatively predict measured data of the phenomena I am trying to understand (e.g., what is the precise amount of time it takes a sunflower to fully mature and offer its seeds and oil?). I might thus refine my model(s), generally increasing its complexity (e.g., by adding more variables), until it again proves its utility, allowing me to better tackle the problem at hand, and so on. This iterative modeling process is clearly at the core of the scientific method itself, and hence in no small way is responsible for the enormous fruits yielded by the scientific and engineering enterprises. Still, even at best case, saying that we have a "complete" understanding of *any* complex physical phenomena in nature should be taken with a grain of salt. Appropriate parsing of the problem at hand into smaller manageable pieces that can then be quantitatively modeled represents a true modern "art form."

Strictly speaking, are such models "true"? ("Truth," in this context, means objective reality; the real state of things.) They are not. After all, models are artificial, utilitarian constructs. Models are either valid or invalid, depending on whether their underlying assumptions are satisfied or not, and whether they accurately capture the reality they were built to mimic. As W.C. Randels once famously asserted, "The essence of engineering is knowing what variables to ignore" [3].

5.6 Electrons and holes

Great. Semiconductors have a unique crystalline structure that produces energy bands separated by a forbidden bandgap, the magnitude of which is especially conducive to building transistors. How does this help? Well, we said that the key to semiconductors lies with being able to easily and controllably manipulate their resistivity in various interesting ways. Let's now combine that notion with our energy band model. To do this we must talk a little about electrons and "holes" present in those energy bands. Figure 5.15 shows the conduction and valence bands of a semiconductor, but now with electrons present (shown schematically obviously – there are a bunch!). As we will see, temperature plays a pivotal role in distributing the electrons among the bands (see sidebar on temperature and temperature scales). One can prove in quantum mechanics that, at absolute zero, 0 K (see Geek Trivia sidebar), the valence band is completely full of electrons (meaning all of its available electron states are occupied), and the conduction band is completely empty (there are tons of electrons states, but no electrons). Furthermore, we can also prove that a filled energy band cannot conduct current. Reminder: At the most basic level, current is simply a moving charge ($I = C/s$, charge per unit time), or more generally, $J = qnv$, where J is the current density (in amps per square centimeter), q is the electron charge (1.602×10^{-19} C), n is the electron density (the number of electrons per cubic centimeter), and v is the average electron velocity (in centimeters per second). No current results if (a) there are no electrons present ($n = 0$), or (b) the electrons present cannot move ($v = 0$). It may seem surprising that no current can be generated by a

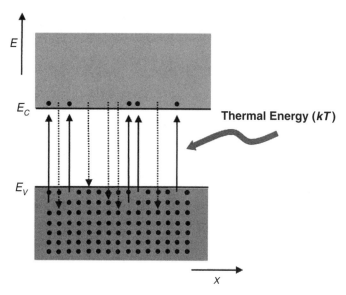

Figure 5.15 Illustration of thermal excitation of electrons from the valence band to the conduction band at temperature (T), followed by their return back to the valence band.

filled energy band, as there are tons of electrons present. Once more we have to invoke a quantum-mechanical explanation – an electron can move in nature only if there is a quantum state available for it to move into [the sacred "Pauli exclusion principle" states that no two fermions (an electron is a fermion) can occupy the same quantum state]. In this case, all electron states are filled and thus $J_{\text{valence band}} = 0$ at $T = 0$ K). At the same time, empty bands obviously have no charge to move, and thus $J_{\text{conduction band}} = 0$ at $T = 0$ K. Moral: Only partially filled energy bands conduct current. This would seem to make a semiconductor pretty dull – at absolute zero, it just sits there, having an infinite resistivity (i.e., it is a perfect insulator) because no current can flow!

It gets more interesting, however, when we warm up the semiconductor, say to "room temperature" (by convention, 300 K = 27°C = 80°F). This is equivalent to saying that we are adding "thermal energy" (kT) to the system (see sidebar discussions on temperature and energy). Adding thermal energy can now "boost" an electron in the valence band up to the conduction band (because there are empty electron states present above E_C). If you stop and think for a second, this fact should greatly surprise you. Why? Well, at 300 K, $kT = 0.0259$ eV (26 meV), and we know that the bandgap of silicon, for instance, is 1.12 eV at 300 K, or 1,120 meV. How can it be that adding 26 meV to the silicon crystal is sufficient to boost Mr. Electron across a bandgap (read: forbidden barrier) that is 43 times larger in magnitude (1,120/26)? Intuitively, this seems impossible. Answer: This is clearly a probabilistic phenomenon (heck, it's quantum mechanics, a world of probability!), meaning that only some small fraction of the electrons can actually make it from the valence band to the conduction band. We can calculate the number of "lucky" electrons with great precision, and from first principles, but this is fairly involved mathematically (if you're curious to know how, see the sidebar "Caffeine Alert #2").

Because the now "thermally excited" Mr. Electron sitting in the conduction is at a higher energy level, and nature prefers to return all systems to their lowest-energy configuration (the "ground state" – this is the way thermodynamics and equilibrium work), after some time t, Mr. Electron will drop back into the valence band. Although this is a dynamic process (always going on), at any given instant in time, there is now a finite density of electrons present in the conduction band (Fig. 5.15). This is important, because there are also plenty of empty electron states in the conduction band, and hence the electron is now free to move if acted on by an external force (e.g., an applied electric field), producing current (music to the ears of any electrical engineer!). Interestingly, in this scenario, the valence band is also no longer full, and hence there are now empty electron states in the valence band, and the electrons there can now also move, similarly producing current (yea!). Pause.

- Moral #1: At finite temperature, both the conduction band and the valence band contribute to the current flowing within a semiconductor, hence determining its resistivity.
- Moral #2: Each time an electron is boosted to the conduction band, it leaves behind an empty electron state in the valence band.

At temperature T there are only a "few" electrons in the conduction band, and hence we can easily keep track of them. Not so for the valence band. There are gazillions of electrons present, and just a very "few" empty electron states. We have a simpler alternative. Instead of counting the vast number of electrons in the valence band, let's instead count only the few empty electron states, because these are the ones that contribute to the current. (Note: Because the electron is negatively charged, an empty electron state is effectively positively charged.) This counting process is depicted in Fig. 5.16. We call the positively charged empty electron states "holes" (imaginative, huh?!), and we can show quite generally that, in terms of current, counting holes is equivalent to counting electrons, and far easier as there aren't many of them. We call the hole density p (number of holes per cubic centimeter). Although it may be tempting to believe that holes are some new type of charged particle capable of carrying current, strictly speaking, this is not true. The hole is really just an artificial construct invented to simplify our complex problem – yep, it's a model! To indicate this, we often refer to the hole as a "quasiparticle." Instead of tracking a gazillion electrons moving in the valence band in response to, say, an applied field, we can instead track just the hole motion because it is mathematically equivalent. In this electron + hole view of energy bands, note that, under equilibrium conditions, n and p are intimately linked – when I create one, I create the other. This dual electron–hole creation is called an "electron–hole pair" (don't worry, their union has been blessed!).

A semiconductor in which $n = p$ is called "intrinsic." Usually intrinsic semiconductors are ones that are in their most purified state, with no intentional

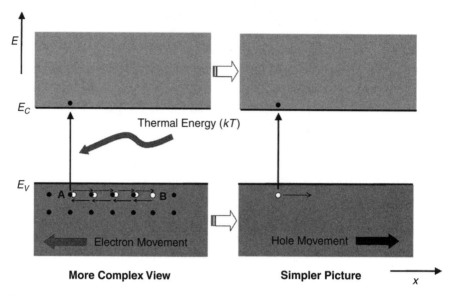

Figure 5.16 Illustration of concept of a "hole" that is free to move in the valence band.

impurities (dopants) present. It shouldn't surprise you that, under equilibrium conditions, $n = p$, and both n and p are functions of both E_g and T. We call this value of $n = p = n_i$ the "intrinsic carrier density." In silicon, for instance, at 300 K, $n_i = 1 \times 10^{10}$ cm^{-3}. Although it may seem as though n_i is a large number ($10^{10} = 10,000,000,000$ pennies is $100 million!), if we convert that dimensionally to cubic microns, that means that, in a transistor-sized 1.0-µm^3 cube of intrinsic silicon at 300 K, there are only 0.01 electrons and holes present (a pretty good insulator). Said another way, to find a single electron and hole in the conduction–valence band of intrinsic silicon at 300 K, I need to have a 100-µm^3 block; 100 times larger, say, than Mr. Transistor. Thus n_i in silicon at 300 K is actually pretty darn tiny in the context of the devices we want to build, and this is in fact important.

Now, under equilibrium conditions, the product of the electron and hole densities can be trivially written as $pn = n_i^2 = $ constant, once E_g and T are fixed. Here, pn is called the "pn product" and the equation is called the "law of mass action" (somewhat obscure sounding, but related to chemical rate equations – the fact that we call it a "law" should clue you in to its importance!). Why is this simple equation so useful? Consider: If I fix the temperature and the semiconductor, then n_i is a constant (i.e., known), and thus if I know n, say, but do *not* know p, I have a trivial prescription for finding it ($p = n_i^2/n$), and vice versa. The law of mass action is thus a clever linking equation between the electron and hole "populations." Know one, you know both.

If we imagine applying an electric field to a sample of an intrinsic semiconductor, we now know that the electrons and the holes can move with some velocity v in the conduction and valence bands in response to that field, generating useful current (they move in opposite physical directions because they are of opposite charge, but both add to the total current). Question: What is the electron's mass in this "transport" process? This might seem like a crazy thing to ask, but it's not. For a fixed applied force, how fast the electron is moving after time t (its speed) depends on its mass (think freshman physics). As we will see, this actually translates into practical speed limits of our transistors (e.g., how fast I can switch from a "1" to a "0" in a digital circuit and ultimately my microprocessor speed). Still puzzled? Consider: I have a BB and a bowling ball. I throw both at the wall. Which gets there first? Answer: The one with least mass. If in doubt, try it! So from a speed perspective, things which weigh less are definitely preferred, all else being equal. Back to the question. What do my electrons and holes in my semiconductor weigh? Easy answer, you say – "they weigh what all electrons weigh, the well-known 'free-electron mass' – $m_0 = 9.11 \times 10^{-31}$ kg, right?" Wrong!

"Why not?," you say. Well, electrons are quantum particles. Energy bands in semiconductors exist only because these quantum particles move in a periodic potential of quantum-sized dimensions. Message: This is as quantum mechanical a problem as you are likely to find! Don't be nervous. It is a pivotal, and frankly rather amazing, fact that we can collapse the entire intense quantum-mechanical nature of this problem of how electrons move in energy bands in semiconductors into a single parameter, the so-called "carrier effective mass" (written m^* by convention, and subscripted with n and p to indicate electron vs. hole, respectively – m_n^* and m_p^*), and then treat the electron or hole moving in the semiconductor as a regular billiard-ball-like particle moving semiclassically (read: Newton's equations), *not* with its normal mass, but with this new quantum-mechanical effective mass ($F = m^*a$). The carrier effective mass is not actually a mass at all in the strictest sense, but instead is a mathematical function of the local curvature of the energy band structure (see Caffeine Alert #1) where the electron sits. Because of the 3D nature of real crystals, m^* is actually a nine-element effective mass "tensor" with explicit directional dependencies, but we don't need to worry about such subtleties. Luckily, m^* happens to have units of mass. This simplification is clearly only an approximation to reality (yep, a model!) and can be violated. It is known as the "effective mass approximation," and were this dramatic simplification from full quantum mechanics to simple Newton's equations not possible, building transistors would likely be an intractable problem. It is important in the grand scheme of things. So exactly how different is the mass of Mr. Electron moving in a semiconductor from its free-electron mass? For most semiconductors, not very. For example, for silicon at 300 K, $m_n^* = 1.18m_0$ and $m_p^* = 0.81m_0$, a difference of only about 20%.

So ... now we know about electrons and holes in energy bands. Excellent. Pause.

- Moral #3: At finite temperature, electrons moving in the conduction band and holes moving in the valence band are responsible for the total current I measure in response to an applied voltage (hence resistivity), such that $J_{\text{total}} = J_n + J_p$.
- Moral #4: The intrinsic carrier density ($p = n = n_i$) is a key semiconductor parameter and is fixed once E_g and T are fixed. Such parameters are often

called "material parameters," because they are constants determined by the material in question. As we will see, the law of mass action ($pn = n_i^2$) is of great utility for semiconductor calculations.

- Moral #5: Electrons and holes in semiconductors move with a "carrier effective mass" (m^*), which embodies the profound quantum-mechanical complexities of the semiconductor band structure and is different from the free-electron mass.

Temperature and Temperature Scales

From an intuitive perspective, temperature is simply a measure of "hotness" or "coldness," and evolution has finely tuned the human body for both maintaining and sensing temperature. One might logically wonder why. In short, the physical properties of virtually all materials depend intimately on temperature, including phase (solid, liquid, gas, or plasma), density, solubility, sound velocity, vapor pressure, and electrical conductivity. In addition, chemical reaction rates depend exponentially on temperature, and temperature controls both the type and the quantity of thermal radiation emitted from an object. Said another way, everything that defines life as we know it depends on temperature.

Microscopically, temperature is the result of the motion of the particles that make up a substance, and an object's temperature rises as the energy of this motion increases. Formally, temperature is defined in thermodynamics by using fundamental definitions of entropy and heat flow, but clearly for quantifying temperature one needs a temperature "scale" with which to compare temperatures with some known value, and there have historically been many such temperature scales invented. For our purposes, the so-called "absolute" or thermodynamic temperature scale is preferred, and is measured in "Kelvins" (K) – pet peeve: *not* degrees Kelvin (°K), despite the many abuses that can be readily located in the literature. The Kelvin and Celsius temperature scales are uniquely defined by two points: absolute zero (0 K) and the triple-point temperature of water under standard conditions (solid, liquid, and gas all coexist at the triple point). Absolute zero is defined to be precisely 0 K and −273.15°C. Absolute zero is where all kinetic motion in the particles comprising matter ceases and is at complete rest in the "classical" sense (myth alert: quantum mechanically, however, all motion need not cease at 0 K). Thermodynamically, at absolute zero, matter contains no heat energy. The triple point of water is defined as being precisely 273.16 K and 0.01°C. This dual focal point (absolute zero and triple point) definition for a temperature scale is useful for three reasons: (1) it fixes the magnitude of the Kelvin unit to be precisely 1 part in 273.16 parts of the difference between absolute zero and the triple point of water; (2) it establishes that a 1 K increment has precisely the same magnitude as a 1° increment on the Celsius scale; and (3) it establishes the difference between the Kelvin and Celsius scales' focal points as precisely 273.15 K (0 K = −273.15°C and 273.16 K = 0.01°C). In science and engineering, $T = 300$ K is often chosen as "room temperature" for calculations (it's a nice round number), although at 80.33°F, it is actually fairly balmy.

For most of the world, the Celsius scale is the everyday temperature scale of choice, in which 0°C corresponds to the freezing point of water and 100°C corresponds to the boiling point of water (at sea level). In the Celsius scale a temperature difference of 1° is the same as a 1 K temperature difference, so the scale is essentially the same as the Kelvin scale, but offset by the temperature at which water freezes (273.15 K), making it very intuitive for relating real-life temperatures to those in scientific and engineering calculations. In the United States, the Fahrenheit scale is still widely used (go figure). On this scale the freezing point of water corresponds to 32°F and the boiling point to 212°F. The following useful formulae for converting between the Kelvin, Celsius, and Fahrenheit scales are

$$T(K) = T(°C) + 273.15, \qquad (5.1)$$

$$T(°C) = 5/9[T(°F) - 32], \qquad (5.2)$$

$$T(K) = 5/9[T(°F) + 459.67]. \qquad (5.3)$$

Geek Trivia: The Quest for Absolute Zero

From basic thermodynamics, one can prove that it is physically impossible to attain 0 K, but the race toward absolute zero has been afoot for some time now. The current record is 500 pK ($500 \times 10^{-12} = 0.000,000,000,5$ K), set by MIT researchers in 2003 using a "gravitomagnetic trap" to confine atoms. For reference, the coldest temperature recorded on Earth was $-128.6°F$, in Vostok, Antartica, on July 21, 1983, and the hottest temperature recorded on Earth was $+134°F$, in Death Valley, CA, on July 10, 1913. Stay tuned – with global warming this latter record seems doomed.

Caffeine Alert #2 (Warning – Read at Your Own Risk!): Calculating Carrier Density

Let's begin by asking a simple question. Exactly how many electrons are there in the conduction band of semiconductor X at temperature T? Wow! That feels like a tough problem. It is! Conceptually, though, we can state a solution *approach* quite easily, and it is a nice example of what is an exceptionally elegant (I would say beautiful!) solution to an exceptionally difficult problem. Don't worry, we'll save the math for another day – it is the approach that I am after, because it is nicely illustrative of the powerful interplay among mathematics, physics, and engineering for solving complex problems.

Here's the thought process. Please appreciate its appealingly simple and intuitive nature. For an electron to be present in the conduction band, there must be a quantum state for it to be in. There indeed exist a certain number of allowed quantum states in the band, but in point of fact, the density (number of states per unit volume) of those states depends on the electron energy itself. Let's call this the "density of states" function [$g_C(E)$]. Now, if there is a quantum state at energy E that is available to the electron, then there is some finite probability that the electron is in that state. That probability itself depends on the energy. Let's call that state-occupancy probability the "Fermi–Dirac distribution" function [$f(E)$]. The product of $g_C(E) f(E)$ thus counts an electron. That is, at energy E_1, g_C times f says "at energy E_1, there is a state, and that state is indeed occupied by an electron." Bamm. We've counted electron #1. Do this again at energy E_2. And again at E_3. Now, if we add up all of those g_C times f "hits" as we move upward in energy within the band, we will add up all of the electrons present. QED.[1] That total, that sum,

(continued)

---

[1] Acronym for the Latin phrase, *quod erat demonstrandum*, literally "this was to be demonstrated," supposedly first used by Euclid in his geometrical proofs.

is identically n, what we are after, and can be expressed mathematically as

$$n = \sum_{\text{band}} g_C(E) \, f(E). \tag{5.4}$$

Fine. Because in principle we are adding our hits up over a large number of energies, to make this mathematically simpler we take the "limit" as the delta between the energies gets vanishingly small and convert the summation into an integral (hint: Here is a good reason to study calculus) over the band, beginning from the bottom of the band (E_C) and going up to infinity (again, chosen only for mathematical simplicity – we will in fact count all of the electrons after going only a "few" electron volts into the band). Thus,

$$n = \int_{E_C}^{\infty} g_C(E) \, f(E) dE. \tag{5.5}$$

You see? Simple. Elegant. Slight problem – what are $g_C(E)$ and $f(E)$? Well, that is a tougher problem, but the answers can be deduced pretty much from "first principles" (meaning, from basic physics). Those derivations are beyond the scope of this discussion, but not beyond the capabilities of a bright senior-level undergraduate student. The answers are

$$g_C(E) = \frac{1}{2\pi^2} \left\{ \frac{2m_n^*}{\hbar^2} \right\}^{3/2} \sqrt{E - E_C} \tag{5.6}$$

and

$$f(E) = \frac{1}{1 + e^{(E - E_F)/kT}}. \tag{5.7}$$

As can be seen, both g_C and f depend on energy, as expected. The only newcomer to the party is E_F. This is a parameter called the "Fermi energy" (named for physicist Enrico Fermi), and it is defined to be that energy at which the electron-occupancy probability of the state is identically 1/2 (equally likely to be filled or empty). The shapes of these two functions, and their product, are illustrated in Fig. 5.17. Finding n is thus equivalent to finding the area under the $g_C(E)f(E)$ curve in the top right part of the figure. Thus we have a calculus problem before us of the form

$$n = \frac{1}{2\pi^2} \left\{ \frac{2m_n^*}{\hbar^2} \right\}^{3/2} \int_{E_C}^{\infty} \sqrt{E - E_C} \, \frac{1}{1 + e^{(E - E_F)/kT}} dE. \tag{5.8}$$

Sad to say, this definite integral cannot be solved in closed form. Booo. Hiss. Still, with some relevant physical approximations, a closed-form answer results, which is given by

$$n = 2 \left\{ \frac{m_n^* kT}{2\pi \hbar^2} \right\}^{3/2} e^{(E_F - E_C)/kT} = N_C \, e^{(E_F - E_C)/kT}, \tag{5.9}$$

where N_C is called the "effective density of states," a constant.

One could follow exactly the same procedure for finding the hole density in the valence band, resulting in

$$p = 2 \left\{ \frac{m_p^* kT}{2\pi \hbar^2} \right\}^{3/2} e^{(E_V - E_F)/kT} = N_V \, e^{(E_V - E_F)/kT}. \tag{5.10}$$

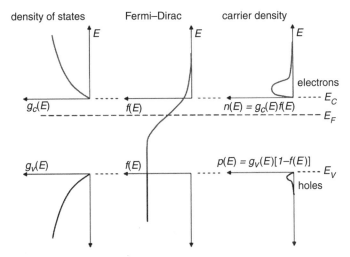

Figure 5.17   Illustration of process of calculating the electron density in the conduction band of an intrinsic semiconductor at temperature T.

A casual glance at the equations for n and p should raise an eyebrow. They are simple, yes. BUT, without independent information on what value E_F is, I cannot arrive at a numerical answer for the electron density, say 37 electrons/cm^3 or 1.5×10^{17} electrons/cm^3. Note, however, some magic here – n by itself requires E_F, and p by itself requires E_F, BUT, if I take the product of p times n, then

$$pn = N_C N_V e^{(E_F - E_C)/kT} e^{(E_V - E_F)/kT} = N_C N_V e^{(E_C - E_V)/kT} = N_C N_V e^{-E_g/kT} = n_i^2, \tag{5.11}$$

because $E_C - E_V = E_g$. Wah-lah! We are back to the law of mass action. The *product* of p and n does NOT require knowledge of the Fermi level, and hence we are now in business. Not so bad, huh?!

5.7 Doping

Intrinsic semiconductors, although convenient for defining some important concepts, don't actually do too much for us if we want to build real transistors. We can now find n and p, yes, but we need a way to *change* n and p. This is where "doping" comes in. In doping, we intentionally introduce impurities into the crystal, precisely changing the relative densities of electrons and holes (hence the sample resistivity). Doped semiconductors are said to be "extrinsic" and are the workhorse of the micro/nanoelectronics world.

Let's consider silicon as our test case. In one type of doping, we introduce a Group V impurity (e.g., arsenic, As) into the (Group IV) host (Si) (refer to Fig. 5.3). Assuming the impurity exactly replaces the silicon atom, the fact that the dopant is a Group V element means that there is one additional valence electron per atom that will be left over when the four covalent electron bonds are made to neighboring Si atoms (which need only four). In this case, the extra

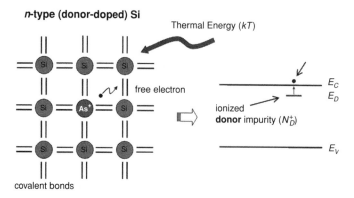

Illustration the donor-doping process. In this case the sample of Si is doped with an *n*-type impurity (As), which with sufficient thermal energy is ionized, producing a positively charged donor impurity and a free electron.

electron will be "loosely bound" (in energy) to the dopant, and with only the application of the thermal energy of the lattice (at 300 K, $kT = 26$ meV), that extra bond is "broken" and the electron is released to the conduction band, where is it free to move in response to an applied field (Fig. 5.18). The dopant is said to be a "donor" impurity, because it "donates" an electron to the conduction band and can be written as a "chemical" reaction equation:

$$N_D \rightarrow N_D^+ + n. \tag{5.12}$$

The post-donation impurity is now positively charged (it gave up an electron), as illustrated in Fig. 5.18. The now electron-rich semiconductor is said to be "*n*-type."

In the other type of doping, a Group III impurity is used (e.g., boron, B), which "accepts" an electron from the valence band, effectively producing a hole, which is similarly free to move (Fig. 5.19). This type of dopant is called an "acceptor" impurity, and produces a hole-rich, "*p*-type" semiconductor according to

$$N_A \rightarrow N_A^- + p. \tag{5.13}$$

Not surprisingly, both the donor- and the acceptor-doping processes can be conveniently represented on our energy band diagram (Figs. 5.18 and 5.19). A semiconductor that has both *n*-type and *p*-type impurities present is said to be "compensated."

What makes a good dopant? Well, clearly the dopant must be soluble in the semiconductor, such that with a "reasonable" temperature (not too low, not too high), the dopant impurity can be dissolved into the host crystal, and it can then migrate to a lattice site where it can seamlessly replace the silicon atom. In addition, though, it is vital that the "ionization energy" (the energy required for "breaking" the bond of the extra electron) of the dopant be at least comparable to the thermal energy at 300 K, such that all of the donors, say, donate their electrons to the conduction band. In arsenic, for instance, the ionization energy is 54 meV. If the ionization energy is too large, only a fraction of the impurities I put in will give me electrons, which is obviously undesirable. The toy model used for

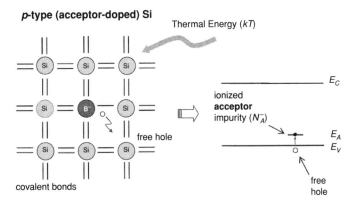

p-type (acceptor-doped) Si

Figure 5.19 Illustration the acceptor-doping process. In this case the sample of Si is doped with a p-type impurity (B), which with sufficient thermal energy is ionized, producing a negatively charged acceptor impurity and a free hole.

understanding the doping process in semiconductors is actually the Bohr model, the most simplistic of the atomic models with at least some level of quantum-mechanical sophistication. It works surprisingly well and can be used to easily estimate the ionization energy of most dopants to within a factor of three or so.

Let's play with some numbers to illustrate the power of doping in changing carrier densities (hence resistivity). Electrons in an *n*-type (donor-rich) semiconductor are called "majority carriers," whereas holes in an *n*-type semiconductor are called "minority carriers" (vice versa for *p*-type material). Why? Consider a sample of silicon at 300 K that is doped with 1×10^{17} cm^{-3} donor impurities. Question: What are the resultant majority and minority carrier densities? Well, electrons are the majority carrier (it is *n*-type) and, with good approximation, n_n (the subscript indicates electrons in an *n*-type sample) equals the electron density contributed by the doping process, or $n_n = 1 \times 10^{17}$ cm^{-3}. Holes are the minority carriers, and we can get the hole density from the law of mass action, and thus $p_n = n_i^2 / n_n = 1 \times 10^{20} / 1 \times 10^{17} = 1{,}000$. Observe – by introducing some impurities, I am able to change the carrier populations from $n = p = n_i = 1 \times 10^{10}$ in intrinsic (undoped) silicon to $n_n = 1 \times 10^{17}$ and $p_n = 1 \times 10^3$. Clearly electrons are in the (vast) majority and holes are in the (sparse) minority, hence their names. In this case the ratio of electrons to holes has been changed profoundly, from 1/1 to 1×10^{14}, a factor of $100{,}000{,}000{,}000{,}000\times = 100$ trillion! And we can do this trivially by simply controllably introducing impurities into the semiconductor! Read: POWERFUL tool.

There is one final concept we need to address in doping, and that is the notion of "charge neutrality." In most EE problems, when I say "charge" you would say, "oh, there is only one type of charge – electrons." Not so with semiconductors. In principle, for any given extrinsic semiconductor, there are actually *four* types of charge that may be present: (1) electrons (*n*), (2) holes (*p*), (3) ionized donor impurities (N_D^+), and (4) ionized acceptor impurities (N_A^-). Electrons and holes are clearly mobile (they can produce current), whereas the donor and acceptor

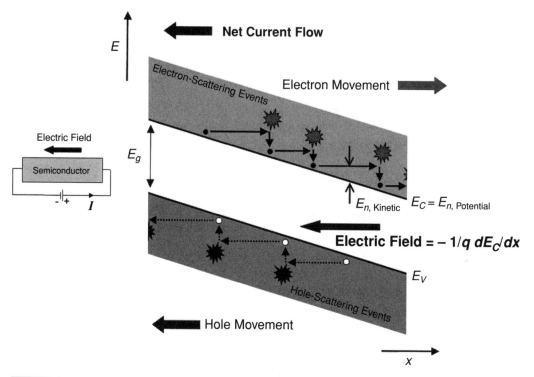

Figure 5.20   Application of an electric field to a semiconductor, resulting in "band-bending." A net movement of both electrons and holes in response to that field is produced, subject to the operative scattering processes, and generating a net current flow from right to left.

impurities are stationary. However, as we will see, under certain conditions, those fixed impurities, although they cannot contribute directly to current flow, can create their own internal electric field! The *net* charge density (ρ_{semi}) is the sum of the four, or

$$\rho_{\text{semi}} = q(p - n + N_D^+ - N_A^-) \qquad (5.14)$$

For our preceding n-type silicon example, $N_D^+ = 1 \times 10^{17}$ cm^{-3}, $N_A^- = 0$, $n = 1 \times 10^{17}$ cm^{-3}, and $p = 1,000$, and thus $\rho_{\text{semi}} = 1000 - 100,000,000,000,000,000 + 100,000,000,000,000,000 - 0 = 0$ (careful: not 1,000, try it on a calculator). When $\rho_{\text{semi}} = 0$, the sample is said to be "charge neutral," and "charge neutrality" is satisfied. For a uniformly doped semiconductor sitting on your desk, the net charge density is thus zero, even though there is a vast amount of charge present in it (100,000 trillion electrons/cm^3 to be precise). This is why if I touch my finger to the sample of silicon sitting on my desk I don't get instantly electrocuted! There is a tremendous amount of charge present, but it is balanced positive to negative, effectively nulling-out the net charge. The sample is "charge-neutral" – go ahead, it's safe to pick up!

Let me torque the thumbscrew slightly. Brace yourself! Poisson's equation, one of Maxwell's equations, is nature's way of mathematically relating the net charge density and the electric field (or electrostatic potential Ψ, aka the voltage), and is given in three dimensions by

$$\nabla \mathcal{E} = \frac{\rho_{\text{semi}}}{\kappa_{\text{semi}} \, \epsilon_0}, \tag{5.15}$$

where κ_{semi} is the dielectric constant of the semiconductor relative to vacuum (11.9 for silicon) and ϵ_0 is the permittivity of free space (8.854×10^{14} F/cm). Don't worry about the math; the point is this: If the net charge density is zero, Poisson's equation guarantees that the electric field is zero, and thus the electrostatic potential is constant ($\mathcal{E} = -\nabla \Psi$). This is equivalent to saying that the energy band diagram is constant (flat), because $E_C = -q\Psi$. Said another way, if E_C and E_V are drawn flat in the energy band diagram for a particular situation, there *cannot* be an electric field present. Conversely, if the energy bands are *not* flat, then an electric field *is* present, and this condition is known affectionately as "band-bending" (Fig. 5.20). The magnitude of that induced electric field in one dimension can be calculated from the slope (mathematically, the derivative) of the band, namely, $\mathcal{E} = -1/q(dE_C/dx)$. As we will see, creative band-bending is at the very heart of transistor operation.

5.8 Drift and diffusion transport

5.8.1 Carrier drift

Wonderful. We now have energy bands, electron and hole densities, and ways to easily manipulate those carrier populations over a wide dynamic range, by means of doping, for instance. Now that we have carriers present in partially filled bands, enabling them in principle to move, we need to ask a couple of simple questions: What magnitude of current actually flows in response to a stimulus? And what does that current depend on? Appreciate that it is the current flowing in a semiconductor device that we will use to do useful work, say build a microprocessor for a computer or a radio for a cell phone.

Interestingly enough, there are in fact two ways for carriers to move in semiconductors; two mechanisms for "carrier transport" in semiconductors if you will: "carrier drift" and "carrier diffusion." Drift is more intuitive, so let's start there. The fundamental driving force in drift transport is the applied electric field (i.e., put a voltage across a sample of length l). Imagine a sample of doped silicon at 300 K, which has a finite density of electrons and holes present (dependent on the doping density, which is known). Now imagine applying a voltage V across the sample. The question of the day: If I put an ammeter in the circuit, what current flows in response to that applied voltage, and what system variables does it depend on? Tough question!

If we apply a voltage across the sample, an electric field is induced ($\mathcal{E} = V/l$), and the electrons and holes will move (drift) in response to that "drift" field in the partially filled conduction and valence bands, respectively. Imagine zooming in on the motion of a single hole in the valence band as it moves (Fig. 5.21). The

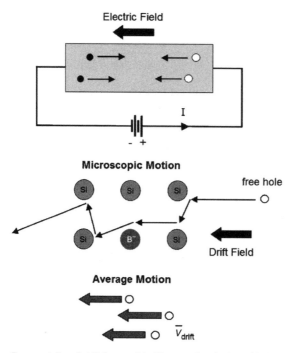

Figure 5.21   Representation of drift transport inside a semiconductor, with the electrons and holes moving in response to the applied electric field. A microscopic view of the hole drift motion (the vector length in each case is proportional to its velocity) is also shown, together with a simplified picture that assumes that all holes move with an average drift velocity.

hole will microscopically follow a very circuitous route through the material in response to the drift field. It will start and stop, speed up and slow down, and even change directions abruptly as it moves along in the field, depending on what it confronts in its path. Anything that impedes the transport of carriers is called a "scattering event" (this is illustrated within the energy band diagram in Fig. 5.20). Think of scattering as a carrier "collision" process. A carrier gains kinetic energy (velocity increases) from the applied field, and then scatters, reducing its velocity (energy). There are many types of scattering in semiconductors, but the two most important are "lattice scattering," the interaction of the carriers with the thermally induced vibrations of the crystal lattice, and "impurity scattering," the interaction of the carriers with the charged dopant impurities. If we now imagine a sea of carriers moving in this drift field (recall: with a boron doping level of 1×10^{16} cm^{-3} there are 1×10^{16} cm^{-3} holes – a bunch!), we can invent an "average drift velocity" ($\bar{v}_{drift}$) for the moving sea of holes (in statistical physics this is called an "ensemble average").

The net result is important. In response to an applied drift field, and in the presence of the various scattering processes, (a) the average drift velocity of the carriers depends on the magnitude of the applied drift field (hence the force driving their motion), and (b) the maximum average velocity of the carriers is finite, and is called the "saturation velocity" (v_s). In lightly doped silicon at 300 K, for instance, v_s is about 1×10^7 cm/s, 223,694 mi/h, about 1/3,000 the speed

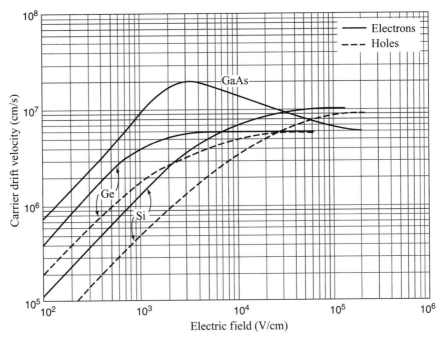

Figure 5.22 Electron and hole drift velocities as functions of the applied electric field for Si, Ge, and GaAs at 300 K.

of light in vacuum (read: mighty darn fast!). The so-called "velocity-field" characteristics of Si, Ge, and GaAs are shown in Fig. 5.22. You will note that, for electrons in silicon, it takes an applied electric field of about 20,000 V/cm to reach saturation velocity. This feels like a pretty large field. It is! Here's the example I like to use to illustrate big vs. small fields. If you walked into a room, and saw a cable with a bright yellow sign saying "DANGER – 20,000 V!" pasted on it, would you walk up, stand in a puddle of water, and bring your finger to within 1 cm of it (this is in fact equivalent to a 20,000-V/cm electric field)? As I like to say, you might, but you'd only get to do it once! Clearly, 20,000 V/cm is a very large field, but amazingly it is a field that is trivial to induce in a semiconductor. Consider: Let's apply 2 V across a transistor only 1 μm long. Congratulations, you have just applied a field of 20,000 V/cm! (Check the math.) Moral: In the micro/nanoelectronics world, small voltages across nanoscale distances generate huge fields, and hence VERY fast carriers. KEY finding.

You will also note from Fig. 5.22 that at "low" electric fields (say below 1,000 V/cm for electrons in silicon), the average drift velocity depends linearly on the drift field. Given this functional relationship, for convenience we define a new parameter known as the "low-field carrier mobility" or just "mobility" (μ_n or μ_p, measured in units of cm^2/V s), such that $\bar{v}_{\text{drift}} = \mu\mathcal{E}$. The mobility is very useful because (a) it collapses an enormously complex piece of carrier-scattering physics into a single parameter (much like the effective mass), and (b) it can fairly easily be measured. Figure 5.23 shows electron and hole mobility data at 300 K as functions of doping. As can be seen, for silicon, electrons are faster than holes,

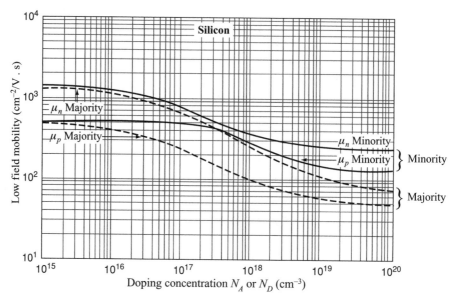

Figure 5.23   Low-field electron and hole mobilities as functions of doping concentration at 300 K.

all else being equal, and this will have important circuit design implications, as we will see.

Okay. With those bare-bones ideas in place, we can rather easily derive the drift current density that flows in response to the applied field. Consider the toy model illustrated in Fig. 5.24 (we'll choose holes in a p-type semiconductor for simplicity). I will use an intuitive step-by-step "dimensional" derivation to get the current [1].

- Step (1) $\bar{v}_{\text{drift}}\, t$ (cm): All holes this distance back from A will cross the plane in time t
- Step (2) $\bar{v}_{\text{drift}}\, tA$ (cm^3): All holes in this volume will cross A in time t
- Step (3) $p\, \bar{v}_{\text{drift}}\, tA$ (# of holes): Number of holes crossing A in time t
- Step (4) $q\, p\, \bar{v}_{\text{drift}}\, tA$ (C): Charge crossing A in time t
- Step (5) $q\, p\, \bar{v}_{\text{drift}}\, A$ (C/s): Charge per unit time crossing A (i.e., current)
- Step (6) $q\, p\, \bar{v}_{\text{drift}}$ (A/cm^2): Drift current density crossing A
- Step (7) $q\, \mu_p\, p\, \mathcal{E}$ (A/cm^2): Drift current density written in terms of mobility

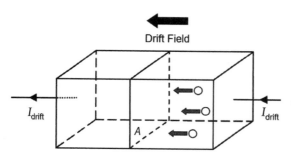

Figure 5.24   Model for determining the drift current that flows in response to the applied electric field.

Sha-zam! Thus, the electron and hole drift current densities, in three dimensions (they are vector quantities), are given by

$$\vec{J}_{n,\text{drift}} = q\,\mu_n\,n\,\vec{\mathcal{E}}, \tag{5.16}$$

$$\vec{J}_{p,\text{drift}} = q\,\mu_p\,p\,\vec{\mathcal{E}}, \tag{5.17}$$

$$\vec{J}_{\text{total,drift}} = \vec{J}_{n,\text{drift}} + \vec{J}_{p,\text{drift}} = q(\mu_n\,n + \mu_p\,p)\,\vec{\mathcal{E}}. \tag{5.18}$$

We come full circle now. Armed with this definition of total drift current density, we can connect our semiconductor resistivity with the electron and hole densities, hence how it is constructed (doped). Because

$$\vec{J} = q(\mu_n\,n + \mu_p\,p)\,\vec{\mathcal{E}} \tag{5.19}$$

and

$$\vec{J} = \frac{1}{\rho}\,\vec{\mathcal{E}}, \tag{5.20}$$

we thus find that

$$\rho = \frac{1}{q(\mu_n\,n + \mu_p\,p)}. \tag{5.21}$$

Observe: Give me the doping and I will calculate the resistivity. For example, most of the world's integrated circuits are build on 8–10-Ω cm p-type silicon, which we now can easily see corresponds to about 1–2×10^{15} cm^{-3} boron doping (check it!).

5.8.2 Carrier diffusion

If you asked most people who know a little about electricity what actually produces current in materials, they will inevitably say, "well, electrons move in response to a voltage according to $V = IR$." Fine in most cases, but NOT for semiconductors. This is only 1/4 of the truth! We already know that we have both electron drift current and hole drift current (two sources), because carrier motion in *both* the conduction band and the valence band contribute to current flow. But that is only 1/2 of the truth! It is an amazing fact that in semiconductors, even with zero field applied (no voltage), current *can still flow*. Sound magical?! It isn't. This "new" type of current flows by diffusive transport, aka "diffusion." Although perhaps less familiar to most people, diffusion is actually firmly entrenched in your normal life experience. Consider: I am in the front of the classroom, and you walk in carrying a sausage biscuit (morning class!), and sit down in the back row. Within a few moments, the smell of your biscuit has my mouth watering. How?

Well, imagine zooming in to a microscopic view of the air molecules in the room (you know, a little oxygen, a lot of nitrogen, and a few trace molecules, hopefully nothing toxic!). Because of the thermal energy present (yep, same kT),

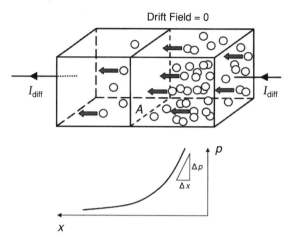

Drift Field = 0

Figure 5.25 Model for determining the diffusion current that flows in response to the carrier concentration gradient.

all of the air molecules are whizzing back and forth at very high speeds, bouncing off one another; massive numbers of collisions – it's a chaotic mess. Now, at one end of the room inject a few "great-sausage-smell" molecules into the mix, and see what happens. Well, all of that chaotic motion will tend to spread out the "great-sausage-smell" molecules into the entire volume of the room. Two minutes later my mouth is watering (hopefully your Mom taught you something about sharing!). Note, it is NOT the air currents present in the room doing this spreading; the same result will occur in a controlled environment with no "stirring" of the air. In short, entropy is at work here, the great leveler of the universe. Said another way, for very deep reasons, nature much prefers the lowest-energy configuration of any system, the so-called "equilibrium state" (see sidebar discussion for a reminder). In essence, nature sees a large concentration of "great-sausage-smell" molecules at one end of the room, and says, "hey, let's equilibrate the system (lower its energy) by spreading out all of the 'great-sausage-smell' molecules uniformly over volume V." Ergo, the "great-sausage-smell" molecules "diffuse" from point A to point B.

Here is the million-dollar question: What is the fundamental driving force in diffusion? Give up? It is the position-dependent variation in particle *concentration* from point A to point B. More precisely (yep, think math), in semiconductors it is the carrier concentration *gradient*, ∇n or ∇p, that drives the diffusion currents. Remember, if charge moves, by whatever means, current will flow. Said another way, if I can induce a variation in n or p such that it is not constant in position in the crystal, its gradient function is nonzero, and current will flow, *even with no applied drift field present*. Think about this – it is rather remarkable.

Sadly, deriving an expression for diffusion current density is a bit more involved than for drift current, and slightly beyond us here, but let me point you in the right direction. Consider Fig. 5.25, which shows a p-type sample with no field applied, but has an induced hole gradient present. By Fick's first law of diffusion for particles (be it electrons, holes, or even "great-sausage-smell" molecules), we can write

$$\vec{\mathcal{F}} = -D\,\nabla\eta, \tag{5.22}$$

where $\mathcal{F}$ is the particle "flux" (particles/cm^2 s), η is the particle concentration or density (number of particles/cm^3), and D is the so-called "diffusion coefficient," simply a proportionality constant dependent on the physical context (dimensionally, $[D] = \text{cm}^2/\text{s}$). Applying Fick's law to semiconductors, we have

$$\vec{J}_{p,\text{diffusion}} = -q\,D_p\,\nabla p \tag{5.23}$$

for the holes, and

$$\vec{J}_{n,\text{diffusion}} = q\,D_n\,\nabla n \tag{5.24}$$

for the electrons (the difference in sign is just due to the negative charge of the electron). Finally, combining our drift and diffusion transport mechanisms, for both electrons and holes, we have

$$\vec{J}_{n,\text{total}} = \vec{J}_{n,\text{drift}} + \vec{J}_{n,\text{diffusion}} = q\,\mu_n\,n\,\vec{\mathcal{E}} + q\,D_n\,\nabla n, \tag{5.25}$$

$$\vec{J}_{p,\text{total}} = \vec{J}_{p,\text{drift}} + \vec{J}_{p,\text{diffusion}} = q\,\mu_p\,p\,\vec{\mathcal{E}} - q\,D_p\,\nabla p, \tag{5.26}$$

with a total current density of

$$\vec{J}_{\text{total}} = \vec{J}_{n,\text{total}} + \vec{J}_{p,\text{total}}. \tag{5.27}$$

Not surprisingly, these two equations are known as the "drift–diffusion transport equations." Clearly we have four potential sources for current flow in semiconductors – pretty neat, huh?! Are all four of these transport mechanisms relevant to real semiconductor devices? You bet! In a MOSFET, for instance, drift transport typically dominates the current, whereas for *pn* junctions and bipolar transistors, diffusion transport dominates.

One other interesting tidbit. Given this final result for current flow, a logical question presents itself (for you deep thinkers anyway!). Given that drift and diffusion represent two fundamentally different means nature has for moving charge in a semiconductor, is there some linkage between the two transport mechanisms? There is! As first proved by Einstein, one can show at a very deep level that

$$\frac{D_n}{\mu_n} = \frac{kT}{q} \tag{5.28}$$

for electrons and

$$\frac{D_p}{\mu_p} = \frac{kT}{q} \tag{5.29}$$

for holes. That is, drift and diffusion are linked by the respective mobility and diffusion coefficients and connected by our most fundamental parameter in the universe – kT. Slick, huh?! These are the so-called "Einstein relations." Appealingly simple.

Geek Trivia: Einstein's Relation

If you are a student of history, you might be curious as to how Einstein arrived at this pivotal result. Hint: It had nothing to do with semiconductors! In 1905, the so-called *annus mirabilis* of physics, Einstein published "On the motion of small particles suspended in liquids at rest required by the molecular-kinetic theory of heat" [5], the result of his study of Brownian motion as a means for examining the basic underpinnings of Boltzmann's kinetic theory of heat. The oh-so-simple Einstein relation elegantly falls out of that analysis and applies generally to any "dilute gas" (in our case, electrons and holes in semiconductors). You can also derive it directly from Eq. (5.25) by assuming the total current is zero (i.e., equilibrium). Dare ya!

Equilibrium

The term "equilibrium" is used to describe the unperturbed state of any system in nature. Under equilibrium conditions, there are no external forces acting on the system (fields, stresses, etc.), and all "observables" (things we can measure) are time invariant (don't depend on time). Knowledge of any given system's equilibrium state is vital because it forms the simplest "frame of reference" for the system. It is also the lowest-energy configuration of the system, and hence nature will strive to return the system to equilibrium when we perturb it by any means. Being able to theoretically (and experimentally) understand any system (semiconductor or otherwise) in equilibrium thus gives us a powerful physical basis for ascertaining the condition of the system once a perturbation is applied [1].

When we declare that a semiconductor is in equilibrium, for instance, we mean that no external electric or magnetic fields are applied, no temperature gradient exists, and no light is shining on the sample. More formally, equilibrium mandates that $J_{total} \equiv 0$. That is, drift and diffusion transport processes are balanced, such that no net charge transport occurs. As can be formally proven, this further guarantees that the Fermi level (E_F in our energy band diagram; refer to Fig. 5.17) is position independent (constant in x, spatially "flat").

5.9   Generation and recombination

Excellent. Semiconductors contain energy bands, and the electron and hole densities in those bands can be varied controllably by means of, say, doping. We can also move those electrons and holes around at high speed by means of either drift or diffusion (or both) to create current flow. Fine. What else? Last topic (I promise!). If I take a semiconductor out of its equilibrium state, by whatever means, what tools does nature have at its disposal to reestablish equilibrium? Good question!

In general, when a semiconductor is perturbed from equilibrium, either an excess or a deficit of local charge density (electron or hole or both) is inevitably produced relative to their equilibrium concentrations. Nature does not take kindly to this, given that it represents a higher-energy configuration, and will promptly attempt to rectify the situation and restore equilibrium. How? Well, carrier generation–recombination (G/R for short) is nature's order-restoring mechanism in semiconductors. G/R is the means by which either a local carrier density excess or deficit is stabilized (if the perturbation remains) or eliminated (if the perturbation is subsequently removed). Formally, "generation" is *any* process by which electrons and holes are created, whereas "recombination" is *any* process by which electrons and holes are annihilated. Because nonequilibrium conditions prevail in all semiconductors devices while in use, clearly G/R processes play a pivotal role in shaping how they are designed, built, and utilized.

Intuitively, we can describe in a nutshell how transistors are used to spawn a Communications Revolution.

- Step 1: Build Mr. Transistor.
- Step 2: Said transistor sits comfortably in equilibrium twiddling its thumbs.
- Step 3: We take said transistor (kicking and screaming) momentarily out of equilibrium.
- Step 4: Mother Nature doesn't take kindly to Step 3 (or us!).
- Step 5: Mother Nature steps in and pulls G/R out of her pocket to restore said transistor to equilibrium.
- Step 6: Unbeknown to Mother Nature or said transistor, we farm the current generated by the G/R restoration processes to do some useful work!
- Step 7: Presto – Communications Revolution!

Have we "fooled Mother Nature"?[2] Perhaps. Afterall, we humans are pretty darn clever. Still, nothing comes totally for free. Read on.

So how do G/R processes actually work? Let's revisit our energy band model. The two most important G/R processes are illustrated in Figs. 5.26 and 5.27: "band-to-band G/R" and "trap-center G/R." We know what energy bands are, but what about "traps"? All crystals contain some finite, albeit tiny, density of defects or imperfections. If those imperfections can communicate with Mr. Electron or Mr. Hole, they are affectionately referred to as "trap" states [meaning: they can capture (trap) and hold onto an electron or hole from the conduction or valence band for some amount of time t_T and then release it]. The band-to-band G/R processes are straightforward. If we need to create an electron to remove a deficit, an electron is excited directly from the valence band to the conduction band. Conversely, the band-to-band recombination event effectively annihilates the electron. In trap-center G/R, a trap state located at energy E_T between E_C and E_V is involved, and the G/R process is now two stepped. That is, to create an electron, the electron is first captured by the trap from the valence band and then released by the trap to the conduction band.

---

[2] Older readers may recall the "It's not nice to fool Mother Nature!' margarine commercials from the 1970s.

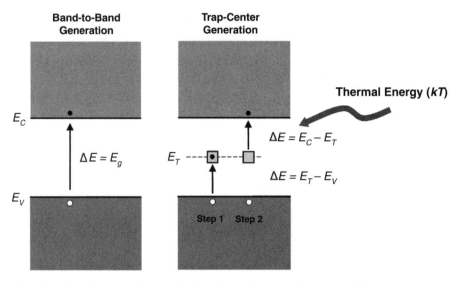

Figure 5.26   Illustration of band-to-band and trap-center thermal generations of an electron and a hole.

Clearly energy exchange (read: nothing is for free) is involved in G/R processes, and, as can be seen in Figs. 5.26 and 5.27, ΔE to first order is equal to the bandgap energy E_g. If, for instance, an electron is annihilated, where does the energy go? (Please don't feel tempted to violate conservation of energy – it has to go somewhere!) In recombination events, direct bandgap materials (see Caffeine Alert #1) favor the production of photons (light) as the dominant energy-release mechanism, whereas indirect bandgap materials favor the production of phonons (lattice vibrations – heat). Not surprisingly, the fundamental differences behind which semiconductors make the best electronic vs. photonic devices originate here. Silicon, for instance, cannot be used to create a good diode laser for a DVD player. This is why.

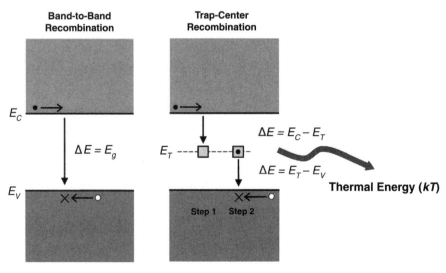

Figure 5.27   Illustration of band-to-band and trap-center thermal recombinations of an electron and a hole.

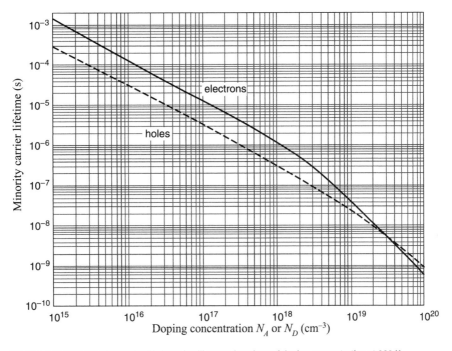

Figure 5.28 Electron and hole minority carrier lifetimes in silicon as functions of doping concentration at 300 K.

For analyzing semiconductor devices, knowing the *net* rate of recombination U (number of carrier annihilations per second per unit volume) is important. Although the solution to this fundamental semiconductor problem is beyond us here (solved first by Shockley and co-workers), with sufficient approximation a very simple answer results:

$$U_n \approx \frac{\Delta n}{\tau_n}. \tag{5.30}$$

That is, the net rate of electron recombination in a p-type material, say, is approximately equal to the "excess" minority electron density ($\Delta n = n_p - n_{p0}$, where the 0 subscript represents the equilibrium value), divided by a newly defined parameter called the "minority carrier G/R lifetime" or just "lifetime" (τ_n). The carrier lifetime is an intuitive concept. It measures the average amount of time it takes an excess electron in the conduction band to recombine with a hole and thus be annihilated. Clearly, τ_n depends on many variables (the energy of the G/R traps, their density, the trap-capture probabilities, the doping level, the bandgap, temperature, etc.), but importantly, it can be fairly easily measured. Figure 5.28 shows minority electron and hole lifetime data for silicon at 300 K, as functions of doping level. Good rule of thumb – in silicon at 300 K, lifetimes are in the range of microseconds (10^{-6}) for reasonable doping levels. The observed decrease in lifetime with increasing doping should not surprise you. If I create an extra electron, it will take less time to recombine if lots of holes are present (high p-type doping). That is, Mr. Electron doesn't have to wait long to find Ms. Hole and consumate their suicide pact!

$$\frac{\partial n}{\partial t} = G_n - U_n + \frac{1}{q} \nabla \cdot J_n$$

$$\frac{\partial p}{\partial t} = G_p - U_p - \frac{1}{q} \nabla \cdot J_p$$

$$J_p = q \, \mu_p \, p \, \mathscr{E} - q \, D_p \, \nabla p$$

$$J_n = q \, \mu_n \, n \, \mathscr{E} + q \, D_n \, \nabla n$$

$$J = J_n + J_p$$

$$\nabla \cdot \mathscr{E} = \frac{\rho}{\kappa \in_0}$$

$$\rho = q \, (p - n + N_D^+ - N_A^-)$$

$$\ldots \mathbb{A}\text{nd } \mathbb{T}\text{he } \mathbb{T}\text{ransistor } \mathbb{W}\text{as } \mathbb{B}\text{orn.}$$

Figure 5.29   The semiconductor equations of state.

5.10   Semiconductor equations of state

Finally, we come to the end of this modestly weighty chapter (okay, fine to now breathe a sigh of relief!). Cheer up; you've just conquered the toughest chapter in the book! Hooray! Energy bands, electrons and holes, drift and diffusion transport, and now G/R processes. This is actually all one needs to solve virtually any semiconductor device problem. Trust me! I close this chapter by bringing together all of our equations into one place. Shown in Fig. 5.29 are the so-called semiconductor "equations of state," our Maxwell's equations if you will, which embody all of the physics we need to tackle any electronic or photonic problem. You will recognize the drift–diffusion equations, Poisson's equation, and charge neutrality. The first two equations are known as the "current-continuity equations," and are simply charge-accounting equations that couple the carrier dynamics (time dependence) and G/R processes to the currents. If you do not speak vector calculus, don't be dismayed. If you happen to, appreciate the beauty. With these seven equations, the transistor was indeed born!

References

1. R. F. Pierret, *Semiconductor Device Fundamentals*, Addison-Wesley, Reading, MA, 1996.

2. B. L. Anderson and R. L. Anderson, *Fundamentals of Semiconductor Devices,* McGraw-Hill, New York, 2005.

3. R. M. Warner Jr. and B. L. Grung, *Transistors: Fundamentals for the Integrated-Circuit Engineer,* Wiley, New York, 1983.

4. R. de L. Kronig and W. G. Penney, *Proceedings of the Royal Society*, Vol. A 130, p. 499, 1931.

5. A. Einstein, *Annalen der Physik*, Vol. 17, pp. 549–560, 1905.

6 Widget Deconstruction #2: USB Flash Drive

Photograph of a refreshing USB Flash Drive (courtesy of CustomUSB).

Memosphere *(meh' moh sfir), n:*
> *The part of the sky one wistfully searches,*
> *eyebrows furrowed,*
> *while trying desperately to recall*
> *something you know you know.*

Sniglets. You know, any word that doesn't appear in the dictionary . . . but should. Memosphere. A frustratingly opaque memosphere can inexorably lead to that universal condition known as a "senior moment," or for the young at heart, the infamous "brain fart"; something you know you know, but just can't quite seem to pull out of the memory banks. Relax, we've all been there. Ahhh, memory, that essential human attribute, and miracle of the human brain (see Geek Trivia sidebar on human memory capacity). How is it that I can instantly materialize a visual image or even movie clip (and thus virtually relive) of myself playing in the dirt with my then-giant bright yellow Tonka dump truck as a 5-year-old; or the instant on New Year's Eve, 1981, when I knelt down and asked my wife Maria to marry me; or the precise moment my son Matt was born 25 years ago . . . and yet, I couldn't tell you what I had for dinner last Wednesday night, or what shirt I wore three days ago, even if you paid me $1,000! The baffling miracle of memory.

> ### Geek Trivia: What is the Memory Capacity of the Human Brain?
>
> Humm…lots?! General estimates of human memory capacity are many and varied, typically ranging from 10s of TB to perhaps 1,000 TB, and are arrived at by use of various neuron-counting schemes. Here's a commonly cited example [1]: The average human brain contains between 50 billion and 200 billion neurons, each of which is connected to perhaps 1,000 to 100,000 other neurons through something like 100 trillion to 10 quadrillion synaptic junctions. Each synapse possesses a variable electrochemical triggering threshold that gets smaller as the neuron is repeatedly activated (a likely component of the memory storage sequence). Now, if we assume that the triggering threshold at each synapse can assume, say, 256 distinguishable levels, and that there are 20,000 shared synapses per neuron, then the total information storage capacity of the synapses of the brain would be of the order of 500 to 1,000 TB. Yep, LOTS! Clearly, equating functional memory storage capacity with simple estimates from numbers of neurons and their sophisticated synaptic linkages is very likely overly simplistic. How we store and process both short- and long-term memories remains a largely unsolved problem. Still, you get the idea. The human brain is a pretty amazing memory device! If only we could make it behave a little better and figure out how to use all of its available capacity!

Just as memory is a remarkably complex and subtle defining feature of what it means to be human, memory is also essential to all electronic objects (read: virtually all of modern technology). Yep, no memory, no Communications Revolution. Consider: Our computer operating systems (basically computational instruction sets) and various software packages we intend to use must reside in memory, so that when we power up our machine it comes to life, the programs are loaded, and the computer is then ready to perform useful work. While the computer is engaged in doing its thing (say, playing a game), pieces of information must also be temporarily stored until needed and then retrieved and passed along (at blazing speeds). And of course our myriad files, word documents, powerpoint presentations, digital pictures, telephone numbers, etc., etc., all must be stored in some form of memory, somewhere, somehow. Not surprisingly, electronic memory comes in many forms and types. In this chapter we will deconstruct a slick and quite new memory gadget.

6.1   With a broad brush

The *USB flash drive*, aka the "thumb drive" or the "memory stick," is the new-kid-on-the-block memory toy (Fig. 6.1). A small, cheap, appealingly easy-to-use memory device for toting files or photos or other electronic softcopy "stuff" around. Think for a second. When was the first time you remember seeing a flash drive? When did you own your first one? (Feel free to consult the memosphere!)

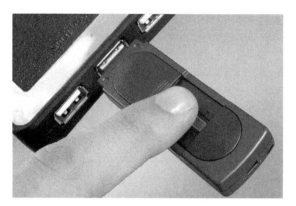

Figure 6.1   The ubiquitous thumb drive (courtesy of CustomUSB).

Humm . . . Answer? Definitely not long ago. The first flash drive came on the market in 2000, and only in the past couple of years have flash drives become truly indispensable and ubiquitous. Folks have them on their key chains, on lanyards around their necks, thrown into their car glove compartment, at the bottom of their handbags. Heck, when I walk into a meeting, the first thing I do is whip out my trusty flash drive and load my powerpoint presentation onto the local laptop connected to the projector. Ready to roll. Your slick report has grown to 100 megabytes (MB) and is now too big to e-mail? No worries; just drop it on a flash drive and take it along with you in your shirt pocket.

What began as a pretty whimpy-sized (the first one only held 8 MB!) and slightly awkward memory option for quite a few bucks is suddenly quite a deal. I just paid $20 for a 8-gigabyte (GB) flash drive last week. Okay, it was on special, but still, a 8-GB flash holds quite a bit of anything for an object so appealingly tiny and so trivial to use even grandma doesn't have to ask for instructions!

Not surprisingly, now that flash drives have taken off, flash-drive-as-fashion-accessory is the new rule of the day. Sleek artsy flash drive? Sure! Figure 6.2. A flash drive to go with your fancy wood desk set? You bet! Figure 6.3. A flash drive for the outdoor enthusiast? Roger that! Figure 6.4. A flash drive for the MD in training? No problem. Figure 6.5. Okay . . . a flash drive for the devout

Figure 6.2   A more visually appealing thumb drive, with an artsy feel (courtesy of CustomUSB, *http://www.customusb.com*).

Figure 6.3 A thumb drive for the fashion conscious (courtesy of CustomUSB).

Figure 6.4 A thumb drive for the outdoor enthusiast (courtesy of CustomUSB). (See color plate 16.)

Figure 6.5 A thumb drive for your favorite physician (courtesy of CustomUSB).

Figure 6.6 There is even a thumb drive for the faithful! (courtesy of CustomUSB).

religious?! Sad to say, yep! Figure 6.6. Techno-geek gadget as fashion statement? What gives? – isn't that an oxymoron? What's the world coming to?! Something so new and so exciting that it gets put into a Swiss Army knife is definitely worth knowing a bit about. Read on!

USB flash drives are actually NAND-type flash-memory data-storage devices integrated with a USB connector (more on all this mouthful in a moment). Because there are actually no moving parts in USB flash drives, they really are not "drives" in a strict sense, but have retained that name from floppy disk *drive* days, an item they have essentially pushed off a cliff into extinction. See the historical sidebar discussion on the invention of the USB flash drive.

Historical Aside: Who Really Invented the USB Flash Drive?

Not surprisingly, several companies lay claim to the invention of the flash drive (can you say patent infringement!). Singapore-based Trek Technology was the first company to actually sell a flash drive, in early 2000 (trade name, *Thumb-Drive*; go figure!), and obviously filed patents on the idea beforehand. M-Systems (subsequently acquired by SanDisk in November of 2006) actually began working on a flash drive in 1998 (tradename, *DiskOnKey*; IMHO much less catchy), and even registered a web domain *www.diskonkey.com* on October 12, 1999, suggesting an intended sales path. Yep, they filed patents. Enter IBM. An IBM invention disclosure (RPS8-1999-0201, dated September 1999; recall: an invention disclosure is NOT a patent) by Shmueli *et al.* appears to actually be the earliest known document to describe a flash drive. Turns out IBM and M-Systems teamed up, and in late 2000 M-Systems was the first to sell a flash drive (for IBM) in North America, an 8-MB "USB Memory Key," on December 15, 2000.

You can see the writing on the wall. A miffed Trek Technology claimed it was the first to have the idea and promptly filed suit against its competitors (now including more than just M-Systems), who then countersued. Ahhh, yes, lawyers. Final result? Newsflash! (pun intended) – in August of 2006, Trek Technology won the patent infringement lawsuit to stop others from copying their flash drive (without paying a major license fee). The Singapore Court of Appeals confirmed the validity of Trek Technology's original patent, calling it "novel and inventive," and quashed (legalese for squashed, squished?) the plea of Trek's four main competitors: M-Systems, Electec, FE Global Electronics, and Ritronics Components, ordering them to immediately stop selling similar products. According to Trek CEO Henn Tan (who must still be smiling), "When we first introduced the ThumbDrive in early 2000, we believed that this little device was set to change the way consumers across the world would store and transport information and data." Indeed it has; flash drives, in only a few short years, have become as almost as ubiquitous as cell phones. Trek Technology holds patents for its ThumbDrive in Japan, Taiwan, South Korea, the United Kingdom, New Zealand, and Singapore. Read: Big bucks, here we come!

USB flash drives are small, lightweight, portable, and instantly rewritable, and they offer major advantages over other portable memory storage devices such as floppy disks and CDs (think wayyyy back! – if you know the difference between a 5.25-in and a 3.5-in floppy you are officially old!). Compared with floppy disks, flash drives are far more compact, much faster, hold significantly more data, and are more reliable against data loss because of their lack of moving parts. Perhaps most importantly, they use a USB connector to interface to the outside world, and USB ports are found on virtually all computers and are compatible with all major operating systems. Buy a USB flash drive from anybody, plug it into any computer, anywhere, anytime, and it will work, no questions asked. Sweet!

Convenience is big. USB flash drives are sturdy enough to be carried about in your pocket, on your key chain as a fob, on a lanyard, or just tossed into your briefcase like so much loose change. The end of the USB connector protrudes from the flash drive, and it is protected either with a removable cap or by retracting it into the body of the drive. To access the data stored on the flash drive, the drive must be connected to a computer, either by plugging it into a USB host controller built into the computer or into an external USB hub. Flash drives are active only when plugged into a USB connection and draw all of their necessary power from the supply provided by that connection. Presto, you're in business!

6.2 Nuts and bolts

But what exactly IS a USB flash drive? Essentially it is a transistor-based (go figure) NAND flash-memory chip (an IC) interfaced to a USB connector, and then all placed in a single convenient package. Flash memory? USB connector? Hold on, first things first. Let's talk a bit about types of memory in the electronics world, and then we'll work our way to flash memory.

Memory Type	Volatile?	Writable?	Max Erase Cycles	Cost (per Byte)	Speed
SRAM	No (with backup)	Yes	Unlimited	Expensive	Fast
DRAM	Yes	Yes	Unlimited	Moderate	Moderate
PROM	No	Once, with a device programmer	N/A	Moderate	Fast
EPROM	No	Yes, with a device programmer	Limited	Moderate	Fast
EEPROM	No	Yes	Limited	Expensive	Fast to read, slow to erase/write
Flash	No	Yes	Limited	Moderate	Fast to read, slow to erase/write
NVRAM	No	Yes	Unlimited	Expensive	Fast

Figure 6.7 The various types of semiconductor memories and their characteristics.

"Memory," in an electronics context, is some object, some widget, that enables us to selectively store or selectively retrieve, or both, bits of information (ultimately binary "1s" and "0s"). The analog to human memory should be obvious, but of course electronic memory needs to be a bit more robust than our neuron-based contraption! There are basically two broad classes of electronic memory you will come into contact with routinely: (1) hard disk drive memory, and (2) an amazingly diverse set of transistor-based semiconductor memories (Fig. 6.7). Electronic memories can be temporary ("volatile") or semipermanent ("nonvolatile"); they can be only "readable," or they can be or readable and "writable." Hold on, let me clarify.

6.2.1 Hard disk drives

Hard disk drives (aka HDDs, or just "hard drives") are nonvolatile memory devices (unless you drop it and shake the bits off the disk!). Turn off your laptop and pull out the battery (no juice), and the files you saved on your hard disk will stay put until you power it up again. Nonvolatile. We can obviously "read" (retrieve) a file from a hard drive, but, more important, we can also "write" a file to the disk. Repeatedly. Hundreds of thousands of times if necessary. Very handy indeed. How does it work? Well, skipping the fancy physics, it is really pretty simple.

HDDs record (store) digital data by directionally magnetizing a ferromagnetic material, one direction (in geek-speak, polarization) representing a binary "0" and the other direction, a binary "1." The stored bits can then be read back out by electronically detecting the induced magnetization state of the material. A HDD physically consists of a spindle that holds one (or more) flat circular disk called a "platter" (think kitchen), onto which the data are recorded (Fig. 6.8). The platters are made of a nonmagnetic material, say glass or aluminum, that has been coated

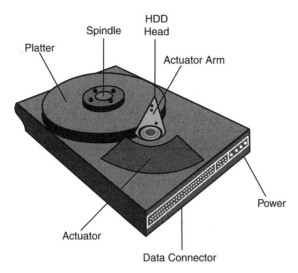

Figure 6.8   Block diagram of a HDD.

with a VERY thin layer (perhaps only a few atomic monolayers) of said magnetic material. Older HDDs used iron oxide as the magnetic material, but current disks use a more efficient, albeit complex, cobalt-based magnetic alloy (e.g., cobalt platinum chromium boron, CoPtCrB). Whew!

HDD platters are spun at exceptionally high speeds to enable very fast read–write times. Data are written to the platter as it rotates past the so-called "read–write head." The HDD head is electronic (yep, uses transistors to do its business) and sits unbelievably close to the fast-spinning platter (Fig. 6.9). The head is used to electrically detect, modify, or both, the magnetization of the material immediately underneath it, and it sits VERY close to the platter as it rotates (10s to 100s of nanometers). An actuator arm moves the heads on an arc (radially) across the platter as it spins, allowing the head to access the entire surface of the platter as it spins. Oh, and a high-end HDD might have many stacked platters running in tandem, each with its own head.

Figure 6.9   Close-up of a read–write "head" hovering over a HDD "platter" (courtesy of International Business Machines Corporation; unauthorized use is not permitted). (See color plate 17.)

The magnetic surface of each platter is divided into many tiny submicron-sized magnetic regions, each of which is used to encode a single binary unit of information (i.e., a digital bit). Each of these magnetic regions is composed of several hundred magnetic "grains," and each magnetic region forms a magnetic "dipole" (think bar magnet) that generates a tiny but highly localized magnetic field. When writing data to the disk, the write head magnetizes a particular region by generating a strong local magnetic field by means of current flow (recall: moving charge, a current, generates a magnetic field). In HDD heads, the read–write elements are separate but in very close proximity on the tip of an actuator arm (see Fig. 6.9). We can then calculate the memory storage capacity of the HDD by multiplying the number of platters by the number of heads by the number of sectors by the number of bytes/sector. Result? Tons of very fast nonvolatile memory!

Some cool facts on HDDs? Sure! In 2007, a typical high-end HDD (think workstation) might store between 160 GB and 1.0 TB of data, have multiple platters that each rotate at 7,200–15,000 revolutions per minute (rpm), and have an aggregate data transfer rate of 1–2 GB/s. This little beast represents a pretty remarkable piece of engineering. Consider: "As an analogy, a magnetic head slider flying over a disk surface with a flying height of 25 nm with a relative speed of 20 m/s is equivalent to an aircraft flying at a physical spacing of 0.2 μm at 900 km/hr. This is what a disk drive experiences during its operation" [4]. The current record for platter memory density is over 100 GB/in^2, and growing! You might logically wonder how this HDD, spinning at thousands of rpm, and with a head located only a few 10s of nanometers away from the platter, remains functional as you carry it around, inevitably jostling it. Answer? VERY carefully designed shock absorbers! Want to get even fancier? How about embedding a MEMS (micro-electro-mechanical system) accelerometer (motion sensor) in the HDD electronics to detect any overly excessive motion (heaven forbid you drop your darn laptop), thereby electrically shutting down the platter rotation in a few microseconds to avoid the now-infamous and quite painfully self-descriptive "head-crash." Very doable. Pretty routine today actually. MEMS meets HDDs. Pretty slick, huh?

Presto – the ubiquitous hard disk drive. Now, how about semiconductor memories? There are many types to choose from. Let's start with the two types found in all computers: RAM and ROM. Huh? Translation: random-access memory (RAM) and read-only memory (ROM).

6.2.2 RAM

Okay – Physics 101. Deep breath. Don't panic. High school physics 101, not graduate school physics 101! Recall: A capacitor is simply an energy storage device. Take two conductors and separate them with an insulator and you have built a capacitor. Now, connect a voltage V to Mr. Capacitor of size C, and a charge of size $Q = CV$ appears on the capacitor (go ahead and check the units for sanity). Said another way, if I place a charge on a capacitor, a voltage must be present, the size of which is determined by the magnitude of the capacitance. Fine. Let's imagine that that voltage across C (1.0 V or 2.5 V or whatever) represents a binary "1," a memory state. Now remove that charge so that the voltage is gone ($V = 0$), and let's call that new state a binary "0." Fine. Now we know that a transistor acts as a simple on–off switch. You'll hear lots more about this in Chap. 8, but take me at my word for now. So how about we make a 2D array of capacitors, comprising "rows" and "columns," and an electronic selection scheme for "addressing" a given memory "bit" in that 2D capacitor array. We now have a scheme for building a memory. At each memory "cell" in the 2D array we now place a transistor (Mr. MOSFET) to isolate the capacitor from the rest of the world. That is, we store (write) a charge on one particular capacitor cell in the 2D array, and then we turn off the transistor so that the charge sits on the capacitor until we need it. We have stored a piece of data. Or, if we want to read the data out, we simply connect the memory cell to a sensing circuit that detects whether charge is present or not. Because we can do this for any of the 2D memeory cells, we have constructed a RAM. There is a slight problem, however. Do you see it? Alas, Mr. MOSFET is not a perfect off switch, but instead leaks charge (current) at a very tiny, but steady, rate (aka, the transistor's "off-state leakage"). Result? The charge placed on the capacitor when we wrote the memory cell will soon leak off, and thus the memory bit is stored for only a short period of time. Bummer. Clearly this RAM is a volatile memory. But … how about we periodically (and before the charge leaks off, thereby changing the desired memory state) "refresh" the cell (read it out and then write it back), and do this dynamically. Say every 100 ns. Result? Dynamic RAM–DRAM. Care to see a block diagram? Sure! Consult Fig. 6.10. What would a transistor-level implementation of DRAM entail? Well, check out Fig. 6.11, which shows a DRAM cell. There is a DRAM "bit line" and a "word line" connected to the source and drain ends of Mr. MOSFET, which are used for x–y cell addressing, and the storage capacitor is formed between two levels of conductive polysilicon in the metalization stack as a classical parallel-plate capacitor. And by the way, this is all done using standard silicon fabrication techniques. Wouldn't it be cool to have patented this simple idea of a "one-transistor" memory? You bet it would. Bob Dennard at IBM Research did just that in 1968. Talk about a semiconductor memory revolution! Why use DRAMs in place of HDDs? Well, they have a MUCH higher memory density, no moving parts to wear out, and can be manufactured on a single piece of silicon, at extremely high volumes, very cheaply. Yep, all the merits associated with the economy of scale of silicon IC production.

Want to get fancier? Easy. If we need even higher DRAM density, instead of using a planer (flat) storage capacitor, we could form the capacitor vertically, resulting in the now-famous and ubiquitous trench DRAM cell (Fig. 6.12). Clever, huh? This type of DRAM can take us to several gigabytes of DRAM on a single silicon die.

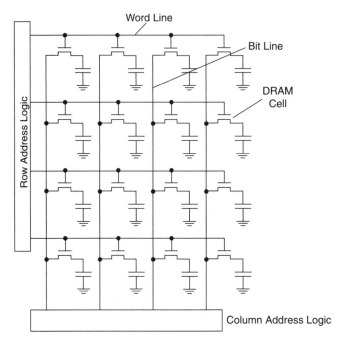

Figure 6.10   A DRAM array consisting of transistors and capacitors.

DRAM Memory Cell

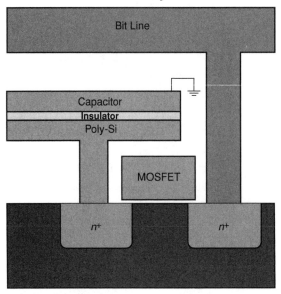

Figure 6.11   A planar DRAM memory cell.

Trench Capacitor DRAM Cell

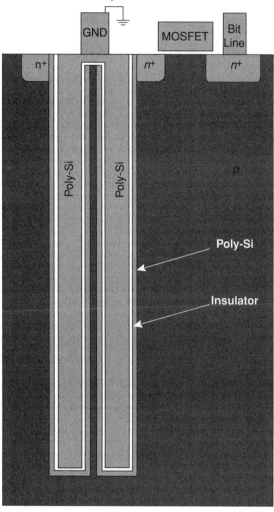

Figure 6.12 A vertical trench–capacitor DRAM memory cell.

Back-of-the-Envelope #1: Counting Electrons in DRAMs

Here's a sobering calculation for your amusement. Let's assume you have a 4-GB DRAM chip, which operates off a 1.5-V supply and has a memory cell capacitance of 5 fF (5×10^{-15} F). All pretty reasonable numbers for modern DRAMs. Question: How many electrons does it take to store a digital bit in that DRAM? Well, $Q = CV$, and there are 1.602×10^{19} electrons/C of charge. So you do the math! You'll find that it takes only 46,816 electrons to store a bit of information! Is this a large number or a small number? Well, turn on your desk lamp for 1 s and assume it draws 1 A of current. $1 \text{ A} \cdot \text{s} = 1$ C of charge, or 1.602×10^{19} electrons! Moral: 47,000 electrons is an unbelievably tiny number.

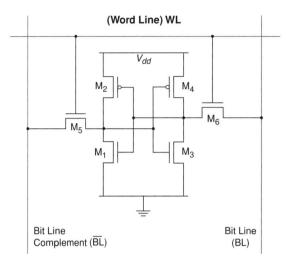

A classical six-transistor SRAM cell.

Even with the appealing simplicity of DRAM, the volatile nature of the stored memory can often be problematic. Logical question – Can we use transistors to create a nonvolatile transistor-based RAM? We can. This is called static RAM, or SRAM. We cannot use the old one-transistor + capacitor trick, but instead must revert to using a transistor-based circuit trick called a "latch" (think of a latch on a door), which will remain in a stable state (door closed, latched!) until we opt to write the memory to a new state. The classical six-transistor SRAM cell is shown in Fig. 6.13, and can be built from six MOSFETs. Slick. Obviously, however, if we remove the power to the memory IC, it reverts to volatile, so some sort of battery backup is needed. Note that there is still a word line and a bit line for addressing the 2D memory cell array, just as in a DRAM. Clearly the memory density of SRAMs is far smaller than for DRAMs, but that nonvolatile feature of SRAM makes it indispensable for most electronic gadgets. Single-transistor, DRAM-like, true nonvolatile RAM (NVRAM) can be built with special magnetic layers used for the memory storage element (sort of a merger between HDDs and DRAM), and this approach is rapidly gaining traction.

6.2.3 ROM

Okay, we now have HDDs, DRAM, and SRAM. But we still need a NAND flash to build a USB flash drive! Next stop? Read-only memory, or ROM, to those in-the-know. Memories in the ROM family are many and varied, but essentially are distinguished by the methods used to write new data into them (called "programming" the ROM), and the number of times the ROM can be rewritten. This classification reflects the evolution of ROM devices from hardwired, to programmable, to erasable-and-programmable. A common feature of all these ROMs is their ability to retain data (hence programs) forever, even during a power failure or when the system is powered off. Hence ROM is the preferred memory type for placing the start-up controls and code in your laptop, say, or your cell phone.

The very first ROMs were fairly primitive hardwired devices that contained a preprogrammed set of data or instructions. The contents of the ROM had to be specified before IC manufacturing began so that the actual data to be stored could be used to hardwire the connections to the transistors inside the ROM by use of the various metalization layers. One step up from this primitive ROM is the PROM (programmable ROM), which is manufactured and purchased in an unprogrammed state. If you examined the contents of an unprogrammed PROM, you would see that the data are made up entirely of "1s" (not very useful). The process of writing your data to the PROM involves a special piece of equipment called a "device programmer" or PROM writer. The PROM writer electrically writes data to the PROM one binary word at a time by applying a voltage to the input pins of the chip. Once a PROM has been programmed in this way, its contents can never be changed, and thus if the code or data stored in the PROM need to be changed, a trash can is the only option. An EPROM (erasable-and-programmable ROM), on the other hand, is programmed in exactly the same manner as a PROM; however, EPROMs can be erased and reprogrammed repeatedly. To erase the EPROM, you simply expose the device to a strong source of ultraviolet (UV) light. An easy indicator that you have an EPROM in your hands is the transparent window in the top of the package that allows the light to reach the transistors. After UV exposure, you essentially reset the entire chip to its initial unprogrammed state. This is getting handy now, although carrying around a strong UV lamp (can you say "tanning bed") is admittedly a bit cumbersome! EEPROMs (electrically erasable-and-programmable ROMs) represent the top of the ROM ladder, and are electrically erasable and programmable. Internally, they are similar to EPROMs, but the erase operation is accomplished electrically, rather than by exposure to UV light. In addition, data within an EEPROM can be selectively erased and rewritten, many times. Once the EEPROM is written, the new data are permanent. How on earth do we electrically erase and program a ROM memory? Very cleverly! In essence, we embed a second gate within Mr. MOSFET and make it electrically floating (not connected to anything). As shown in Figs. 6.14 and 6.15, we can then change the terminal voltages to induce electrons either into or out of the floating gate, changing the properties of the ROM cell transistor (i.e., shift the turn-on or threshold voltage of the device). For the not-so-faint-of-heart, this all happens by quantum-mechanical tunneling. The quantum world in action! If you are curious about the details of how this erase–program mechanism changes the behavior of the transistor, you'll get some more exposure to how MOSFETs actually work in Chap. 8, so stay tuned. The primary trade-off for this improved functionality with EEPROMs is their increased complexity and hence higher cost, although write times are much longer than for DRAMs or SRAMs, preventing you from using EEPROMs for your main system memory, except in very unusual conditions. Drumroll please! Next stop, flash memory!

EEPROM Cell Erase Condition

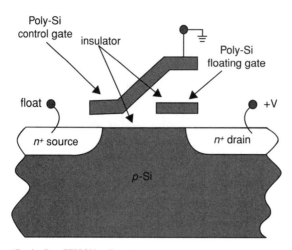

| Figure 6.14 | "Erasing" an EEPROM cell. |

6.2.4 Flash memory

Flash memory is newest gizmo in the memory world and combines the best features of all of the transistor-based semiconductor memories. Flash memories have DRAM-like densities, are very low cost, are nonvolatile, are reasonably fast (for read cycle, not write cycle), and importantly are electrically reprogrammable (each transistor in the flash memory has a floating gate and thus uses the same basic technique as found in EEPROMs for cell erasure and programming). Cool! These advantages are compelling in many instances, and, as a direct result, the use of flash memory has increased dramatically in recent years. Hence the flash-memory cards in your cell phone, your digital camera, and of course our own USB flash drive. Flash-memory cards are everywhere, and of course represent

EEPROM Program Condition

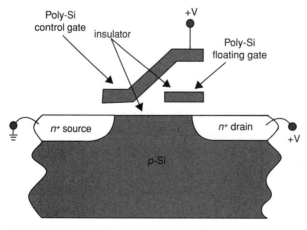

| Figure 6.15 | "Writing" an EEPROM cell. |

NOR Flash

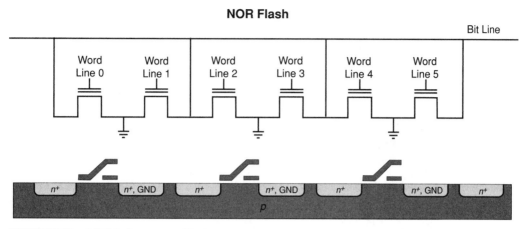

Figure 6.16 A NOR flash-memory architecture.

BIG business. There are actually two major types of flash memory, NOR flash and NAND flash, which differ in two major ways: (1) the internal transistor connections of the individual flash-memory cells are different, and (2) the electronic interface required for reading and writing the memory is different. NAND flash is the simplest and least flexible of the two, but nevertheless offers the highest memory density and ultimately lowest cost, usually a compelling advantage.

NOR flash and NAND flash get their names from the structure of the logical interconnections between memory cells (if you need a refresher on digital logic, see Appendix 3). In NOR flash (Fig. 6.16), memory cells are connected in parallel to the bit lines, allowing cells to be read and programmed individually. The parallel connection of the cells resembles the parallel connection of transistors in a CMOS NOR logic gate (hence the name). In NAND flash (Fig. 6.17), however, cells are connected in series, resembling a CMOS NAND logic gate, preventing cells

NAND Flash

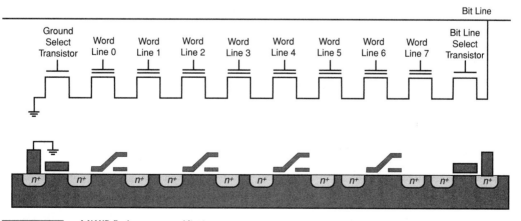

Figure 6.17 A NAND flash-memory architecture.

Figure 6.18 The current NAND flash record holder (2007): A 64-Gbit Samsung IC built using 32-nm CMOS technology (courtesy of Samsung Corporation).

from being read and programmed individually. That is, the cells are connected in series and must therefore be read in series.

One might legitimately wonder how many times a NAND flash can be erased and reprogrammed before it bites the dust. Interestingly, the "endurance" of NAND flash is generally much greater than that of NOR flash (typically 1,000,000 cycles vs. 100,000 cycles), because cell programming and erasure in a NOR flash rely on fundamentally different and asymmetric microscopic charge transport mechanisms (hot electron injection vs. quantum-mechanical tunneling, respectively) to take charge on and off of the floating gate in the transistor, whereas NAND flash utilizes a single (hence symmetric) mechanism (Fowler–Nordheim tunneling).

Figure 6.19 A flash-memory "card" for a digital camera, a close cousin to the thumb drive (courtesy of Sandisk Corporation).

The asymmetric nature of NOR flash programming and erasure increases the rate at which memory cells degrade during use. Clearly, however, for your USB flash drive, 1,000,000 cycles is going to be just fine for most things you will use it for, so don't sweat it!

Some NAND flash trivia? Sure! The current record holder (December 2007) in memory density is Samsung's 64-Gb NAND flash, built in 32-nm CMOS technology (Fig. 6.18). No, you probably wouldn't put this in a 64-Gb USB flash drive for your key chain! More commonly you will see your flash chip tucked safely away under the covers of an IC package. Figure 6.19 shows a common example from a my own digital camera; but, yep, it's the same-old NAND flash-memory chip under the hood.

Great. Now that we know a bit about memory (sorry!), and importantly, what a flash memory actually is, let's have some fun and deconstruct your USB flash drive!

6.3 Where are the integrated circuits and what do they do?

Okay, let's now rip the covers off of Mr. USB flash drive and see what actually lurks inside. As shown in Fig. 6.20, there are eight major components to a USB flash drive, all of which are mounted on a two-sided printed circuit board (PCB), some on the front and some on the back to minimize form factor.

- Component 1, *USB connector:* Clearly a USB flash drive needs a USB connector! USB? Universal serial bus. The original plug-and-play adapter! The details and history behind this important innovation are given in the sidebar discussion. Essentially the USB connector acts as an interface, a gateway of sorts, between the NAND flash-memory chip and the outside world (e.g., your laptop).

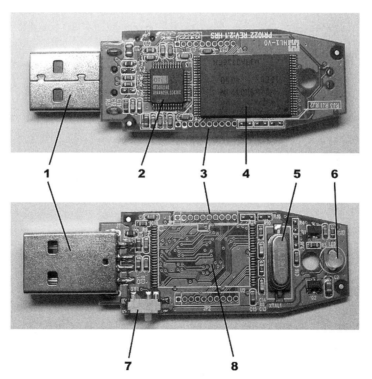

Figure 6.20   Front and back side views of a naked thumb drive, showing the building blocks. Components 1–8 are defined in the text (for reference, component 4 is the NAND flash IC). (Photograph and annotations by J. Fader, 2004.) (See color plate 18.)

- Component 2, *USB mass storage controller IC:* The USB connector it-self needs various electronic control functions for proper communication (signal levels, data timing, power conditioning, etc.), and this is provided by the USB mass storage controller IC – sounds important, but it's pretty routine!

- Component 3, *"Test points":* These electrical pins on the side of the PCB serve only as electrical contact points, "test points," to electrically stimulate and exercise the flash drive during the assembly process. It is obviously much more cost effective to identify any defects or dead ICs on the board prior to completing the assembly sequence and encapsulating the beast, and these test points serve that function nicely.

- Component 4, *NAND flash-memory IC:* No explanation necessary – here is Mr. NAND flash IC, the guts of the USB flash drive.

- Component 5, *Crystal oscillator:* The crystal oscillator is exactly what it says. A piece of quartz crystal designed to vibrate at a very particular frequency (just like in your watch), and then serve as a very stable system digital "clock" (like in a microprocessor) to time the read and erase program sequences, drive the communications between on-board ICs and the external world, provide signal timing for the USB interface, etc.

- Component 6, *LED indicator light:* Ah, a nice merger of photonics and elec-tronics. A semiconductor LED (light-emitting diode) that serves as a handy indicator light that the flash drive is on and functioning.

- Component 7, *Write-protect switch:* A switch to safeguard the contents you just dumped onto the drive. If toggled, this switch will prevent your friends from inadvertently erasing your Ph.D. dissertation to make room for their latest cell phone pics or MP3 files. Can be handy, although a lot of USB flash drives don't have these.

- Component 8, *Space to put a second NAND flash-memory IC!:* For those in the business of making money, consider: How about leaving an extra piggyback slot for a second NAND flash IC to double your memory capacity? Business students pay attention: We can go from a 1-GB to a 2-GB flash drive, at double the selling price, by simply using two (much cheaper) 1-GB NAND flash ICs instead of one (more expensive) 2-GB IC, and keeping all of the old circuitry and board of my original 1-GB flash drive. Ka-Ching!

There you have it. Mr. USB flash drive. Very handy gadget. Quite sophisticated! If you don't currently own one . . . you soon will. One last tidbit before moving on to bigger and better things. So . . . gulp . . . do you just yank out the USB flash drive after transferring files from your laptop and assume all will be well, or do you have to go through that laborious process of shutting it down properly first by means of your Windows "Safely Remove Hardware" function on your laptop? Answer? USB flash drives are designed to "hot swap," so it's fine to just pull it out. Please don't quote me on that!

Figure 6.21 A collection of different types of USB plugs (courtesy of agefotostock.com, © Wolpert).

Figure 6.22 The type of USB connector used in thumb drives (courtesy of agefotostock.com, © B. Krozer).

Companion Gadget #1: USB Drives

The universal serial bus (USB) is a serial bus standard designed to enable easy and robust interfacing of various electronic gadgets. In the context of PCs, USB "ports" and cables (Figs. 6.21 and 6.22) are designed to allow peripherals to be connected to the PC by a single standardized interface socket, enabling "plug-and-play" capabilities and allowing various components to be connected and disconnected without rebooting your machine (aka "hot swapping" – nice idea). Other slick USB features include (1) providing juice to low-power-consumption objects (think flash drives!) without the need for an external power supply (very cute), and (2) allowing different objects (flash drive, camera, CD player, mouse, etc.) to be used without requiring manufacturer-specific, individual software drivers to be installed (the whole notion behind plug-and-play). P.S. If you lived before the emergence of USB you will vividly recall the MAJOR pain associated with connecting *anything* new to your PC. Grrrr. In 2004, there were about 1 billion USB-enabled devices in the world.

(continued)

Companion Gadget #1: USB Drives (*continued*)

A little USB history? Sure! USB was standardized by the USB Implementers Forum (USB-IF), an industry-driven standards body that included, among others, Apple, Hewlett-Packard, NEC, Microsoft, Intel, and Agere. The USB 1.0 (version 1) was introduced in November of 1995. The USB was strongly promoted by Intel (UHCI and open software stack), Microsoft (Windows software stack), Philips (Hub, USB-Audio), and U.S. Robotics. USB 1.1 (read: let's fix things wrong with 1.0 without a complete USB overhaul) came out in September 1998, and soon things took off in a big way. If you own a new computer, you now also probably own USB 2.0 drives (2G USB). USB 2.0 essentially is a higher-bandwidth version of USB 1.1, running at 480 Mbits/s, and was standardized by the USB-IF at the end of 2001. Smaller USB plugs and receptacles for use in handheld and mobile devices (look at the transfer cable on your digital camera for an example), called Mini-B, were recently added to the USB family, and a newer variant, the Micro-USB, was announced by USB-IF on January 4, 2007. USB 3.0? You bet! On September 18, 2007, a USB 3.0 prototype was demonstrated at the fall Intel Developer Forum. USB 3.0 is targeted for $10\times$ the current USB 2.0 bandwidth, or roughly 4.8 Gbits/s (read: pretty darn fast), utilizes a parallel optical cable, and will be backwards-compatible with both USB 2.0 and USB 1.1.

References

1. See, for instance, *http://answers.google.com/answers/*.
2. See, for instance, Wikipedia: *http://en.wikipedia.org*.
3. M. Barr, "Introduction of memory types," *http://www.netrino.com/Publications/Glossary/MemoryTypes.php*.
4. G. C. Hadjipanayis, *Magnetic Storage Systems Beyond 2000*. NATO Science Series II, Kluwer Academic Publishers, New York, 2001.

7 Bricks and Mortar: Micro/Nanoelectronics Fabrication

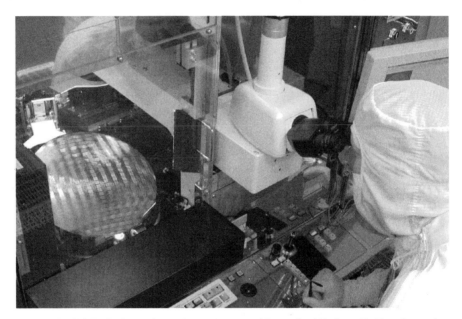

Photograph of IC fabrication equipment in use (courtesy of International Business Machines Corporation. Unauthorized use is not permitted). (See color plate 19.)

If You Have Built Your Castles In the Air
Your Work Need Not Be Lost;
That Is Where They Should Be.
Now Put The Foundations Under Them.

Henry David Thoreau

Modern miracles. Consider once more the facts before us: (1) We have journeyed from a thumbnail-sized single transistor (*uno, eins, ichi, ena, odin, un, yi, en, jeden, egy, moja* – you get the idea) in 1947, to a thumbnail-sized integrated circuit (IC) containing well over 1 billion (1,000,000,000) transistors – all in a span of a little more than 60 years. (2) The manufacturing facilities used to build these ICs require in the range of $2–3B (yep, billions!) of capital investment just to get in the game, and yet even with this GDP-sized investment, lots of

Figure 7.1   A 300-mm silicon wafer of Intel Pentium Centrino Mobile microprocessors (courtesy of Intel Corporation).

companies around the world make lots of money building ICs for a living (recall: over $100B was spent in 2005 for the ICs alone).

All modern silicon ICs begin as beach sand and end as a large IC "wafer" (Fig. 7.1) containing hundreds of identical ICs. Wafer? How about, flattened monster potato chip? Ultimate, Ultimate Frisbee? Really thin flying saucer? No wait, got it – mega Necco wafer! (orange is my favorite, but I do have a penchant for licorice). Currently, these silicon wafers can be as large as 300 mm in diameter (about 12 in). Why so big? Easy; more ICs per wafer, lots more money for about the same effort. A single IC (Fig. 7.2, in this case an Intel Pentium Centrino Mobile microprocessor) is the goal, and it will then be tested for functionality (does it have a heartbeat?), cut up and packaged for consumption, and then sold for big bucks to you and me (Fig. 7.3).

This chapter will show you, gentle reader, how we get from point A to point B – sand to IC; Dark Ages to Communications Revolution. It should be easy to glean, with even a nanosecond's reflection, that putting 1,000,000,000 of anything onto a centimeter-sized piece of something, and expecting all 1,000,000,000 to work, requires some skill. Lots of skill. I would argue that modern IC manufacturing easily represents the most intricate piece of human engineering ever successfully conducted (you know already my affinity for bold claims, but good luck trying to refute it). (P.S. I said human engineering, not nature's engineering – she still has us beat by miles.)

The modern IC begins and ends in the IC fabrication facility, the "wafer fab," and involves seven major types of operations, each of which must be executed with almost unbelievable dexterity (think atomic-scale accuracy):

- *Establish the environment*: the cleanroom, defect density, air handling, chemical purity, deionized water.

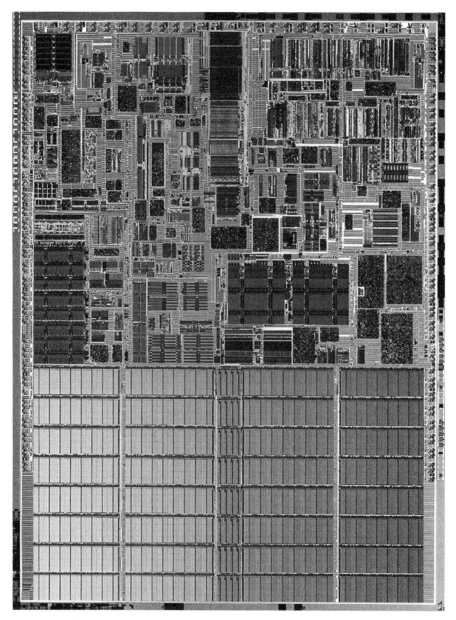

Figure 7.2   Die photo of an Intel Pentium Centrino Mobile microprocessor (courtesy of Intel Corporation).

- *Selectively add materials*: doping, oxidation, thin film deposition, epitaxial growth, ion implantation, metalization, spin-coating photoresist.
- *Pattern the materials*: photolithography.
- *Selectively remove materials*: chemical (wet) etching, plasma (dry) etching, photoresist developing, chemical–mechanical polishing.
- *Selectively redistribute materials*: diffusion, silicide formation, strain engineering.
- *Apply damage control*: annealing, hydrogen passivation, sacrificial layers.

Sand to ICs

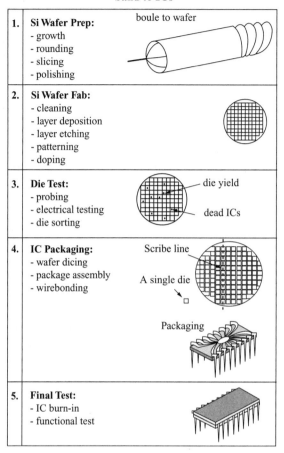

1.	**Si Wafer Prep:** - growth - rounding - slicing - polishing
2.	**Si Wafer Fab:** - cleaning - layer deposition - layer etching - patterning - doping
3.	**Die Test:** - probing - electrical testing - die sorting
4.	**IC Packaging:** - wafer dicing - package assembly - wirebonding
5.	**Final Test:** - IC burn-in - functional test

Figure 7.3   Representation of the entire IC fabrication "process," from semiconductor crystal growth, to wafer fabrication, to packaging of the IC, to final IC test.

- *Admire our handiwork and then package our shiny new IC:* functional test, dicing, packaging, die attach, wirebonding, burn-in, final test.

First we start with a really, really clean, bright, and shiny, 300-mm silicon wafer. A carefully controlled environment is needed to cordon off the IC fab process from a host of undesirable outside influences, and thus IC "processing" happens inside what is known as a "cleanroom." To build Mr. Centrino, we selectively add materials to our wafer by a host of means; we pattern those materials by using essentially photographic techniques; we selectively remove materials we do not need; we redistribute certain materials; we remove any damage we introduce in the process; and finally we package them for consumption. Done. Seem easy? Well, building Mr. Centrino requires thousands of individual fabrication steps and takes several months of 24/7 activity. Not to mention some highly skilled workers to keep things running smoothly, gobs of electricity, the purest chemicals on Earth, and some very, very expensive tools. Curious as to how it's done? Read on.

Figure 7.4 Inside a 300-mm IC manufacturing facility. (Courtesy of International Business Machines Corporation. Unauthorized use is not permitted.)

7.1 The IC fabrication facility (aka "the cleanroom")

The best IC cleanrooms (Fig. 7.4) are roughly 100,000 times cleaner in terms of airborne particulates (dust, bugs, you get the idea) than the best hospital operating room. A cartoon of a simplified cleanroom cross section is shown in Fig. 7.5. The key to cleanliness lies in constant air flow through sophisticated filtration systems. Thus air is continually circulating in the cleanroom at high speed, 24/7, flowing downward through sophisticated filters, out through the "raised floor," and back through the filters, etc. Cleanrooms are classified by "Class" (sorry), and Fig. 7.6 shows the number of allowed particulates of a given size in a cubic foot of air at any given time (as you can see cleanroom classes are historically tied to 0.5-μm particle densities). A modern 300-mm IC production "line" is Class M-1 (yep, off the scale!).

Why so clean? Well, recall the size of Mr. Transistor – 10s to 100s of nanometers. Any particulate that happens to land on the surface of the wafer during a critical processing step will effectively "shadow" Mr. Transistor, preventing the completion of the construction step in question, be it addition, subtraction, or patterning of that particular layer. Read: R.I.P. Mr. Transistor. And one DoA transistor can lead to one DoA IC; and there might be 1,000,000,000 Mr. Transistors on one IC. Cleanliness is next to godliness.

To minimize air-handling volume and hence (seriously expensive) filtration requirements, only the business ends of the various wafer processing equipment

Fabrication Cleanroom Classes

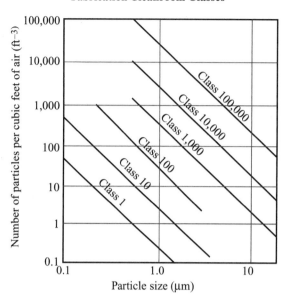

Figure 7.5 U.S. Federal Standard 209E for airborne particulate densities (by cleanroom "Class") in an IC fabrication facility.

open into the cleanroom (see Fig. 7.7). The guts of the equipment are kept in a separate equipment "core" or "chase," and do not require as pristine an environment (maybe only Class 1,000 – still 4,000 times cleaner than you enjoyed during your last knee surgery!). Exceptionally pristine environment for eliminating all transistor-sized foreign objects from the cleanroom? – check.

So what about the people who actually build the ICs and necessarily inhabit the cleanroom? Well, I'm not quite sure how to break this to you, but on a transistor scale we humans are downright filthy creatures, a veritable "Pig-Pen" of Charlie Brown fame, surrounded by a constant cloud of skin flakes, hair bits, bacteria, viruses, chemical residues, and other noxious exfoliants! Yuck. The sodium in your sweat is particularly deadly to Mr. Transistor. Nope, not a pretty sight, sorry. To enable us to even enter the cleanroom, we must therefore get hosed down with

	Particles/ft^3				
Class	**0.1 μm**	**0.2 μm**	**0.3 μm**	**0.5 μm**	**5 μm**
M-1	9.8	2.12	0.865	0.28	
1	35	7.5	3	1	
10	350	75	30	10	
100		750	300	100	
1,000				1,000	7
10,000				10,000	70

Figure 7.6 Particle density as a funcion of particle size for the various cleanroom classes.

The Fab

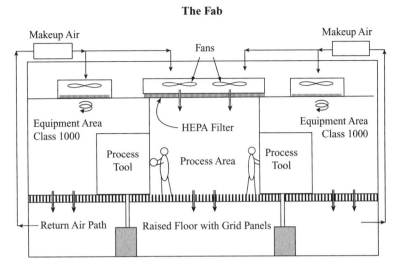

Figure 7.7 The basic construction and air-handling of an integrated circuit fabrication facility.

high-velocity air, and then "suit up" in the infamous cleanroom "bunny-suit," don mask, hairnet (yep, serious hat-hair after work), and latex gloves (Fig. 7.8). The successful manufacturing of IC wafers requires both time and energy, utilizing three shifts of folks a day, seven days a week, 52 weeks a year. Bunny-hopping fab workers traverse all of the dark reaches of the modern IC fab many, many times for the execution of an individual step in the "process flow" at a piece of very

Figure 7.8   The ubiquitous (and infamous) Gor-tex "bunny-suit" used in cleanroom environments to protect our precious ICs from particulate-bearing humans (courtesy of Intel Corporation).

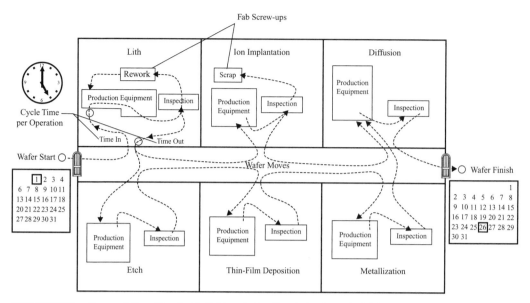

Figure 7.9 Representation of the start-to-finish wafer handling during the wafer fabrication "process."

expensive "production equipment" (aka the "tool"), followed by the inspection of what was just done (to make sure it's right before moving on), and process step rework (if needed), as illustrated in Fig. 7.9. Clearly, minimizing the contact time and handling of the IC wafers by us error-prone humans is generally a good thing (despite its strength, drop a 300-mm wafer and it will shatter into a million pieces), and increasingly, modern IC fabs are highly automated, using robotic "cluster tools" for wafer handling and tool-to-tool transfer (Fig. 7.10). Now that you know the ins-and-outs of the IC fab, let me now take you on a quick tour through the various production steps needed to build (fabricate) an IC and introduce you to the nuances of the various tricks of the trade [5,6].

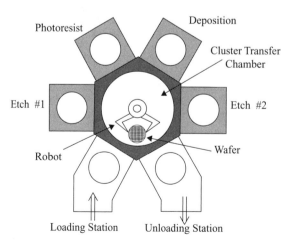

Figure 7.10 Representation of a robotic "cluster" tool used in IC fabrication.

Figure 7.11 Growth of a silicon crystal "boule" from a seed crystal lowered and rotated in a vat of highly purified molten silicon (courtesy of Texas Instruments, Inc.). (See color plate 20.)

7.2 Crystal growth and epitaxy

Let's start with the semiconductor crystals themselves, as they are clearly the most basic building blocks needed to produce Mr. IC. Ever spend time as a kid trying to grow a large cubic salt crystal, or perhaps those tasty "rock-candy" sugar crystals (yum . . .)? Think way back. Yep, you guessed it – I did. My own favorite was the sapphire-blue, copper-sulfate crystal (slightly trickier, but absolutely gorgeous if you get it right). Remember how? Dissolve as much of the material (the solute) in question into your boiling water (the solvent), creating a supersaturated solution once it cools down. Pour a little of the solution in a shallow dish, and allow it to evaporate. A crust of mini "seed crystals" results. Choose a nice one, tie a thread around it, and then lower it into your supersaturated solution. Cover it up, put it in the closet, and one week later, wah-lah! – a beautiful gemlike crystal.

7.2.1 Semiconductor crystals

Giant silicon crystal? Same basic idea. Melt the ultra-pure silicon, lower a seed crystal into the vat, rotate it very slowly to ensure a round crystal, and now, over several days, under very, very tightly controlled conditions, gently "pull" the seed from the melt (Fig. 7.11). This crystal-growth technique is called the Czochralski process (CZ for short – for obvious reasons). The result is a pristine silicon "ingot" or "boule." Think 300 mm in diameter by 6 feet long (Fig. 7.12). By any stretch the largest and most perfect crystals on Earth; more perfect by far than any flawless gem-quality diamond (if only silicon crystals looked brilliant and clear like diamonds – imagine a 12 in × 6 ft rock in your engagement ring!).

Figure 7.12   A 300-mm silicon crystal boule (courtesy of Texas Instruments). (See color plate 21.)

Chemistry in Action: The Purification of Silicon

We live in a silicon world [1]. This statement is literally as well as figuratively true. Silicates, the large family of silicon–oxygen-bearing compounds, such as feldspar ($NaAlSi_3O_6$), beryl [$BeAl_2(Si_6O_{18})$], and olivine [$(MgFe)_2SiO_4$], make up 74% of the Earth's crust. Silicon is the third most abundant element on planet Earth (15% by weight), after iron (35%) and oxygen (30%). One need go no further than the beach and scoop up some sand to hold silicon in your hand and start the crystal-growth process. But how is it purified to the extreme levels needed for producing gazillions of 300-mm silicon wafers? Ahhh... Clever use of chemistry. Low-grade silicon is first produced by heating silica (impure SiO_2) with carbon in a furnace. The carbon pulls the oxygen out of the SiO_2 (in chem-speak, "reduces" it), leaving behind still impure silicon. Next it is chlorinated to yield $SiCl_4$ (or $SiHCl_3$), both of which, importantly (and surprisingly), are *liquids* at room temperature. This silicon "halide" liquefaction process is a clever path, as there are many, many ways to purify liquids. In this case we distill (yep – think moonshine!) the $SiCl_4$ (or $SiHCl_3$) solution multiple times, reaching our final ultra-pure state. Now, finally, we heat the solution in a hydrogen atmosphere, yielding pure silicon ($SiCl_4 + 2H_2 = 4HCl + Si$). We cool it down, bust it into chucks, and throw these pure silicon pieces into our CZ vat, and we are now in business [2].

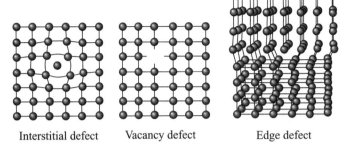

Interstitial defect   Vacancy defect   Edge defect

Figure 7.13   Several common types of "defects" encountered in the growth of crystalline semiconductors.

The key to success in building functional ICs (target: 100% "yield," every one of the ICs on the wafer alive and kicking postfab) is that the crystal imperfections (defects) introduced during the boule growth be kept to miniscule densities. Typical crystal defects include "interstitials" (an extra atom where it does not belong), "vacancies" (a missing atom), and a edge defect or "dislocation" (a row or line of extra or missing atoms), as illustrated in Fig. 7.13.

Following growth, the boule is then sliced into thin "wafers" (see Fig. 7.14) (a 300-mm silicon wafer is about 800 μm thick – a little less than a millimeter) by a diamond saw (the only thing hard enough to cut silicon!) and polished on one side to produce a pristine, mirror-like finish (Fig. 7.15). Postpolished and then chemically cleaned wafers are atomically flat (i.e., < 1-nm variation across the entire 300-mm surface – read: pretty amazing) on the side of the wafer where we will begin the IC fab process. It should be obvious that there are enormous economic advantages to making the starting wafer as large as possible (at fixed yield, more ICs per wafer to sell), and silicon wafer sizes have thus grown steadily since the early 1960s (Fig. 7.16) to their present mammoth size of 300 mm. The

Figure 7.14   Rough-sawed 300-mm wafers from a silicon boule (courtesy of Texas Instruments, Inc.).

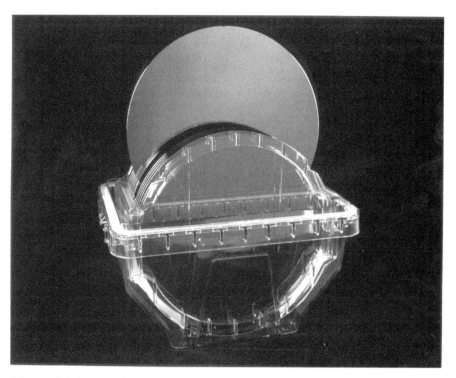

Figure 7.15   A "cassette" of 16 cut and polished 300-mm wafers. (Courtesy of International Business Machines Corporation. Unauthorized use is not permitted.)

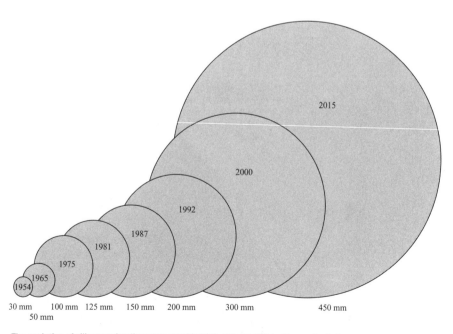

Figure 7.16   The evolution of silicon wafer diameters used in high-volume silicon IC manufacturing.

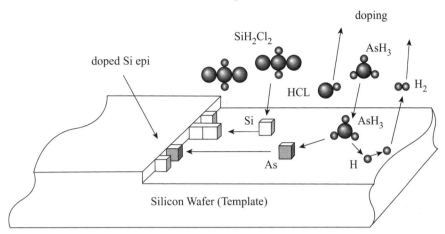

Silicon Epi Growth

Figure 7.17 Illustration of the growth process of conventional arsenic-doped high-temperature silicon epitaxy.

ability to grow near-defect-free large silicon wafers is actually rather unique in the grand scheme of all semiconductors and is a major contributor to the dominance of silicon-based manufacturing. Clearly one also must also increase the wafer-size capability of the fab tools during this evolution, without losing the cross-wafer manufacturing control, and this is one reason why modern wafer fabs run into the billions of dollars.

7.2.2 Epitaxy

In addition to defect-free semiconductor wafers, we also typically require additional silicon crystal grown directly on top of the starting wafer (i.e., a second layer of pure silicon crystal that is atomically aligned – defect free – with the underlying "host" crystal wafer). This extra silicon, extended crystal if you will, is called *epitaxy*, or *epi* for short, and can be grown either doped or undoped (intrinsic), depending on the need. Silicon epi can be grown by means of either a high-temperature (think $1,100°C$) process or a low-temperature (think $500–600°C$) process. The former is fairly routine today, whereas the latter remains quite challenging and requires a sophisticated (read: multi-$M) toolset. The correlation between epi growth temperature and the ease of producing high-quality epi makes intuitive sense – the higher the temperature, the more readily the silicon atoms can "find" where they need to be on the surface of the host lattice and incorporate themselves into the crystal matrix without creating a defect. The chemistry of a high-temperature n-type silicon epi growth process is illustrated in Fig. 7.17.

So why would we want to produce epi at very low temperatures and risk introducing defects into the wafer? Good question! One answer: to enable "bandgap engineering" (recall the idea of energy bandgaps in Chap. 5). As depicted in

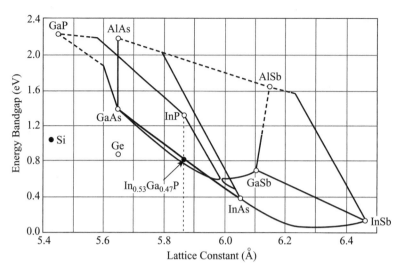

Figure 7.18 Energy bandgap as a function of lattice constant for the major semiconductors. It can be seen that whereas Si and Ge are not lattice matched, AlAs is lattice matched to GaAs, as is In$_{0.53}$Ga$_{0.47}$As to InP.

Fig. 7.18, the energy bandgap more or less depends reciprocally on lattice constants in semiconductors. Being able to selectively change bandgaps *within* a given device, often multiple times, or even as a function of position, has proven to be an exceptionally powerful lever for engineering the properties of both electronic and photonic devices to suit our myriad needs (e.g., to speed up a transistor or change the color of a laser diode). We can grow a semiconductor alloy of a given composition to achieve a desired energy bandgap. Inevitably, however, we must embed that resultant alloy in the device, and this requires that we grow it as epi and seamlessly mate it to the host crystal. Low-temperature epi is usually mandated because we do not want to disturb the underlying structures or materials we have already assembled our device with.

Consider the following technologically important example as a test case. A silicon germanium (SiGe) alloy can be grown on a silicon host wafer to practice silicon-manufacturing-compatible bandgap engineering for high-speed transistor design (this is my personal research field). Unfortunately, as illustrated in Fig. 7.19, such a SiGe alloy is not "lattice matched" to the silicon crystal (that is, they have different lattice constants – in this case the lattice constant of Si = 0.543 nm, whereas that of Ge is 0.566 nm, a difference of about 4%). This inherent "lattice mismatch" between epi and wafer is often the case in practical bandgap engineering. Intuitively, this atomic mismatch doesn't sound like a good thing, and it's not, because the lattice in the epi we are trying to grow (say a SiGe alloy with a 25% Ge mole fraction) does not line up with the lattice of the crystal-growth template of the wafer. Not only that, but we would like to grow the epi at temperatures below say 600°C, where the growth process is inherently more challenging (believe it or not, 600°C is actually very low as semiconductor processing temperatures go!). The natural consequence is that we can easily introduce extended dislocation defects at the epi growth interface during fabrication

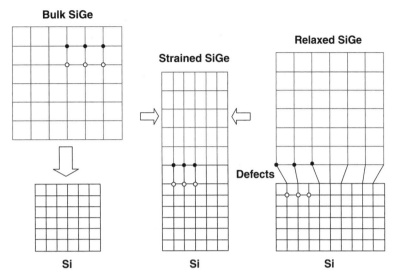

Figure 7.19 Illustration of the heteroepitaxial growth process for depositing one semiconductor (here, a SiGe alloy) of a given lattice constant onto a host semiconductor crystal (here, silicon) of a different lattice constant. The defect-free "strained layer epitaxy" (center) is the desired result.

(Fig. 7.20), which from the top side are suggestive of a cacophony of train tracks at a switching station (Fig. 7.21) – read: not a good thing for a high-yielding IC! In actuality, it took people over 30 years to go from the basic idea of SiGe bandgap engineering to its successful implementation in manufacturing, clearly representing an especially challenging materials growth problem. It turns out, though, that growing defect-free SiGe epi on large Si wafers is fairly routine today, provided you have the know-how and build some very sophisticated equipment for doing it.

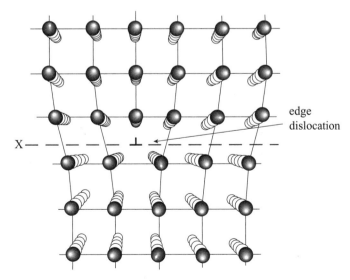

Figure 7.20 Illustration of an "edge dislocation" formed at the growth interface during an epitaxial growth process, which can then "thread" its way to the crystal surface, forming an extended defect.

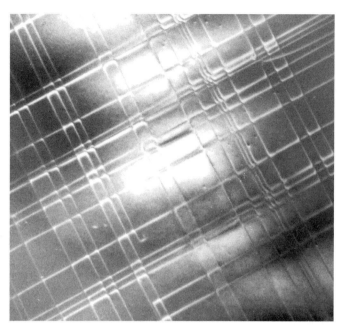

Figure 7.21 Top-view image of a dislocation array produced in a failed epitaxial growth process (here, a SiGe alloy was epitaxially deposited onto a 200-mm silicon wafer).

When done right, the epi SiGe alloy is squeezed (compressively strained) during its atom-by-atom growth sequence, "forcing" the film to adopt the underlying Si lattice constant and producing defect-free nanoscale material in the resultant "epitaxial strained SiGe alloy" (see Fig. 7.22). A host of properties beneficial to device engineering results. Many different types of nanoscale, defect-free "pseudomorphic" epitaxial alloys (see sidebar) are in practical use today, lurking inside this or that electronic or photonic device. Clearly, however, it is not hard to imagine that the amount of strain that can be accommodated in pseudomorphic epi films is finite, and thus if we get greedy by either (a) putting to much Ge in the film (too much mismatch) or (b) growing too thick a film (too much strain), we will exhaust nature's patience (formally, exceed the thermodynamic stability criterion), and the whole structure will "relax" back to its natural unstrained state, generating tons of extended defects in the process (read: instant death to any IC). Knowing that precise compositional boundary (% Ge + SiGe film thickness) that nature will allow us to march up to, but not cross, is clearly of great practical importance in successful bandgap engineering.

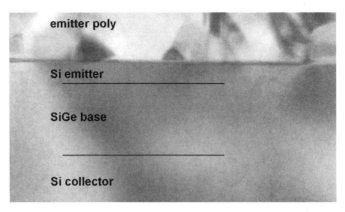

emitter poly

Si emitter

SiGe base

Si collector

Figure 7.22 Cross-sectional view of a defect-free nanoscale SiGe strained epitaxy deposited on the surface of a 200-mm silicon wafer. (Courtesy of International Business Machines Corporation. Unauthorized use is not permitted.)

7.3 Doping: Diffusion, implantation, and annealing

Okay. We now have a semiconductor crystal-construction platform (either a single-crystal semiconductor wafer or an epitaxial layer grown on a wafer, or both) to build our transistors on. For reasons discussed at length shortly, we now need to be able to precisely tailor the doping in the crystal to alter its resistivity; and not just locally change its doping type (p vs. n), but also its concentration ($N =$ dopants/cm^3) as a function of depth into the crystal [this is often called the "doping profile," $N(x)$]. Finally, we need to be able to controllably locate our doping at a particular place on the surface of the crystal. That is, we want to draw shapes and patterns of doping into the surface of the crystal (i.e., let's put doping here in this particular shape, but not over there).

Once that doping is safely on the surface of the crystal, we need to then redistribute it within the volume of the crystal. This dopant redistribution accomplishes two major things: (1) we can precisely tailor the doping profile $N(x)$ to suit our needs (e.g., to change the properties of Mr. Transistor), and (2) we can electrically "activate" the dopant impurity during the redistribution process. Oh yeah? Recall: For the donor impurity to give up its electron to the conduction band and make itself available for current flow (i.e., fulfill its destiny as a donor), the impurity atom must replace a silicon atom in the crystal lattice.

Historically, this dopant introduction and redistribution process was accomplished in a very high-temperature (think 1,000°C – pretty darn warm) "diffusion furnace" (Figs. 7.23 and 7.24). In this process, the silicon wafers are placed in a quartz "boat" inside a quartz "diffusion tube," heated to a very high temperature (literally glowing red-hot), and then the selected dopant in a gaseous form (e.g., phosphine gas for phosphorus n-type doping) is allowed to flow into the tube. At temperatures this high, the dopant has sufficient energy to migrate (diffuse) into the crystal, in a process known as "solid-state diffusion" (note that nothing is melted here) – the dopant atoms worm their way (diffuse) into the solid by one

Figure 7.23   A set of vertical diffusion furnaces (courtesy of Intel Corporation).

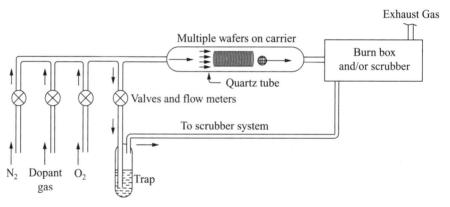

Figure 7.24   Representation of a dopant diffusion furnace.

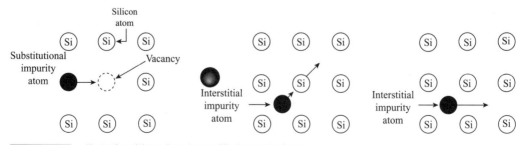

Figure 7.25   Illustration of the various dopant diffusion mechanisms.

of two basic means: (1) "substitutional" diffusion (the dopant moves via vacant silicon lattice sites); or (2) "interstitial" diffusion (the dopant moves between the silicon atoms in lattice) (Fig. 7.25). When you determine that the dopants are properly located on the silicon lattice sites, turn down the temperature and they are effectively frozen into place. Perfect.

Caffeine Alert #1 (Warning – Read at Your Own Risk!): The Mathematics of Diffusion

As intimated in Chap. 5, diffusion is diffusion, whether for electrons diffusing into a semiconductor to generate a current flow, or a dopant impurity diffusing into the crystal at high temperature, and the mathematics of diffusion begins and ends with Fick's laws. Fick's first law of diffusion (in one dimension) states that

$$\mathcal{F} = -D\,\frac{\partial N}{\partial x}, \tag{7.1}$$

where $\mathcal{F}$ is the dopant atom "flux" (dopants/cm^2 s), N is the dopant concentration (number of dopants/cm^3), and D is the "diffusion coefficient," a proportionality constant dependent on the physical context of the problem that is typically experimentally determined. A general continuity equation (just as for electrons and holes) relates the rate of increase in dopant concentration over time (time derivative) to the divergence of the dopant flux, which in one dimension can be written as

$$\frac{\partial N}{\partial t} = -\frac{\partial \mathcal{F}}{\partial x}. \tag{7.2}$$

Combing these two equations results in Fick's second law of diffusion (1D):

$$\frac{\partial N}{\partial t} = D\,\frac{\partial^2 N}{\partial x^2}. \tag{7.3}$$

This lean-and-mean partial differential equation can typically be solved with a separation-of-variables technique, subject to the appropriate boundary conditions. The two most practical boundary conditions are (1) "constant-source diffusion" (the surface dopant concentration is held fixed at the surface of the crystal during diffusion), and (2) "limited-source diffusion" (a fixed quantity of dopants is placed at the surface and is not replenished). Solving these two problems results in the classical time-dependent diffusion profiles:

$$N(x, t) = N_0\,\mathrm{erfc}\left(\frac{x}{2\sqrt{Dt}}\right) \tag{7.4}$$

for case (1) (erfc is the so-called "complementary error function" from probability theory fame), and

$$N(x, t) = \frac{2N_0\sqrt{Dt/\pi}}{\sqrt{\pi\,Dt}}\,\exp\left[-(x/2\sqrt{Dt})\right]^2, \tag{7.5}$$

for case (2), which is the famous Gaussian distribution function. Both functions look quite similar in shape, as shown in Figs. 7.26 and 7.27, and their time-dependent evolution at fixed temperature is fundamentally driven by the Dt product. As an important aside, rapid-thermal-processing (RTP) systems succeed in suppressing dopant diffusion during high-temperature annealing by minimizing the so-called "Dt product" in these functions.

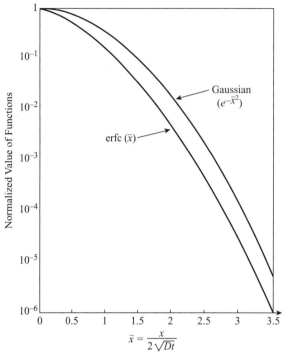

Figure 7.26 A comparison of the complementary error function and Gaussian function dopant profiles during solid-state diffusion.

For early IC manufacturing, solid-state diffusion was the preferred means of doping crystals, but it is in fact rarely used today to build state-of-the-art transistors like you might find in your PC or cell phone. Why? At the extreme temperatures needed for effective solid-state diffusion, it becomes nearly impossible

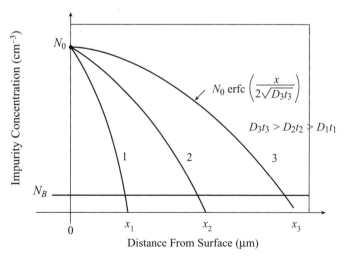

Figure 7.27 Doping profile evolution as a function of varying Dt for a diffusion process.

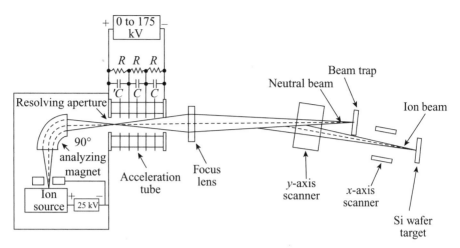

Figure 7.28 Representation of an ion implantation system (inspired by [3]).

to control the location and shape of the doping profile to the levels of precision needed to build modern transistors. Consider: If I want to put 100,000,000 transistor on a 1-cm^2 piece of silicon for a microprocessor, it is not hard to imagine that the lateral and vertical dimensions of its doped regions must be very, very, very shallow and very, very, very close together, without touching (think precise control over 100s of nanometers). This is impossible to accomplish in standard high-temperature diffusion processing.

So how do we do it? The answer may appear wacky, but essentially we build a compact particle accelerator to speed the dopant atoms to a fraction of the speed of light, and then slam them into the surface of the crystal! At such speeds the atoms literally penetrate the surface and become embedded in the crystal. To accelerate the dopants, we first turn them into ions (strip an electron or two off) so that they become charged and thus easily accelerated by a high voltage. Such a beast is called an "ion implanter" (okay, not so imaginative), and the basic idea in fact dates back to Shockley and friends. Clearly, although the idea is appealingly simple, the implementation is not. Still, by the early 1980s commercial ion implanters were in use, and today they represent the gold standard for precision doping of transistors in ICs (Fig. 7.28).

Not so fast though. Intuitively, ion implantation feels like an awfully violent process (billiard ball ion + cannon + speed of light + aimed at the surface of my perfect crystal = big mess!). Yep, very violent indeed. In practical ion implantation, the impinging dopant ions actually "amorphize" (turn amorphous, destroy the lattice structure) the crystal surface, a seemingly harebrained idea after going to all that trouble to create a perfect 300-mm-diameter crystal! Hence successful ion implantation can be achieved only if we figure out a clever way to repair ("anneal") the implantation damage and restore the perfect crystallinity of the semiconductor sample, now with implanted impurities safely on silicon lattice sites and electrically active. Nicely, nature is reasonably kind in this regard. If we simply expose the postimplanted damaged crystal to high temperatures in an

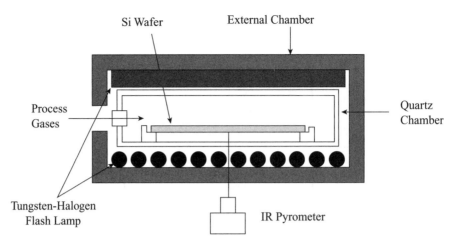

Figure 7.29 A RTP system for annealing and diffusion (inspired by [3]).

"annealing furnace," the crystal surface regrows nicely, removing all implantation damage and activating the dopant impurities we implanted. But wait a second. Didn't we turn to ion implantation for doping to avoid having to expose the doped crystal to high temperatures?! Yep. Bummer. We have to be a bit more clever. It turns out that, although we do in fact need high temperatures to anneal the implant damage, if we can apply those high temperatures for only a very, very short time interval, the dopant impurities simply do not have time to move around very much. The answer to our problem is thus the so-called "rapid-thermal-processing" (RTP) system (Fig. 7.29), again a workhorse in modern IC manufacturing. In a RTP system, the wafer is surrounded by a tungsten–halogen "flash lamp" (just what it sounds like), which can raise the temperature of the

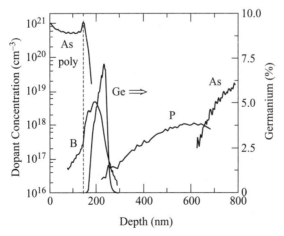

Figure 7.30 Measured 1D "doping profile" through the center of a real transistor (in this case a bandgap-engineered SiGe bipolar transistor), showing both doping concentration and germanium fraction as functions of depth into the silicon crystal. The dashed line indicates the original surface of the silicon wafer. (Courtesy of International Business Machines Corporation. Unauthorized use is not permitted.)

wafer exceptionally quickly and in a highly controlled manner (think 950°C for 5 s!). RTP annealing gets the repair job done without moving the dopants around. We're in business!

With ion implantation and RTP at our disposable, the precision with which doping profiles can be engineered in modern transistor design is quite remarkable. Figure 7.30 shows a typical doping profile [$N(x)$] of a high-speed transistor. This device has arsenic (As) n-type doping, boron (B) p-type doping, and phosphorus (P) n-type doping; varying in concentration from about 1×10^{16} cm^{-3} to 1×10^{21} cm^{-3}, all over a distance of about 800 nm (< 0.1 μm, 0.1 millionths of a meter)! [This bandgap-engineered transistor also uses SiGe epi embedded inside the device for speed enhancement; hence the Ge(x) region is also shown.]

7.4 Oxidation and film deposition

In addition to the precise introduction of dopants into the crystal, we need to be able to add new layers. Why? Well, for various reasons. Those added layers on top of the crystal might play an active role (e.g., the gate oxide in Mr. Transistor – more on this in the next chapter), or act as an ion implantation or diffusion barrier (to control exactly where our dopants end up), or even serve as structural scaffolding for the buildup of a given device. We have two basic means at our disposal for putting new layers down: thermal oxidation of the crystal itself (think rust!), and film deposition by various chemical means. As previously mentioned, the fact that silicon trivially oxidizes to form one of nature's most perfect dielectrics is arguably its single most important attribute contributing to its total dominance of the Communications Revolution.

So how exactly do we cause silicon to rust? Recall Chemistry 101. If I take a hunk of silicon crystal, expose it to pure oxygen, and add a little heat to make it all go, silicon dioxide ("oxide" for short) results. In chem-speak,

$$\mathrm{Si} + \mathrm{O}_2 \quad \rightarrow \quad \mathrm{SiO}_2. \tag{7.6}$$

This would be called a "dry" oxide. When I say "add heat," we might need 800°C or 900°C to make it go quickly enough to be practical. Even so, if we simply leave a pristine silicon wafer on the shelf, at room temperature, a "native oxide" of maybe 1–2 nm will result. If you are in a hurry, you could alternatively expose silicon to heated water vapor (steam) and do the same thing as in dry oxidation, but much more quickly, resulting in a "wet" oxide, according to

$$\mathrm{Si} + 2\mathrm{H}_2\mathrm{O} \quad \rightarrow \quad \mathrm{SiO}_2 + 2\mathrm{H}_2. \tag{7.7}$$

In practice, dry oxides generally offer the best quality (e.g., higher breakdown strength) whereas wet oxides are quick-and-dirty. Part of the quality associated with dry oxides lies in the minimization of imperfections (often called "dangling bonds") at the growth interface between the SiO$_2$ and the silicon crystal that can adversely impact a transistor's properties or long-term reliability (more on

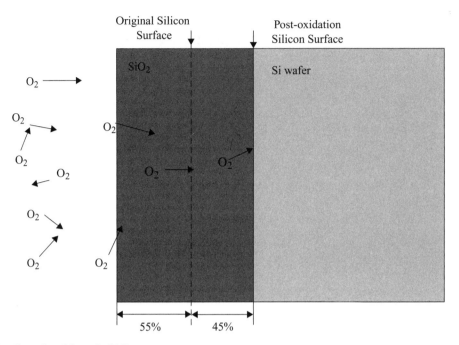

Original Silicon
Surface

Post-oxidation
Silicon Surface

SiO₂

Si wafer

O₂

55%   45%

Figure 7.31   Illustration of thermal oxidation process.

this later). As you might imagine, if I wanted to optimize the time required for growing an oxide, I might start with dry oxidation to establish a pristine growth interface, then switch to wet oxidation to grow the thickness I need quickly, and then at the end switch back to dry oxidation to ensure the top surface of the oxide is again pristine. This is a common fabrication path taken in building Mr. Transistor.

A dry oxidation process is illustrated in Fig. 7.31. Just from the basic chemistry we can see that silicon is actually consumed in the production of oxide, and thus we effectively eat up our precious crystal to make oxide (the final oxide layer will be about 55% above the original silicon surface, and the silicon crystal surface now 45% below where it started; you get it: 100 − 55). Read: Not always a good thing, esepcially if we need fairly thick oxide layers (say more than a few hundred nanometers).

For such thick oxides, or even other materials we might need in our toolkit, we can engage creative chemistry to "deposit" layers (aka "films") on top of the wafer. This is typically accomplished with the ubiquitous "chemical-vapor deposition" (CVD) system. Such systems are often run at reduced (below atmospheric) pressures, and thus are known as LPCVD (low-pressure CVD) tools. LPCVD is a very simple concept (Fig. 7.32): Load your wafers into a furnace that can be controllably heated, and then have a means to flow in (and exhaust) the appropriate gases (chemical vapors) needed to create the film you are trying to deposit. Presto!

As an example, one means to deposit (in fab lingo, "put down") oxide is to use dichlorosilane at about 900°C (there are typically several available chemistries

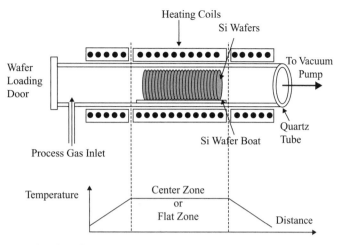

Figure 7.32 Illustration of a LPCVD system (inspired by [3]).

for any given desired film, each with their own set of pros and cons). This reaction in chem-speak is

$$SiCl_2H_2 + 2N_2O \rightarrow SiO_2 + 2N_2 + 2HCl. \qquad (7.8)$$

Yep, we produce hydrochloric acid as a by-product! One might also use tetraethyl-orthosilicate (TEOS on the street, for obvious reasons):

$$Si(OC_2H_5)_4 \rightarrow SiO_2 + \text{various by-products.} \qquad (7.9)$$

Although these oxides are generally inferior electrically to thermal oxides, one can put them down easily and quickly, with excellent "conformality"; meaning, they can be deposited on both horizontal and vertical surfaces at similar rates, and can thus both cover steps in the crystal surface and fill cavities in the surface, often indispensable attributes.

If we wanted to deposit polysilicon (poly) on the wafer, we can also do that with LPVCD by using silane, according to

$$SiH_4 \rightarrow Si + 2H_2. \qquad (7.10)$$

We might even choose to dope the polysilicon as we deposit it [e.g., n-type using phosphine (PH_3) – nasty stuff that supposedly smells like almonds, but so wickedly toxic you have to wonder how anyone actually knows what it smells like and lived to tell!]. Alas, silane is extremely explosive (and toxic). If you happen to ever visit a real fabrication facility you will note that the 50-ft-tall, large white silane cylinders are conveniently removed a few hundred feet from the main building and placed in what are affectionately known as "bunkers" – you get the idea! Exceptionally sensitive gas detectors and all sorts of safety measures for things like silane and phosphine handling are a dime-a-dozen in the fab, for obvious reasons (okay, $100,000 a dozen!). Polysilicon is very commonly used

Figure 7.33   An automated wet-etching tool (courtesy of Intel Corporation).

in fabrication when we do not need a perfect silicon crystal, but still require a doped silicon layer to serve as a good conductor, and it can be put down directly on oxide, for instance, to isolate it electrically from the silicon crystal.

Finally, an alternative dielectric to oxide is silicon nitride ("nitride" for short), which we can deposit with LPCVD by using silane plus ammonia:

$$3SiH_4 + 4NH_3 \quad \rightarrow \quad Si_3N_4 + 12H_2. \qquad (7.11)$$

Nitride is not as good an insulator as oxide, but importantly it has very different etching (next topic) properties and thus is often essential in the fabrication process.

7.5 Etching and polishing

Awesome! Crystal-construction platform – check. Means to dope said crystal – check. Way to sequentially grow and deposit new materials on said crystal – check. Now we need a way to selectively remove those materials ("etch" them), either from the crystal layer itself or from the various layers we just took pains to deposit. As we will see, etching will prove indispensable for selectively patterning layers as we build up Mr. Transistor from scratch.

There two basic ways to etch deposited materials: "wet etching," using a bath of some liquid etchant to eat away the material in question (this is intuitive), and "dry etching," using a reactive "plasma" to whisk away the material in question (this is perhaps less intuitive). A basic chemical-etching system is not much to look at (Fig. 7.33), but is conceptually very simple in its operation: (1) Create a vat of very pure etchant for the material in question; (2) lower your wafer containing the deposited material into said vat; (3) let the etchant eat away said material for time t; (4) remove wafer, and then rinse it off. *Finito*. Of course, controlling the precise amount of material that is etched can be a fine art (it might only be a few microns, or even 10s of nanometers), and doing it uniformly across a 300-mm wafer with 500,000,000 features to be etched is an even finer art. Still, the process is fairly routine today.

As an example, a very common etchant for oxide is buffered hydrofluoric acid (BHF). I wouldn't advise dipping your fingers into it! BHF is an especially nice etchant because it attacks oxide but does not attack silicon. Hence, if I lower my wafer with oxide deposited on top into a vat of BHF, the oxide will etch until it reaches the silicon and then stop, no matter how long it stays in the bath (amazingly, called an "etch stop"). In this case BHF is said to have good etch selectivity to silicon (doesn't etch it). If I further pattern a protective layer on top of the oxide in a particular shape (this is called an "etch mask") that does not etch in BHF (an organic photoresist would be a good example of a etch mask on oxide – see next section), then I can use BHF to "transfer" that patterned shape onto the surface of the wafer (meaning: BHF etches the oxide down to the silicon everywhere except where I put the protective etch mask – instant pattern transfer).

If I instead wanted to etch silicon (or polysilicon), for instance, I might use potassium hydroxide (KOH). KOH has good selectivity to both oxide and nitride, with silicon "etch rates" of approximately 2 μm/min. An oxide layer underneath a polysilicon layer could be used as an etch stop in this case. Interestingly, KOH also attacks different silicon crystal directions at different etch rates. For example, the differential etching ratio for <100>:<111> directions (don't panic – simply two different directions in the crystal lattice, one at a 45° angle to the other) can be as high as 400:1; meaning, if I open a pattern in an oxide etch stop on the surface of a silicon wafer and simply expose the wafer to KOH, I can trivially etch deep V-shaped groves into the silicon surface. If I instead use a wider rectangular-shaped pattern, I can actually control the etch depth independent of the etch time, creating a self-limiting deep etch (quite clever and useful). As we will see in a later chapter, such etching techniques form the basis of so-called "bulk micromachining" used in MEMS/NEMS technology.

So why wet vs. dry etching? Well, it relates to the resultant "etch profile" desired. As illustrated in Fig. 7.34, wet etching typically has an "isotropic" etch profile (etches the same in all directions), whereas dry etching can be easily tailored to create a number of useful "etch profiles," from isotropic to "anisotropic" (etches in only one direction) to etch narrow but deep "trenches" into the various materials. Anisotropic etching becomes especially important as the shapes to be etched get finer and finer and more closely packed, because wet etching tends to undercut the etch mask, "blowing up" or "ballooning out" shapes during etching, hence limiting how close two shapes to be etched can be placed to one another. This fundamentally determines the size of Mr. Transistor and hence how many transistors per IC can be used (read: a big deal).

Etch Profile

Type of Etch	Sidewall Profile	Visual
Wet Etch	Isotropic	
Dry Etch	Isotropic	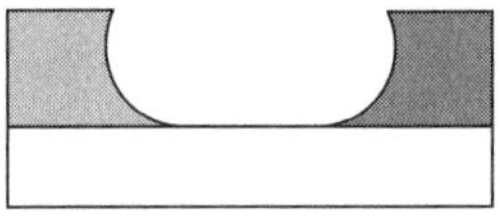
	Anisotropic	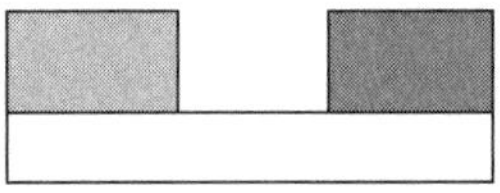
	Anisotropic (tapered)	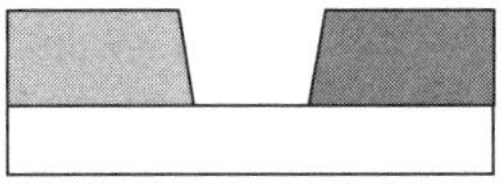
	Silicon Trench	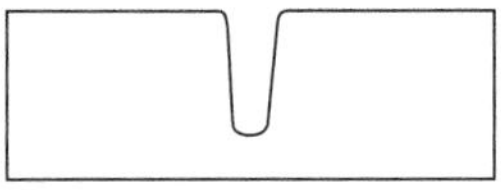

Figure 7.34   The sidewall etch profiles for both wet- and dry-etching processes.

Dry etching is performed with a reactive "plasma." Oh yeah?! Think back to high school chemistry. Got it? Plasma – fourth state of matter (solid, liquid, gas, *plasma*). A plasma is essentially a gas that has had energy applied to it in some form (typically electrical), stripping the valence electrons from the gas molecules, and thus creating an "ionized" gas (literally a gas of ions). Plasma in action? – think fluorescent light; or, on a grander scale, the *aurora borealis* (see Geek Trivia sidebar). In our case, one of the nice features of plasmas is that if we choose the right starting gas to ionize, the resultant plasma can be very chemically reactive – read: a great etchant. Excellent.

> ### Geek Trivia: *Aurora Borealis*
>
> The *aurora borealis* is the haunting glowing lights sometimes seen in polar night skies. At northern latitudes, this phenomenon is known as the *aurora borealis*, named for Aurora, the Roman goddess of the dawn, and Boreas, the Greek name for the north wind. The *aurora borealis* is often called the "northern lights" because they are visible only in northern skies, typically from September to October and March to April. In the southern polar regions, we instead have the *aurora australis*; australis meaning "of the South" in Latin. *Auroras* of either type are caused by the collision of energetic charged particles (e.g., electrons) that come from the solar wind or coronal storms and then are subsequently trapped in

the Earth's magnetic field (magnetosphere), with gas atoms found in the Earth's upper atmosphere (often at altitudes above 80 km), creating glowing plasmas. Light emitted during an *aurora* is often dominated by emission from an oxygen plasma, resulting in either a ghostly greenish glow (at 557.7 nm) or a sunrise-like dark-red glow (at 630.0 nm).

In a typical plasma-etching system (Fig. 7.35), an appropriate etch gas is introduced into the chamber, and is then ionized by use of RF power (high-frequency EM energy) at 13.56 MHz (no real magic on the number; it is set by the FCC). The energetic plasma ions then bombard the surface, suitably protected by an etch mask in various regions we do not want to etch. The ions bond with the target host atoms to be etched and are whisked away and exhausted out of the chamber (Fig. 7.36). Because we can also apply a dc electric field to the plasma that the ions will respond to, we have easy directional etch control in the plasma process, allowing us to produce sharp anisotropic etch profiles. Such a plasma-etching system is often called a "reactive ion etching" tool, or RIE for short. Typical RIE etch gases include SF_6 and O_2 for organic films, CCl_4 and CF_4 for polysilicon, C_2F_6 and CHF_3 for oxides, etc. More sophisticated RIE tools used today for etching very deep (100s of microns), high-aspect-ratio (deep but narrow) trenches include inductively coupled plasma (ICP) tools, a workhorse in the MEMS/NEMS field.

The final commonly used material removal technique in microelectronics is called chemical–mechanical polishing (CMP) and represents an interesting combination of both chemical etching and mechanical abrasion. Although a relative latecomer to the fab bag of tricks (1990s), CMP is now ubiquitous in modern IC manufacturing, particularly in back-end-of-the-line (BEOL) planarization

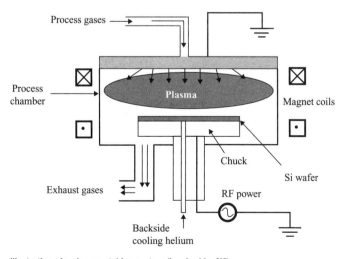

Figure 7.35   Illustration of a plasma-etching system (inspired by [3]).

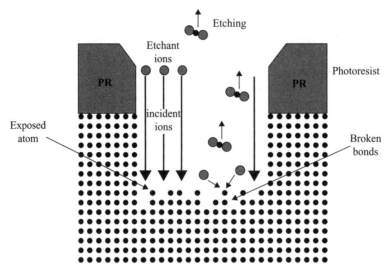

Figure 7.36 Illustration of the plasma-etching process.

(to make flat). The need is obvious. As we build up the many levels of metal/dielectric/metal/dielectric needed for interconnecting modern ICs (this could be 8–10 levels), we continue to deposit many additional new layers. If the starting surface for that BEOL process is not perfectly flat, it strongly affects the density of the metal wires that can be placed on any given layer. The question is one of how best to flatten (planarize) the surface of an entire 300-mm wafer without harming what is underneath. Neither wet etching nor dry etching will do the planarization trick needed here. Answer: CMP! In CMP, a chemical "slurry" is used as an etchant or abrasive, and the wafer is turned upside down, slowly rotated, with backside pressure pushing the wafer down against the CMP "pad" to effectively

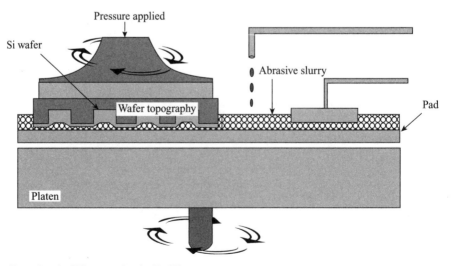

Figure 7.37   Illustration of a CMP system (inspired by [2]).

grind away the surface topography (much like optical lens grinding and polishing). Material removal rates are generally slow, in the 100-nm/min range, and can be controlled precisely by varying the slurry flow and chemical pH, applied pressure, rotational speed, and temperature. CMP is also used to polish the business end of silicon wafers to achieve the atomic flatness needed for beginning wafer manufacturing. It still surprises me that CMP, arguably the most brute-force semiconductor processing technique, works so well, but alas, it does! A schematic of a CMP system is shown in Fig. 7.37, and if you refer to Fig. 7.42 in Section 7.7, which shows the BEOL interconnect layers of a modern 90-nm CMOS microprocessor, you can clearly see the beautifully planarized metal/dielectric layers, the singular footprint of CMP. So much for etching and polishing. Now on to shape patterning.

7.6   Photolithography

The essential step in all of microelectronics fabrication is the pattern transfer process previously alluded to. That is, how do we go from a rectangular shape of dimension $x \times y$ on my computer screen where I am designing Mr. Transistor to implementing that identical shape in material z on the surface of my silicon wafer? That pattern transfer process is called "photolithography" or simply "lith" to those in-the-know. A modern IC might require upward of 30 distinct lith steps, and, given the scale of lith involved in building say a microprocessor, which might require patterning 300,000,000 rectangles that are 90 nm on a side on a single IC in a single lith step, clearly lith is *the* most delicate and hence costly processing step used in semiconductor manufacturing (by far). In fact, we often classify the fab facility by its lith capabilities: Building A is a 90-nm fab; Building B is a 180-nm fab, etc.

So what's the deal with lith? In short, to fabricate an IC, thin films (think fractions of a micron) of various materials must be deposited onto the wafer to act as etch stops, ion implantation or diffusion barriers, conductors, insulators between conductors, etc. Those thin films typically need to be patterned into particular shapes on the wafer, or holes ("windows") must be etched into the films to either allow dopants through or to make an electrical contact between two different layers. These wafer patterns are first designed on a computer, and then are transferred to a photolithographic "mask" and subsequently onto the surface of the wafer. Lith uses an exceptionally elegant photoengraving process much like that used in conventional film photography (I assume at least a few of you still own a nondigital camera!). The mask patterns are first transferred from the mask onto a light-sensitive organic material called a photoresist ("resist"), and then either dry or wet etching is used to transfer the final pattern from the resist to the thin film in question.

Each lith step (pattern transfer sequence) is itself made up of a number of individual processing steps, as illustrated in Fig. 7.38. Let's imagine that I have

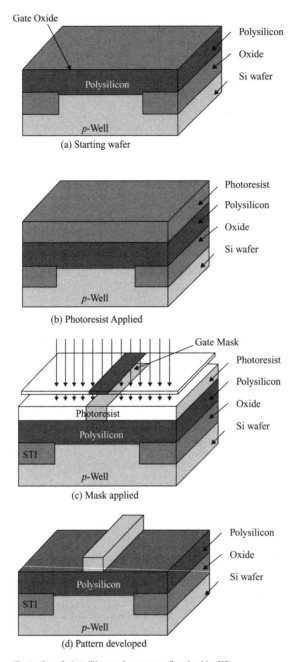

(a) Starting wafer

(b) Photoresist Applied

(c) Mask applied

(d) Pattern developed

Figure 7.38 Illustration of photolithography process (inspired by [3]).

a rectangle I want to create in a polysilicon layer on Mr. Transistor. Here's how I would use lith to go about this:

- Deposit the poly by using LPCVD across the entire wafer. This would be called a "blanket film deposition".
- Apply a chemical primer (think paint) or adhesion layer to the blanket poly film to help the resist stick to the thin film. Typically hexamethyldisilazane (HMDS for obvious reasons) is used.

- Spin on the resist (literally rotate the wafer at high (1,000–5,000) rpm's and drop on the liquid resist so that it "spin-coats" the whole wafer to a given well-controlled thickness of a fraction of a micron to several microns, depending on the lith step).
- Cook the resist at low temperature (maybe 80°C – this is called a "soft-bake").
- Apply the mask by using a lith exposure system.
- Align the mask to the wafer (this obviously gets tricky if I have to align this mask to a previously patterned layer, to a fraction of a micron resolution).
- Expose the wafer to light. Wherever the shape is *not* present on the mask will be exposed to the light, chemically altering its polymer structure.
- Remove the mask.
- Soft-bake the wafer again to solidify the chemical changes to the resist.
- "Develop" the resist (think photography), using a wet etch to dissolve all of the resist not exposed to the UV light (burning off the resist with plasma etching in oxygen – "resist ashing" – is also used).
- Cook it once more at a higher temperature ("hard-bake") of maybe 150°C to toughen the resist to plasma etching.
- Get it under the microscope and make sure everything went smoothly ("lith inspection"). Rework if needed; else, lith is done.

Whew! Twelve separate steps just to get the image from the mask onto the resist! With the intended pattern now safely transferred into the resist, we can expose the wafer to a plasma etch to remove the poly everywhere but under the resist pattern, and then use a different wet etch (or plasma) to remove the final resist. Now take a final peak under the microscope to ensure all is well. Look okay? Good. The image is now completely transferred from the mask into the poly. Clever, huh?!

A word or two more about the masks used to expose the resist. First I design my rectangles for my IC on the computer, where I can easily manipulate them (remember – there could be 300,000,000 of these submicron-sized objects for a given mask layer for a single IC, and maybe 25 different layers needed!). Next, I feed those rectangles as electronic data to a "pattern generator" that replicates them on a "reticle." The reticle is essentially a copy of what was on my computer screen, except that it is generally made 5× or 10× larger than the original drawn images (i.e., it is a magnified version of the design). A "step-and-repeat" camera (well named) is then used to replicate the reticle on a master lith mask over and over again, such that many identical copies of the reticle (one for each final IC) are included on the mask. For instance, on a 200-mm mask, we can place 1,200 copies of a 5 mm × 5 mm IC [4]. The final lith mask is essentially a piece of high-quality glass onto which a thin metal coating (typically chrome) has been applied, which can now serve as a pristine master photographic plate for use in lith (the chrome patterns are also defined through use of a lith process). One such mask is needed for each lith step. A complete set of state-of-the-art masks (again, there might be 25) can easily cost $1,000,000 and are easily the most expensive

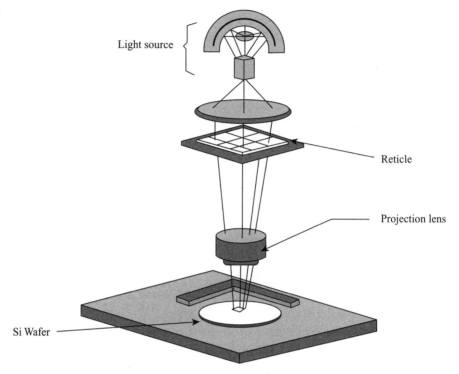

Light source

Reticle

Projection lens

Si Wafer

Figure 7.39   Representation of an optical projection lithography system (inspired by [3]).

article needed to produce an IC. You can start to see the utility of the lith process. It is, by design, massively parallel, such that tons of ICs are built up together – the ultimate assembly line efficiency. This economy of scale is the only way to build cheap ICs (think Moore's law), even though the individual processing steps are very complicated and costly.

So how then is the master mask used to produce a pattern in the resist? Well, we use an optical projection lithography system (Fig. 7.39). Sound expensive? It is! Frighteningly complicated in practice, but conceptually quite simple. A wafer is covered with resist, and then the mask is held above the wafer while UV light is shone through the mask and a series of lenses that focus the images sharply onto the wafer, exposing the resist to the light. Why UV? Well, recall from Chap. 2 the relation between the wavelength of the light and the object we are trying to see. If I want to expose a 1-μm^2 object in the resist, I better use light with a wavelength much smaller than 1 μm (read: smaller than visible wavelengths of 0.4 to 0.7 μm); otherwise it will have fuzzy, washed-out edges (read: lith disaster). The typical light source in a state-of-the-art optical lith system is a Kr-F excimer laser, which can generate a pure and intense deep-ultraviolet (DUV) wavelength at 193 nm. Clever mask games can then be played to go to even smaller geometries using optical projection (e.g., phase-shifting masks – a topic for another day). Beyond that, for patterning shapes down to, say, 10s of

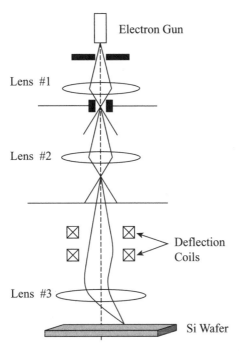

Electron Gun

Lens #1

Lens #2

Deflection
Coils

Lens #3

Si Wafer

Figure 7.40 Representation of an electron-beam lithography system (inspired by [3]).

nanometers, one has no option but to move to much shorter wavelengths still for resist exposure, and the most common choice is to use electron-beam (e-beam) lithography, as depicted in Fig. 7.40 (recall the discussion of using electrons for imaging in Chap. 2). Interestingly, there is no lith mask *per se* needed in e-beam lith, which might appear to be a huge cost advantage over optical lith. Alas, no, because the e-beam "write time" is far longer than that for optical systems, producing a lith-throughput bottleneck (the electron beam has to literally trace out each shape to be patterned, whereas in an optical system the whole die is exposed at one time). Still, e-beam lith is the gold standard for ultra-fine-line nanoscale fabrication and is in use worldwide, especially in the research arena. If you take an even casual glance at the ITRS roadmap, which projects a 20-nm feature size by about 2014, e-beam lith will have to fairly quickly become a mainstream IC manufacturing tool with the requisite throughput.

Geek Trivia: *Photoresist*

Interestingly enough, good ole William Shockley had his hand in the application of photolithography to semiconductor manufacturing [7]. In the early 1950s, one of Shockley's highly capable assistants was one Jules Andrus, who just happened

(continued)

Geek Trivia: *Photoresist* (*continued*)

to be a commercial artist by profession. In 1954, Shockley had a "crazy" idea about applying photoresist used in commercial film photography to pattern shapes in transistors, and he sent Andrus to Eastman Kodak (think Kodak camera film) in Rochester, NY, on a fact-finding mission. A prototype lith technique was quickly and successfully applied to transistor fab, culminating in the famous Andrus patent, U.S. Patent 3,122,817, filed February 4, 1963, and issued March 3, 1967.[1]

Enough of the history – so how exactly does photoresist work? Most polymers (think plastics, with long strands of linked carbon-based molecules), when exposed to UV light in the range from 200 to 300 nm, form "radical species" that can result in polymer cross linking and increased molecular weight, altering their physical and chemical properties such that what once easily dissolved in chemical *A* no longer dissolves in chemical *A* (that is, its so-called dissolution rate is changed). Whereas the automotive and aerospace industries focus a ton of research on inhibiting such light–polymer reactions in order to extend the lifetime of plastic coatings on components, the semiconductor industry actively exploits these physical changes for the photoresist used in lithography. A "negative" resist becomes insoluble in the resist developer after exposed to UV light, whereas a "positive" resist becomes soluble (i.e., I can either leave a hole in the resist where my intended image is or leave the image, depending on what is needed for fabrication).

One very common positive UV photoresist is based on a mixture of Diazonaphthoquinone (DNQ) and novolac resin (a phenol formaldehyde resin). DNQ inhibits the dissolution of the novolac resin. On UV exposure, however, the dissolution rate increases to even beyond that of pure novolac. DNQ–novolac resists are developed by dissolution in tetramethyl ammonium hydroxide in water. DUV resists are typically polyhydroxystyrene-based polymers with a photoacid generator used to provide the requisite solubility change.

7.7 Metallization and interconnects

All of these sophisticated wafer fab techniques would mean little at the end of the day if we didn't have a clever way to connect all of the pieces together. That is, it is one thing (a pretty big thing actually) to build 300,000,000 transistors on a centimeter-sized silicon die. But it is quite another to connect all of those 300,000,000 buggers together in some useful fashion (e.g., to build a microprocessor to serve as the brains of your laptop). This is where IC "interconnects" come into play (think fancy point-to-point wiring). Figures 7.41 and 7.42 show top-down and cross-sectional views of state-of-the-art interconnects used in microprocessors. In the 1970s, only two layers of interconnect were typically

---

[1] Given all that I have told you about Shockley's personality, it is truly amazing that Shockley's name is not on this patent!

Figure 7.41 Top view of the copper metalization of a modern IC. (Courtesy of International Business Machines Corporation. Unauthorized use is not permitted.) (See color plate 22.)

used; today, 8–10 layers, and growing. Simply put, with 300,000,000 transistors to connect up, this veritable rat's nest of wires can be handled only by many independent layers. As a reminder, a centimeter-sized modern microprocessor literally has several MILES of micron-sized metal wires connecting up all its internal guts.

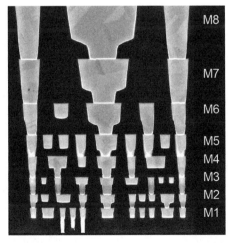

Figure 7.42 A cross-sectional view of the eight different layers of metal interconnects on a 90-nm CMOS technology. From the bottom of metal layer 1 (M1) to the top of metal layer 8 (M8) is about 20 μm, 20 millionths of a meter (courtesy of Intel Corporation).

With only a moment's reflection, you can convince yourself that these interconnects are going to have to be made from metals. Why? Well, remember RC time constants from high school physics? Never fear, let me remind you! Imagine that I have a capacitor C that I want to charge up through a resistance R to voltage V. That is, I want to move the voltage signal from point A to point B. How long does that take? Well, from basic physics, this is an exponential charging process (determined by the differential equation governing the process – don't be frightened, no need to pull it out), and the charge on the capacitor q builds (current flows) in time t (establishing the voltage on the capacitor) according to

$$q = CV \left(1 - e^{-t/RC}\right) \tag{7.12}$$

RC clearly has dimensions of time (the exponential function is dimensionless). Hence, the smaller RC is, the faster we can charge the capacitor to the needed charge or voltage (this might represent our logical "1" for a digital bit moved from point A to point B). In fact, we can easily show that within one "RC time constant" the charge on the capacitor reaches ($1 - e^{-1} = 63\%$) of its final value. Got it? Good.

Now, let's assume that the length of the wire connecting two transistors is fixed, and hence the capacitance associated with the wire itself is fixed. Then, the smaller the R is of our wire, the faster Mr. Transistor can talk to his neighbor (smaller RC time constant). For a given geometry of the wire in question (length × width × height), the smaller the resistivity of the wire material used to fab it, the smaller R will be. Materials with the smallest resistivity? Yep – metals.

Okay, so what metal? Historically, aluminum (Al) has dominated the IC interconnect business until very recently. Why? Nice and low resistivity (2.65 µΩ cm: recall, a typical starting silicon wafer in IC fab might be 10 Ω cm) and relatively easy to both deposit and etch. So how do we build the interconnect? First we blanket deposit a thin layer of Al across the entire wafer and then use lith to pattern the intended wire shape, followed by either wet or dry etching to remove the Al everywhere except where we want it. Need two layers of interconnect? Fine; now cover that first Al wire with chemical-vapor-deposited dielectric, CMP it to planarize the surface, and now use lith again to pattern a metal-to-metal contact point (called a "via"), etch the via, fill it up with another metal [typically tungsten (W), to form a W "stud"], deposit the second metal layer, pattern it, and etch it. Done. We have Metal 1 (M1) for "wire routing," Metal 2 (M2) for independent wire routing, and the ability to connect M1 and M2 with a via, as needed. Figure 7.42 has M1 through M8 and also shows metal-to-metal vias, along with clear evidence of CMP.

How do we deposit the requisite Al? Evaporation is the classical path for most thin-film metal deposition. It may sound strange, but in essence, we put our wafer inside a vacuum chamber, pump it down, and then use an electron "beam" to heat up the metal to its vaporization temperature (i.e., create a vapor of metal atoms), and then the metal atoms deposit themselves onto the surface of said wafer. We rotate the wafers during the evaporation process to ensure a uniform thickness

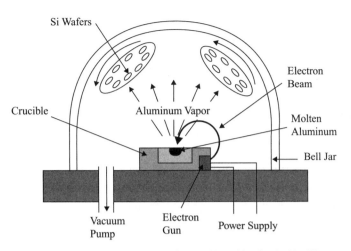

Figure 7.43 Representation of an e-beam evaporator for metal deposition (inspired by [3]).

(Fig. 7.43). Done! For metals that may not evaporate well (e.g., W), we can use an argon ion beam (recall the "ion cannon" in Star Wars *The Empire Strikes Back*!) to "sputter" metals onto wafers. A little more violent, but the same idea.

It turns out that state-of-the-art interconnects are actually built with copper (Cu), not Al – observe the copperish color of the metals in Fig. 7.41 (color plate 22). Why? Well, it is all a matter of resistivity. Al has a resistivity of 2.65 $\mu\Omega$ cm, but Cu has 1.70 $\mu\Omega$ cm, the best of all easily accessible metals for IC manufacturing. Although this may seem like a modest improvement, it's not. During the 1990s microprocessor designers already began to bump up against "interconnect delay" limits [8]. In essence, if the RC time constant of the wires in the "critical delay paths" is too large, and I am continuing to run Mr. Microprocessor faster and faster with each technology scaling advance, eventually I reach a point where the IC speed is limited by the wiring delays and not the transistors, and hence Mr. Microprocessor just doesn't get any faster in going from scaling node A to scaling node B (say 130 nm to 90 nm) – major bummer since you just plopped down $5B for the 90-nm fab! Solution? Well, you might rightly say, "let's just make the Al wires thicker, to lower their total resistance." You'd be right, up to a point. The drawback is that, the thicker the Al wires get, the farther apart they have to be placed (a density hit), and the more problems with planarization I have to deal with. Better solution? Change metals in my interconnects. For the same wire length and geometry, I can get a 36% improvement in interconnect delay just by moving from Al to Cu. See Head Scratcher #1 for a discussion of gold interconnects.

Improved Cu resistivity is the upside (and its a big upside). The downside is that it is VERY difficult to either wet or dry etch Cu, and this effectively held up progress on Cu interconnects until the late 1990s. Interestingly enough, to get around this, one actually electroplates Cu to a seed layer, rather than depositing it and etching it away as done with Al. In electroplating (an ancient process), an electrically conducive "seed layer" is deposited and patterned, and then the wafer is immersed into the Cu plating solution and a dc bias applied between the Cu solution and the seed layer, resulting in Cu deposition (plating) on only the patterned seed layer. At the state-of-the-art, we use a technique called a "dual-Damascene process," which allows us to create both plated Cu interconnect and Cu-to-Cu via (repeatedly) within the same fab sequence. The 90-nm BEOL shown in Fig. 7.42 is done this way.

One final thing regarding interconnects. One of the simplest ways to obtain a decent wire directly on the silicon wafer surface is to deposit a metal on the silicon and then give it a gentle bake to cause a chemical reaction between the metal and silicon, producing a "silicide." For example, titanium (Ti), cobalt (Co), and tungsten (W) are all common silicides found in IC manufacturing ($TiSi_2$; $CoSi_2$; WSi_2). Silicides typically have relatively poor resistivity compared with that of pure metals (e.g., $TiSi_2 = 25$ $\mu\Omega$ cm), and they do consume silicon when they react; but if used in only short or noncritical interconnects, they can prove to be highly advantageous and are thus commonly used, especially INSIDE transistors to minimize resistive parasitics.

Head Scratcher #1: *The Price of Gold*

Recently during a lecture I asked my students to tell me a disadvantage associated with using gold for metalization in IC manufacturing. A hand shot up – "it is too expensive!" Angry buzzer sounds – nope, but thanks for playing! My interest piqued, I told the class that I would give some extra credit to the first student to solve the following problem [9]. Assuming we want to deposit a 1.0 cm long by 1.0 μm wide by 1.0 μm thick gold wire onto a silicon wafer, what would be the cost of the gold required, assuming current market price? Answer: 4.2 microcents per wafer! (4.2×10^{-8} dollars). Read: Cost is not a big deal. Said another way, I could fab 23,809,523 such wafers before spending a buck for the gold! Not a hard problem – give it a try! Question: So let me then ask you, dear reader – What would in fact be a disadvantage of using gold for metallization in silicon? Hint: It has to do traps and carrier recombination (refer to Chap. 5).

7.8 Building Mr. Transistor

Whew! Finally we have arrived. Our fab toolkit is now complete, and we can actually build Mr. Transistor and connect him up to do something useful. Consider Figs. 7.44 and 7.45, which show cross-sectional images of modern high-speed transistors. The first is a 130-nm SiGe heterojunction bipolar transistor (HBT) used in high-speed communications ICs, and the second is a 90-nm field-effect transistor (MOSFET) used in microprocessor applications. Both look pretty complex, and indeed they are. Still, they can be constructed by use of the basic fab techniques previously outlined. Disclaimer: Building the transistors shown would actually require 25–30 lith masks and hundreds of individual processing steps

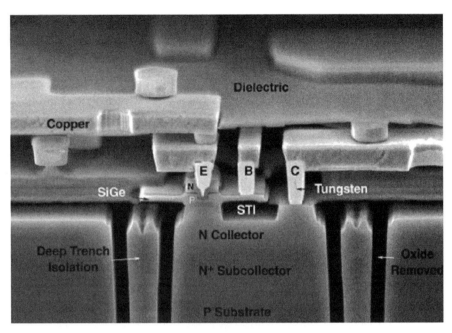

Figure 7.44 Decorated cross-sectional view of a modern bandgap-engineered SiGe bipolar transistor (Courtesy of International Business Machines Corporation. Unauthorized use is not permitted.) (See color plate 23.)

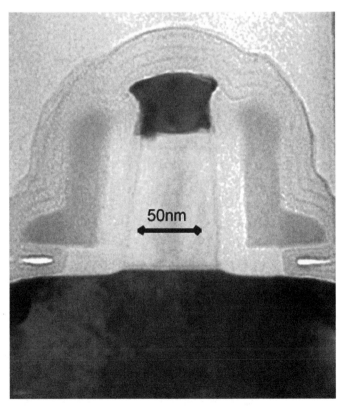

Figure 7.45 A cross-sectional view of a FET from a 90-nm CMOS technology. The active transistor region is 50 nm, 50 billionths of a meter (courtesy of Intel Corporation).

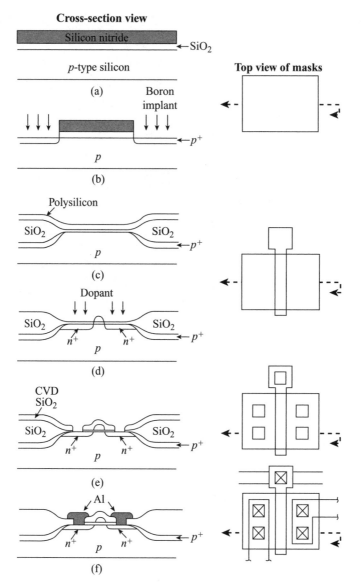

Figure 7.46   Simplified cross-sectional view of the step-by-step contruction of a MOSFET. The lithography masks are shown to the right (inspired by [4]).

(the fab "recipe" might take up half of this book!), requiring say 8–10 weeks of 24/7 effort by lots of folks in a state-of-the-art multi-billion-dollar 300-mm facility. Read: What I am after is simply showing you a "bare-bones" path to transistor fab, and this is intended only to give you a realistic feel for how all this stuff happens. Real life is always complicated!

So...let's build a MOSFET! Figure 7.46 shows a dual cross-sectional + top-down view of the simplest possible transistor (a MOSFET), as a step-by-step construction sequence. Figure 7.47 lists the "process flow" (what actually happens when) of the sequence, together with the required lith mask levels. This MOSFET fab sequence would be called a "five-mask process" (requires five

Fabrication Flow

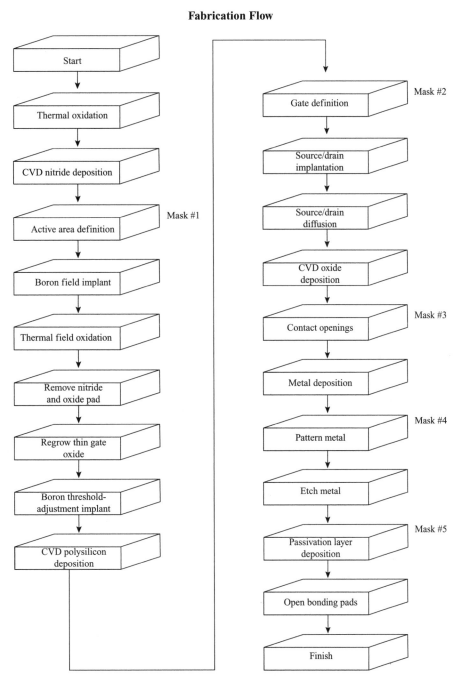

Figure 7.47   Simplified step-by-step fabrication flow for building a MOSFET (inspired by [4]).

lith masks to make it). This is the sort of MOSFET that good universities will actually allow students to go into the lab and learn how to build themselves during a one-semester course in micro/nanoelectronics fabrication. Nothing fancy, and the lith will be coarse (maybe 5 μm), but it will function as a transistor and nicely illustrates how fabrication works. Education in action.

CMOS Cross-section

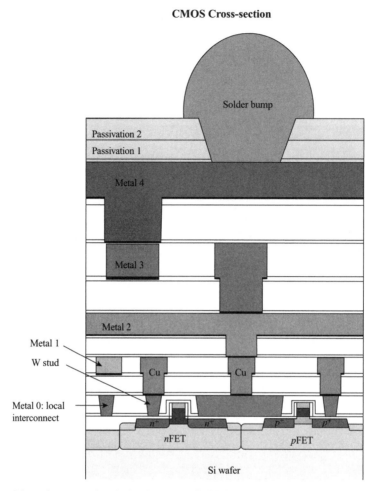

Figure 7.48 Schematic cross-section of a four-layer-metal CMOS IC (inspired by [3]).

Basic Steps to Building Mr. Transistor:

- Deposit dielectrics on the wafer surface, pattern them with lith + etching, and form the thick "field-oxide" isolation by thermal oxidation.
- Deposit n-type poly using LPCVD, and pattern the layer with lith + etching (this will be the MOSFET "gate")
- Use ion implantation + rapid thermal annealing to form heavily n-type doped regions (this will be the MOSFET "source" and "drain" regions).
- Deposit CVD oxide on the wafer, pattern the contact openings, and etch them.
- Evaporate Al onto the wafer, and then use lith to pattern the metal (M1) and etch it.

Done! You could of course now begin to add additional metal interconnect layers on top of this first one, as needed, each obviously requiring additional masks and processing.

Figure 7.48 shows a realistic (to scale) schematic cross section of a four-metal-layer CMOS technology (with both n-type and p-type MOSFET transistors needed to build CMOS electronic circuits – next chapter's topic). Observe that most of the "overlayer" superstructure of a real CMOS technology resides in the interconnects. In a current-day eight-level metal 90-nm CMOS technology, the distance from the silicon surface to the top metal is in the range of 15–20 μm, compared with the transistor thickness of maybe 0.5 μm. Figure 7.48 also shows the final steps for IC fab. Once the final top metal layer is in place, we deposit protective "passivation" layers (e.g., Si_3N_4) on the top surface of the wafer, use lith to define and then etch "bond pad" windows, and finally, depending on the IC packaging required, follow that by deposition of PbTi "solder bumps" (basically 100-μm-sized balls of solder, sometimes called "C4 balls" – you'll see why these are handy in a moment). Mr. IC is now complete. *Finito!* Piece of cake, right? Congratulations! You've just built your own transistor!

7.9 IC packaging: Wirebonds, cans, DIPs, and flip-chips

At this stage, Mr. IC (containing 300,000,000 Mr. Transistors), although dressed up and pretty, is still in wafer form (Fig. 7.1) – Read: not especially useful for putting into your laptop or cell phone! IC packaging comes next; postfab fab, as it were. When I say IC "packaging," think of a container into which I place Mr. IC that (a) protects him from the world (e.g., our grubby fingers), and (b) provides an electrical means to get power (the applied voltage to run the circuit) and signal lines (carrying "1s" and "0s") on and off Mr. IC cleanly. This is actually pretty scary to think about. A modern microprocessor might literally have 100s of input–output (I/O) pins for an IC, which is a centimeter on a side. Not surprisingly, IC packaging has evolved over time in its sophistication to handle the growing numbers of I/Os, and today IC package types are many and varied. I'll introduce you to some of the more common ones.

First things first though. We need to go from silicon wafer to silicon "die," that centimeter-sized silicon "microchip" (often just "chip"). How? Simply use a circular saw running at high rpm's and literally saw up the the wafer into pieces (put them on a piece of blue two-sided sticky tape first so they don't fly all over the room). What kind of saw do we use? You guessed it – a diamond saw (one of the very few things that can cut silicon cleanly and at high speed). First, however, we will electrically test each IC (remember, there may be 100s per wafer) for functionality and identify with an ink-blot marker which ones work and which ones are dead (aka the IC yield). Yield for a mature 90-nm technology should be in the >90% range (if you intend to stay in business). The functional IC die is then picked up by a robot and placed (literally referred to as the "pic-and-place" tool, go figure) into said package, attached with epoxy (uninspiringly called "die attach"), wirebonded (more on this in a second), carefully sealed, typically with a rather boring cured black plastic, then retested to make sure all is well, and finally sent to Company X for use in Gizmo Y (Fig. 7.49). To my mind it is a little sad that black plastic is used to seal IC packages, because this hides Mr. IC's

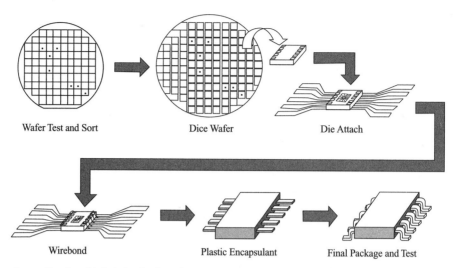

Figure 7.49 Process flow from fabricated wafer through completed IC package (inspired by [4]).

good looks from lurking eyes, and hence most folks just don't have a clue what glory actually lies under the package cover. Sigh . . . Okay, the black plastic does also protects Mr. IC from any harmful light that might cause problems, but alas, it just isn't very aesthetically appealing (I'll freely admit I'm a sucker for those see-through cell phones and clocks that show all the guts).

Some common package types? Well, for a single (discrete) transistor, you might use a "can" package (no need to explain – look at Fig. 7.50). For 6–40-pin ICs I might use a dual-inline package (the famous and now-ubiquitous DIP), also shown in Fig. 7.50. More sophisticated packages such as those you might find in

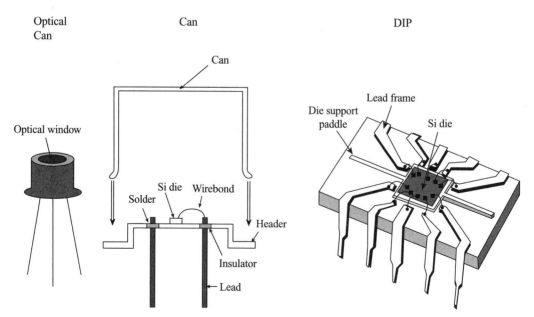

Figure 7.50 Schematic cut-away view of two popular types of packages used in low-integration-count ICs (with wirebonds shown): (a) the TO (transistor outline) "can" and (b) the "dual-in-line package" (DIP) (inspired by [1]).

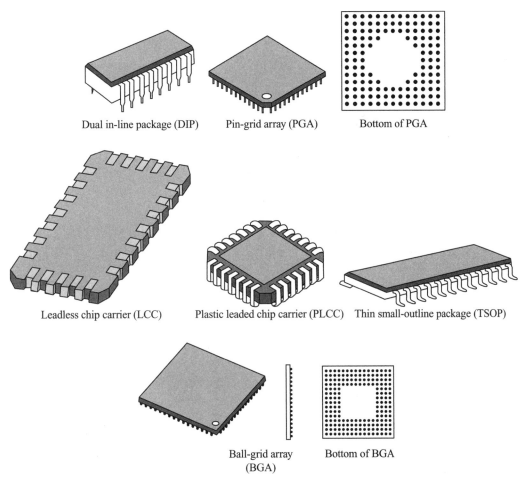

Dual in-line package (DIP)   Pin-grid array (PGA)   Bottom of PGA

Leadless chip carrier (LCC)   Plastic leaded chip carrier (PLCC)   Thin small-outline package (TSOP)

Ball-grid array
(BGA)   Bottom of BGA

Figure 7.51 A variety of IC packages that support much higher circuit integration levels. The DIP and pin-grid array (PGA) packages are "through-hole mount" packages (IC's leads stick through the PCB and are soldered on the backside), whereas the rest are "surface-mount" packages (IC is soldered to the surface of the PCB) (inspired by [1]).

your laptop or DVD player or cell phone should you opt to pry off their covers (don't try this at home kids) are shown in Fig. 7.51 and include the leadless chip carrier (LCC), the pin-grid array (PGA), and the ball-grid array (BGA). Remember all IC packages do the same thing (protect Mr. IC + provide electrical connections to the outside world), but these fancier packages allow far higher pin counts than a typical DIP, for instance.

So how does one connect the leads of a package to Mr. IC? In one of two ways, wirebonding or "flip-chip." Wirebonding is far more common today and still rather remarkable. Basically, a sewing machine takes a fine gold wire and fuses one end to a metalized pad on the package and then fuses the other end to the metal bond pad on the edge of the IC that might be only 100 μm × 100 μm in size (often smaller). This process is called "thermocompression bonding," and the steps are illustrated in Fig. 7.52, with the result shown in Fig. 7.53. Look easy? It's not! First of all, the Au wire is only 15–75 μm in diameter! For reference,

Wire Bonding

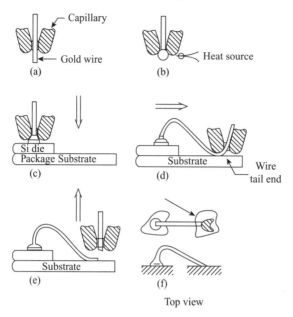

(a)

(b)

(c)

(d)

(e)

(f)

Top view

Figure 7.52 Illustration of the wirebonding process. This case shows "thermosonic ball bonding" (inspired by [4]).

a typical human hair is about 50 μm in diameter – go ahead, pull one out right now and take a look; and then imagine using it in your Mom's sewing machine to sew a vest for a flea! You get the idea. AND, the wirebonding machine is going to execute in the range of 10 bonds per second, 24/7, yielding thousands of IC packages a day. Nope – definitely not easy.

Flip-chip is just what it says. On the die, open the bond pads and then deposit solder balls (Fig. 7.48). Now flip the chip upside down (hence the name), align it to the metal pads on the package, and gently heat it so that it the solder ball fuses to the package (Fig. 7.54). Done. Why bother? Well, several reasons: (1) Using

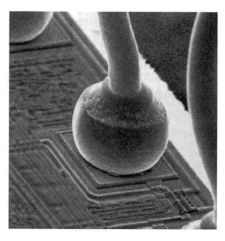

Figure 7.53 Decorated view of a single gold wirebond on the surface of an IC. (Courtesy of International Business Machines Corporation. Unauthorized use is not permitted.) (See color plate 24.)

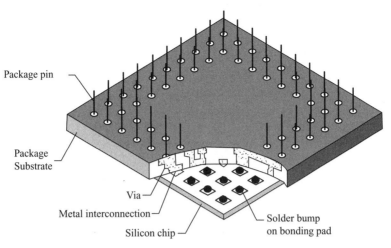

Package pin

Package Substrate

Via

Metal interconnection

Silicon chip

Solder bump on bonding pad

Figure 7.54 Illustration of "flip-chip" package using solder bumps (aka C4 balls) on top of the contact pads (inspired by [4]).

flip-chip, I can populate solder balls literally all over the surface of Mr. IC, not just around the edges, so the pin density can be very high indeed; and (2) flip-chip is especially useful for minimizing the parasitics associated with the connection. That is, in a classical wirebond, the parasitic inductance and resistance of the gold bond wire itself, although tiny, are actually large enough to cause major headaches for high-speed circuits (especially RF circuits – think cell phones). Flip-chip essentially eliminates these parasitics, because the electrical path length is now 100s of microns instead of the several millimeters in wirebonding. Although it took flip-chip packaging some time to catch on because of the complexity involved (you might ask yourself, for instance, how one orients the solder balls on the upside-down die to the pads on the package – think mirrors!), flip-chip represents the gold standard in high-speed, high-pin-count IC packaging.

Okay, so now we have Mr. IC packaged. What's next? Well, finally we want to integrate the packaged IC into useful electronic system, and again we have multiple options. Figure 7.55 shows the mounting of an IC package onto a printed circuit board (PCB). Think of the PCB as a second level of packaging that can now accomodate multiple IC packages and has its own lith-defined wiring to connect up the I/Os of the various packages as required. Many common electronic products (e.g., laptop, cell phone, digital camera, etc.) have a PCB "motherboard," as shown in Fig. 7.56. In this case, ICs can be attached to either one or both sides of the PCB. In this case the IC packages utilize "surface-mount" packaging technology (just what it says). A second packaging approach is called "chip-on-board" (CoB), in which the bare die is attached to the PCB, and then wirebonded, followed by additional passive components (resistors, capacitors, etc.) that might be needed to build the circuit (Fig. 7.57). A logical extension of CoB is the multichip module (MCM), which has various bare die and passive components attached to a packaging "substrate" (Fig. 7.58). Importantly, in MCM, you can also build up the package substrate with multiple layers (even 10–20) of metal

IC Packaging

Leads   Pins

IC package

Printed Circuit Board

Edge connector

PCB

System Assembly   System board

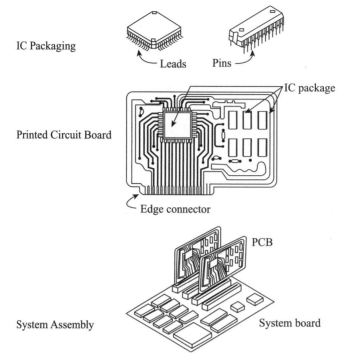

Figure 7.55   Assembly flow from package to board to system (inspired by [4]).

interconnects, such that very high levels of wiring density can be achieved in a tiny package. In MCM, the package substrate serves as its own independent interconnect platform.

This MCM approach can be extended to its logical conclusion (the subject of much present-day research); namely, 3D packaging (Fig. 7.59). 3D packaging has substantial functionality within the package, beyond simple wiring that a MCM might have, and can also accommodate a host of different types of components to build a tiny form-factor complete system. In such a 3D "system-on-package" (SoP), we might integrate a variety of individual IC subsystems, including a microprocessor (itself a "system-on-a-chip"), RF radio(s) for wireless communications, MEMS sensors and actuators, memory, an optical interface (maybe an optical receiver and a laser diode + driver), optical waveguides, and even ICs built from different types of semiconductors (e.g., silicon + Group III-V). Sound

Figure 7.56   Actual PCB with surface-mount IC packages attached (courtesy of International Sensor Systems).

Figure 7.57 Actual CoB package (courtesy of International Sensor Systems).

complicated? It is, and although we are not there yet for commercial systems, clearly the push to integrate more and more electronic–photonic functionality into increasingly sophisticated IC packages will continue to accelerate over time.

7.10 Reliability

Build Mr. Transistor? – check. Use 300,000,000 Mr. Transistors to build Mr. IC? – check. Package Mr. IC for consumption? – check. Off to the races! Still, a simple question remains. How long will Mr. IC remain alive and well and contributing to society once I power up the beast in a real product? Good question! Let's end this chapter on IC fab by touching briefly on reliability issues. In a nutshell, I am interested in the following imminently practical question: How long can I reasonably expect Mr. IC to last, and when he does kick the bucket, how will he pass? – implode, explode, spontaneously combust? How about a postmortem IC death discussion? You got it!

Fact. One of the most compelling attributes of ICs is that they possess unbelievably high reliability. Consider placing 300,000,000 of anything into a centimeter-sized space of something and having them work well as a unit for years and years without complaining. Absolutely remarkable feat. Routine today. Still, as most of us know, ICs do fail, often in very inconvenient ways. I was put into a headlock by this crushing reality not 10 days back, when, in the middle of happily typing along on my laptop I heard a muted "pop" and the infamous "screen of death" stared back at me from my machine. SPEAK TO ME BUDDY! Alas, only deathly silence. A test of your level of addiction to modern technology? Have your cherished laptop (that you haven't backed up recently – shame on you!) meet a unexpected and violent end and then perform a litmus test on

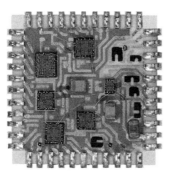

Figure 7.58 Illustration of MCM package, combining several ICs on the same package (courtesy of International Sensor Systems).

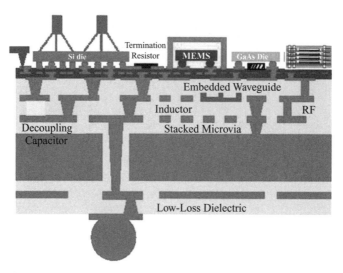

Figure 7.59 Vision for a 3D package composed of many different types of IC subsystems of widely differing form and functionality. This integration approach is known as "system-on-package" (SoP) (courtesy of M. Swaminathan, Georgia Tech).

your resultant emotional state. Zen-like calm? Mild amusement? Head shake and a frown of disappointment? Me? Utter panic! What would I do? How would I live? How much of my manuscript was gone?! My world was at an end. Serious reality check. Me, a high-tech junkie? Guilty as charged.[2]

So let's get down to business. ICs die in one of four major ways [10]: (1) bulk-related failures, (2) surface-related failures, (3) interconnect-related failures, and (4) packaging-related failures. Bulk failures, meaning within the semiconductor crystal itself, are relatively unimportant today, especially for silicon ICs. Pristine starting wafers are the norm, and hence latent defects that might cause trouble downstream are generally few and far between. Surface-related failures, meaning at the interfaces between the semiconductor and the various dielectrics that make up the IC, account for around 35% of all IC failures [10]. Interfaces (e.g., Si/SiO_2) are in general prone to damage from chemical contaminants, charge introduction, or energetic carriers (the infamous "hot-carrier" aging). The most common failure modes of ICs, however, are actually interconnect related (typically between 50% and 60% of all IC failures), and can result in either "short circuits" or "open circuits," either of which can mean death to Mr. IC (a short circuit that allows lots of current to flow from power to ground can be considerably more violent!). The naturally high current densities found in metal interconnects are especially troublesome, because, with technology scaling, wires get smaller and hence current densities tend to rise. At a sufficiently high current density (say > 5–10 mA/μm^2), "electromigration" becomes a dominant failure mode. In electromigration, the electrons in the wire that comprise the current flow can actually push the metal atoms around (aka mass transport), generating an electrical

---

[2] Fortunately I was lucky – they were able to recover my disk, and I endured only about a week of frustration until my new pristine laptop arrived.

"open" over time. Historically this was especially a problem for Al interconnects, but can be mitigated with the addition of small amounts of silicon and copper to the metal layer (obviously less of an issue for full Cu metalization found in state-of-the-art ICs). Finally, packaging-related failures generally involve either wirebond fatigue and failure, or cracking and failure of the die attach between the package and the IC.

So how good is IC reliability? Simple question; complicated answer. We characterize IC reliability by attempting to determine the statistical mean time between failures (MTBF) for the IC in question. Generally, the MTBF of a modern IC might be in the range of a few million hours. Although this sounds great, it clearly is problematic for reliability projections, because directly testing the lifetime takes way too long (want to twiddle your thumbs for 5,000,000 h waiting for a fail?). Instead, we subject the IC to "accelerated life testing" to speed up the failure process (essentially rapid aging).

Most IC failure mechanisms are "thermally activated," meaning formally that they depend exponentially on temperature, according to

$$MTF = MTF_0 \, e^{E_a/kT}, \tag{7.13}$$

where MTF_0 is a constant, E_a is a failure activation energy (typically a fraction of an electron volt, depending on the failure mechanism), and kT is the ubiquitous thermal energy. This suggests that a good way to accelerate failure is to simply heat up the IC and is called "burn-in" (go figure). We might also choose to run the IC at higher than expected current or voltage to force it to fail early. Following accelerated life testing, we then build a mathematical model to capture this accelerated failure behavior and then project backward to a "reasonable" lifetime trend for the IC as it is intended to actually be used. Generally, industry would attempt to ensure a average projected in-use lifetime of at least 10 years (hint: your laptop is not likely to last that long!). Sound easy? It's not. Lots of folks spend lots of time trying to ensure adequate reliability of ICs. Failure mechanisms can be a moving target as technology evolves, and hence trying to anticipate the next-big-problem before it rears its ugly head is a full-time job.

So much for semiconductor fab. Armed with this new knowledge of what Mr. Transistor looks like, let's now get a good feel for how he actually does his thing.

References and notes

1. J. D. Cressler and G. Niu, *Silicon-Germanium Heterojunction Bipolar Transistors,* Artech House, Boston, 2003.

2. R. F. Pierret, *Semiconductor Fundamentals,* Addision-Wesley, Reading, MA, 1996.

3. B. L. Anderson and R. L. Anderson, *Fundamentals of Semiconductor Devices,* McGraw-Hill, New York, 2005.

4. R. C. Jaeger, *Introduction of Microelectronic Fabrication,* Prentice-Hall, Englewood Cliffs, NJ, 2002.

5. H. Xiao, *Introduction to Semiconductor Manufacturing Technology,* Prentice-Hall, Englewood Cliffs, NJ, 2001.

6. M. Quirk and J. Serda, *Semiconductor Manufacturing Technology,* Prentice-Hall, Englewood Cliffs, NJ, 2001.

7. R. M. Warner Jr. and B. L. Grung, *Transistors: Fundamentals for the Integrated-Circuit Engineer,* Wiley, New York, 1983.

8. J. D. Meindl, J. A. Davis, P. Zarkesh-Ha, C. S. Patel, K. P. Martin, and P. A. Kohl, "Interconnect opportunities for gigascale integration," *IBM Journal of Research and Development*, Vol. 46, pp. 245–263, 2002.

9. Ted Wilcox, ECE 3080, Georgia Tech, Spring 2007, solved a related problem.

10. A. B. Grebene, *Bipolar and MOS Analog Integrated Circuit Design,* Wiley-Interscience, Hoboken, NJ, 2003.

8 Transistors – Lite!

Photograph of a packaged IC (courtesy of Science and Society Picture Library). (See color plate 25.)

The Invariable Mark of Wisdom
Is To See The Miraculous
In the Common.

Ralph Waldo Emerson

Ever been tempted to name a law of nature after yourself? No? Well, I'm feeling bold, so hold on tight. *Cressler's 1ˢᵗ Law: "Transistor form and function are inseparable."* Okay, so maybe it's not a law of nature, but it does have a nice ring to it! Care for a translation? Okay. The way we use semiconductor fab to build Mr. Transistor is fundamentally linked to what we actually want to DO with Mr. Transistor. For example, we might want him to act only as a fast on–off switch inside of a microprocessor (e.g., an Intel Pentium, see the figure on the previous page). Embodiment #1. Or, we might instead want to use him to amplify weak radio signals at the input of a cell phone. Embodiment #2. Or, we might want him to serve as a part of a nonvolatile memory element. Embodiment #3. Each transistor just described will look quite different at the end of the day, because the optimal transistor architecture for that particular use is dictated by very different system constraints and driving forces. That is, depending on the end use, the way we build, and importantly, optimize, Mr. Transistor will change; sometimes profoundly.

In the previous chapter we talked about how we humans use quite a bit of cleverness and some rather expensive toys to fabricate useful objects from semiconductors and assorted materials; all executed at almost unbelievably tiny dimensions. Let's now talk about WHY we need to build Mr. Transistor, and, more important, how those beasts actually do their thing. By the way, in common geek parlance we call anything useful built from semiconductors "devices" – admittedly not very inspired naming, but there you have it; a transistor is a semiconductor "device," and the physical understanding and design of transistors falls under the topic known as "device physics" – this is what I do for a living. The present chapter plumbs the depths (okay, scratches the surface) of what transistors are all about: What is their form, how do they function? And why. First things first, though.

8.1 The semiconductor device menagerie

Check this out. There are seven major families of semiconductor devices (only one of which includes transistors!), 74 basic classes of devices within those seven families, and another 130 derivative types of devices from those 74 basic classes [1]. Gulp . . . Linnaeus would have had a field day! Sound like a veritable device zoo? You got that right! That is actually a boatload of devices to know something about, and let me apologize in advance for the myriad of acronyms! Relax, though, we only need three basic devices to do business: *pn* junction diodes, bipolar junction transistors (BJTs), and metal-oxide semiconductor field-effect

The Transistor Food Chain

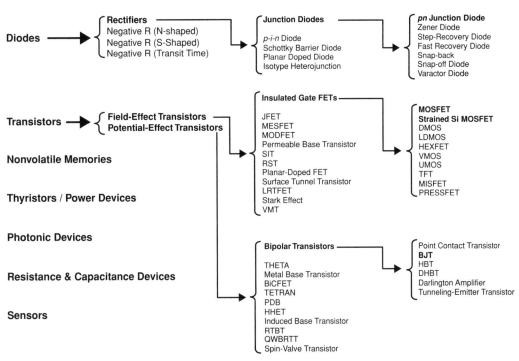

Figure 8.1   The transistor food chain (inspired by [1]).

transistors (MOSFETs). Figure 8.1 illustrates the transistor food chain [1]. Diodes are required in our present discussion because, even though they are plenty useful in their own right, you cannot construct transistors without them.

Interestingly enough, all of our semiconductor device menagerie can be built from a surprisingly small set of tinker toys (Fig. 8.2), including [1]

- the metal–semiconductor interface (e.g., Pt/Si; a "Schottky barrier")
- the doping transition (e.g., a Si p-type to n-type transition; a pn junction)
- the heterojunction (e.g., n–AlGaAs/p–GaAs)
- the semiconductor–insulator interface (e.g., Si/SiO$_2$)
- the insulator–metal interface (e.g., SiO$_2$/Al).

Important derivative structures from these basic building blocks include

- ohmic contacts
- planar doped barriers
- quantum wells.

Needless to say, all of the physics of transistors is bounded by the physics of these basic semiconductor structures (careful – their behavior is not as simple as it may first look!).

The transistor family of semiconductor devices is broken into two major classes: field-effect transistors (FETs) and potential-effect transistors (PETs)

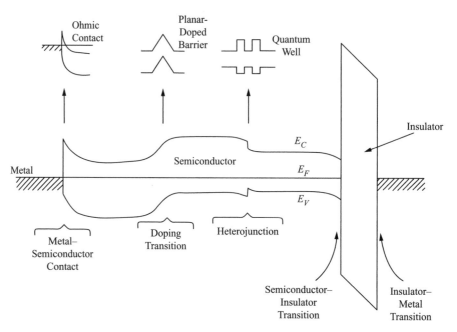

Figure 8.2 The fundamental building blocks of all semiconductor devices: (1) the metal–semiconductor interface (Schottky barrier); (2) the doping transition (a *pn* homojunction); (3) the heterojunction; (4) the semiconductor–insulator interface; and (5) the insulator–metal interface. Important derivative structures include (1) ohmic contacts; (2) planar doped barriers; and (3) quantum wells (inspired by [1]).

[1,2]. Within the class of FETs, we are interested in "insulated gate FETs," and the MOSFETs that are ultimately derived from these (Fig. 8.1). As we will see, MOSFETs are used to build CMOS, and CMOS is the building block of most digital systems (e.g., the guts of microprocessors), hence a principle cornerstone of the Communications Revolution. Read: MOSFETs are pretty darn important in the grand scheme of life.

Within the class of PETs lie all types of bipolar transistors, and our second transistor target: The derivative BJT. You may recall from Chap. 3 that BJTs, although not nearly as common today as MOSFETs in terms of sheer numbers inhabiting planet Earth, were actually the first modern transistors to be demonstrated (albeit the basic idea for a MOSFET actually predates the BJT demo), and are still in wide use in analog and RF systems such as those needed for the global communications infrastructure. Read: BJTs are also pretty darn important.

Field effect, potential effect; what's the difference? Well, the "field effect" in semiconductors was actually defined by Shockley (go figure) to mean the "modulation of a conducting channel by electric fields" [1]. A FET differs fundamentally from a PET in that its conducting channel (where the current is carried) is modulated (changed) capacitively (indirectly) by a transverse (perpendicular to the surface) electric field. Huh? That is, we modulate the characteristics of the conducting channel where all the transistor action lies in a FET by putting an insulator on the channel and then a conductor to form a capacitor. Apply a voltage to the top plate of the capacitor, and we induce a transverse electric field in the

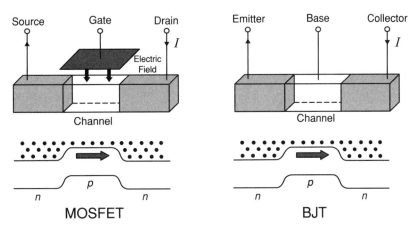

Figure 8.3

High-level commonality between the operational principles of a MOSFET and a BJT.

channel (without dc current flow!), altering the nature of the channel for useful ends. This is not your simple metal–insulator–metal capacitor, but might be, for instance, a heavily doped polysilicon–SiO_2–lightly doped silicon capacitor. Same idea though. In a PET, however, the conducting channel is modulated by *direct* electrical contact (i.e., no capacitor), and hence additional current is involved in the channel modulation process.

Still, at a deep level, FETs and PETs are closely related, as can be gleaned from Fig. 8.3, which shows the most basic MOSFET and BJT. In a MOSFET there is a "gate" control electrode (an electrode is often called a "terminal") that capacitively influences the channel, altering the current flow from the "source" electrode (which acts as a source of electrons) to the "drain" electrode (which drains said electrons). In a BJT, there is a "base" control electrode, which directly influences the channel, altering the current flow from the "emitter" electrode (which emits electrons) to the "collector" electrode (which collects said electrons). Be careful: At this level, MOSFETs and BJTs may look remarkably similar – but refer to Cressler's 1st law. They function very differently and hence have very different forms. Observe also in Fig. 8.3 that both MOSFETs and BJTs have alternating *p*-type and *n*-type regions of semiconductor within them. No coincidence. As we will see, these are the ubiquitous *pn* junctions required for building transistors. No *pn* junctions, no transistors.

8.2 Why are transistors so darn useful?

Simple query: Why do we actually need transistors in the first place? Good question! The answer is best stated generally. Deep breath. *Cressler's 2nd Law: "The universe attenuates all electrical signals."* No way around it. By this I mean, the magnitude of all electrical signals (think "1s" and "0s" inside a computer, or an EM radio signal from a cell phone, or a TV broadcast) necessarily decreases (becomes smaller) as I go from point A to point B; clever manipulation #1 to clever manipulation #2.

8.2.1 Loss

In electronics parlance, we say that an attenuating system exhibits "loss" (nothing to do with grief; think "to make smaller"). Consider two relevant situations involving loss. Scenario #1. Let's say that I have a metal wire of some length L, which, because all metals have a finite resistivity, has an associated resistance R. If I apply a voltage V_{in} to one end of the wire and ground the other end of the wire so that current can flow, transferring the voltage from point A to point B, then some time later the voltage V_{out} at the end of the wire will be given by $V_{out} = V_{in} - IR$, where $V = IR$ is just Ohm's law. Moral? The finite resistivity of materials means that my voltage will definitely be attenuated in going from one point on that wire to another. Is this a problem? Depends. What if my wire length is really, really long (remember, a modern microprocessor has several MILES of internal wiring in even a centimeter-sized chip)? Then I can get substantial "voltage drop" down the wire, and my input voltage V_{in}, which is supposed to be a logical "1" in my digital system, say, will instead produce a output voltage V_{out} that is now so low that it can no longer by interpreted as a logical "1." This is a problem.

Scenario #2. Let's say I want to use my cell phone to send a call and the requisite cell tower is several kilometers away. My cell phone launches the modulated RF carrier omnidirectionally (same in all directions) into space, with my voice encoded on it of course, and to first order, it spreads out spherically as it moves away from me. The resultant EM signal intensity decreases rapidly as it moves away from me (see Geek trivia for a bit more on this important effect). Moral. Propagating EM signals rapidly attenuate. Soon the "signal strength" will become too small to reliably detect by the electronic receiver on the other end (moral: congratulations – you have a dropped call).

Geek Trivia: Inverse Square Law

The intensity of an EM radio signal (or light, or even sound) radiating away from a "point source" (formally, the energy per unit of area radiated is perpendicular to the source) is inversely proportional to the square of the distance from the source ($1/r^2$). Hence an object (of the same size) located twice as far away receives only one-quarter of the energy (over the same time period). More formally, the EM intensity (power per unit area in the direction of propagation) of a spherical EM wavefront varies inversely with the square of the distance from the source; assuming, of course, that there are no additional complexities that are due to losses caused by absorption or scattering. As a practical example, the intensity of Sun's radiation is a cozy 1,370 W/m^2 on the surface of the Earth [1.0 Astronomical Unit (AU) away – aka 93,000,000 miles], but it is 9,140 W/m^2 (mighty darn hot!) on the surface of Mercury (only 0.387 AU away). That is, a 3× decrease in distance results in a 9 × (3^2) increase in the intensity of the EM radiation.

Why $1/r^2$? Simple. The surface area of a sphere is given by SA = $4\pi r^2$ (if you have taken calculus you can easily prove this), so if the point source is at the center of the sphere, the intensity on the surface a distance r away is given by the total radiated power P divided by the surface area of the sphere, or $I = P/A = P/4\pi r^2$. Hence $I \propto 1/r^2$. QED.

Clearly, even non-techno-geeks like photographers and theatrical lighting professionals must at least intuitively wrestle with the inverse-square law when determining the optimal location to place a light source to achieve proper illumination of the subject in question. [Not to offend photographers and other lighting professionals. Their Geek-ness Quotient (GQ) – GQ = 0 means normal Joe clueless of all things technological, GQ = 10 means total techno-nerd saddled with every cool toy and gadget known to man – is for sure nonzero.]

There are actually unintended perks associated with Cressler's 2nd Law. If there were no loss associated with signal transmission, all radio and TV stations within a close frequency band would interfere mightily with one another, yielding a garbled mess for global communications. There would be no way, for instance, to have 90.1 FM (WABE – my beloved local NPR station) in Atlanta and 90.1 FM in Chicago; or Channel 5 on TV in Washington and Channel 5 in New York. Just imagine how useless cell phones would be if all multibillion users were trying to talk on the same few hundred channels of allotted EM spectrum at the same time!

8.2.2 Gain

Well then, what's a fella to do about all this loss? You guessed it, create another law! *Cressler's 3rd Law: "Gain is essential for implementing useful electrical systems."* What is "gain," you say? Just what it says. To make larger. To grow. Yep, opposite of loss. In our case, we want to make either the voltage larger, or the current larger, or both larger; or, at a deeper level, make the electric-field or magnetic-field, or both, vectors in an EM wave larger. THE unique property of transistors that makes them indispensable for the implementation of complex electrical systems is that they possess inherent gain. Push a time-varying voltage signal of magnitude V_{in} into Mr. Transistor, and I get a larger time-varying voltage signal of magnitude V_{out} out of Mr. Transistor. We define the voltage gain (A_V) to be $A_V = V_{out}/V_{in}$. You could do the same thing with current ($A_I = I_{out}/I_{in}$), or even power if I'm driving a "load" ($A_P = \text{Power}_{out}/\text{Power}_{in}$). Clearly, gain (or loss) is dimensionless, but electronics geeks often speak about gain in decibel (dB) units for convenience (refer to sidebar discussion on dB).

In this case, when we present an attenuated input signal to the transistor, it creates an output signal of larger magnitude (i.e., $A_V > 1$), and hence the transistor serves as a "gain block" to "regenerate" (recover) the attenuated signal in question. VERY useful concept. Critical in fact to all useful electronics systems – read: human life. In the electronics world, when Mr. Transistor is used in this manner as a source of gain, we refer to it as an "amplifier." Note: This is exactly what the "amp" on a sound stage does – it takes the small but creatively modulated output signals from say an electric guitar (or a microphone) and amplifies them to a large enough level to effectively drive those honkin' 30-ft speaker towers you are listening to.

Measures of Gain and Loss: Decibels

So what's with all the decibels (dB), anyway? Decibels are officially ubiquitous, so get used to them! Like most things, the decibel is defined for convenience and is simply a logarithmic unit of measurement that expresses the magnitude of a physical quantity relative to a specified (or implied) reference level. Given its logarithmic nature (think powers of 10), both very large and very small ratios are represented in a shorthand way by dB, in a manner similar to that of scientific notation. (Aside: The Richter scale for earthquakes is also a logarithmic scale – a Magnitude 6 quake is 10 times more power than a Magnitude 5 quake, and 100 times more powerful than a Magnitude 4 quake, etc.) Because decibels are given as a ratio, they are a dimensionless unit. Decibels are used in a wide variety of contexts, including acoustics, physics, and of course electronics.

Care for a little history on Mr. dB? Sure! Actually, the dB is not an official SI (Système International d'Unités, the International System of Units) unit, although the International Committee for Weights and Measures (Comité International des Poids et Mesures, or CIPM) has recommended its inclusion in the SI system. Following the SI convention, the d is lowercase because it represents the SI prefix *deci* (one-tenth), and the B is capitalized, because it is an abbreviation of a name-derived unit, the *bel* (see subsequent discussion). The full-name decibel then follows the usual English capitalization rules for a common noun. The decibel symbol (dB) is often qualified with a suffix, which indicates which reference quantity has been explicitly assumed. For example, dBm indicates that the reference quantity is one milliwatt of power. That is, power gain in dBm means how much power we get out relative to 1-mW power in.

A decibel is one tenth of a *bel* (B). What's a *bel*? Well, it was conceived by Bell (detect a bit of arrogance here?!) Telephone Laboratory engineers to quantify the reduction in audio signal strength over a 1-mile (1.6-km) length of standard 1920s telephone cable. The bel was originally called the "transmission unit" or TU (boring), but was renamed in 1923 or 1924 in honor of the laboratory's founder and telecommunications pioneer Alexander Graham Bell. In many situations, however, the bel proved inconveniently large, so the decibel (divide by 10) has become the de facto standard [4]. Cute, huh?

How does all this work in practice? Well, for voltage gain A_V we define

$$A_{V,\mathrm{dB}} = 20 \log_{10} |A_V|, \tag{8.1}$$

where $\log_{10}$ is just the base-10 (common) logarithm. Translation: Voltage gain in dB is just 20 times the base-10 logarithm of the absolute value of the dimensionless voltage gain ratio. You could also obviously reference that to some standard voltage, say 1 mV, if you'd like. Example: If the voltage gain is 100 (V_{out} is 100 times larger than V_{in}), then $A_{V,\mathrm{dB}}$ is just 40 dB (log base-10 of 100 = 2 because $10^2 = 100$, and thus 20 times 2 = 40 dB). So saying the voltage gain is 40 dB is absolutely the same as saying 100. Or a gain of 0 dB = 1 (affectionately called "unity gain"). Seem crazy? It's not. A gain of 100,000,000,000 is simply 220 dB. And if the voltage gain increases by 20 dB, then I immediately know I have a 10× increase in gain. Or, a 6-dB decrease gives me a 2× decrease, etc. Hence decibels are especially convenient for relative changes in parameters. How about loss expressed in dB? Well, if the gain is 0.1 (V_{out} is 10 times smaller than V_{in} – i.e., loss) then the gain is −20 dB, and the negative indicates we have loss (try it!).

Not only can Mr. Transistor serve as a wonderful nanoscale-sized amplifier, but importantly he can also be used as a tiny "regenerative switch"; meaning, an on–off switch that does NOT have loss associated with it. Why is this so important? Well, imagine that the computational path through your microprocessor requires 1,000,000 binary switches (think light switch on the wall – on–off, on–off) to implement the digital binary logic of said computation. If each of those switches contributes even a teensy-weensy amount of loss (which it inevitably will), multiplying that teensy-weensy loss by 1,000,000 adds up to unacceptably large system loss. Read: Push a logical "1" or "0" in, and it rapidly will get so small during the computation that it gets lost in the background noise. Computation aborted; Mr. Microprocessor can do nothing worthwhile for us. If, however, I implement my binary switches with gain-enabled transistors, then each switch is effectively regenerative, and I can now propagate my signals through the millions of requisite logic gates without excessive loss, maintaining their magnitude above the background noise level. Bamm! Mr. Microprocessor suddenly got very useful.

So how do we control whether Mr. Transistor is being gainfully employed (pun intended) as an amplifier or as a switch? You asked for it! *Cressler's 4th Law: "Circuit environment determines the functional nature of transistors."* By circuit environment, I mean the collection of other transistors, resistors, capacitors, and inductors (and of course wires to hook them all up) that are used to build the electronic circuit in question. As we will see, the simplest digital switch can be built with only two transistors, and nothing else. A decent amplifier, however, might need a transistor or three and a handful of RLC "passive" components. Change the way all those objects are connected (wired together, literally), and the functional nature of the transistor changes: voltage amplifier; regenerative binary switch; memory element; you name it. You can see the downstream implications. Turn a sea of very clever humans armed with trillions of dirt-cheap gain-producing transistors loose (recall from Chap. 1: In 2006, there were 10,000,000,000,000,000,000 – 10-billion billion – transistors), and what do you get? A revolution!

Okay, quick review. (1) Mr. Transistor can serve in one of two fundamental capacities: amplifier or regenerative switch. (2) Amplifiers and regenerative switches work well only because Mr. Transistor has the ability to produce gain. So then, where does transistor gain come from? Ah, good question! Fascinating answer! Read on. Yep – Now is the time to tell you how *pn* junctions, BJTs, and MOSFETs actually do their thing. Brace yourself.

8.3 The *pn* junction

Virtually all semiconductor devices (both electronic and photonic) rely on *pn* junctions (aka "diodes," a name that harkens back to a vacuum tube legacy) to function. This is especially true for transistors. Moral: To understand the idea of

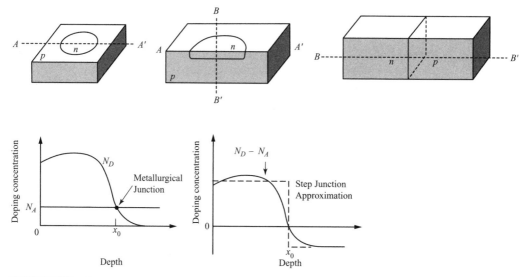

Figure 8.4 Conceptual top-down and cross-section views of *pn* junctions, together with their doping profiles and step junction approximation.

gain in transistors, you first need to understand a little about *pn* junctions. Here goes.

8.3.1 What is it?

We'll start with the simplest embodiment of a *pn* junction – the *pn* "homojunction." Homojunction just means that within a single piece of semiconductor (e.g., silicon) we have a transition between *p*-type doping and *n*-type doping (e.g., *p*–Si/*n*–Si). The opposite would be a *pn* heterojunction, in which the *p*-type doping is in one type of semiconductor (e.g., *p*–GaAs), and the *n*-type doping is within another type of semiconductor (e.g., *n*–AlGaAs), to form a *p*–GaAs/*n*–AlGaAs heterojunction. More on heterojunctions later. As shown in Fig. 8.4, there are several ways to build *pn* junctions. I might, for instance, ion implant and then diffuse *n*-type doping into a *p*-type wafer. The important thing is the resultant "doping profile" as you move through the junction [$N_D(x) - N_A(x)$, which is just the net doping concentration]. At some point x, $N_D = N_A$, and we have a transition between net *n*-type and *p*-type doping. This point is called the "metallurgical junction" (x_0 in the figure), and all of the action is centered here. To make the physics easier, two simplifications are in order: (1) Let's assume a "step junction" approximation to the real *pn* junction doping profile; which is just what it says, an abrupt change (a step) in doping occurs at the metallurgical junction (Fig. 8.4). (2) Let's assume that all of the dopant impurities are ionized (one donor atom equals one electron, etc.). This is an excellent approximation for common dopants in silicon at 300 K.

Does the *pn* junction sound simple? It is and it isn't. Trivial to make, yes; but a decidedly subtle and complicated piece of physics when it operates. But

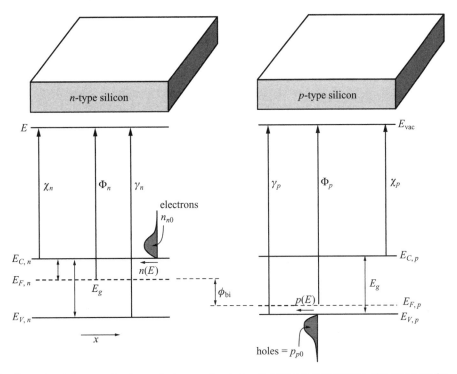

Figure 8.5 Energy band diagrams of n-type and p-type semiconductors isolated and at equilibrium. Also shown are the major carrier populations in the respective bands (electrons for n-type and holes for p-type).

don't be nervous! – remember, we'll just scratch the surface here. Which reminds me ... *Cressler's 5th Law: "All semiconductor devices get more complicated the more closely you examine them."*

8.3.2 How does it work?

Great; now hold on tight! As intimated earlier, the operation of ALL semiconductor devices is best understood at an intuitive level by considering the energy band diagram (introduced in Chap. 5), so let me briefly revisit a couple of key points to refresh your memory. First take a quick peek at Fig. 8.5. Done? Good. Now, consider:

- The energy band diagram plots electron and hole energies as functions of position as we move through a device.
- An n-type semiconductor is electron rich ($n_{n0} = N_D$, the majority carriers) and hole poor ($p_{n0} = n_i^2/n_{n0}$, the minority carriers).
- A p-type semiconductor is hole rich ($p_{p0} = N_A$, the majority carriers) and electron poor ($n_{p0} = n_i^2/p_{p0}$, the minority carriers).
- Those majority electron and hole populations in the conduction and valence bands, respectively, are themselves functions of energy, but they can be easily visualized as a distribution of charge in the respective bands, as shown in

Fig. 8.5 [3]. Formally, the total carrier population is just the area under that distribution curve (a definite integral, if you know a little calculus).

- If the energy band is not constant (flat), then an electric field MUST be present – always. This is called band-bending (for obvious reasons).

- The one new detail we do need is the location in energy of the "Fermi levels" for the n-type and p-type materials ($E_{F,n}$ and $E_{F,p}$). The Fermi level was introduced in Chap. 5 as a prerequisite for actually calculating n and p. (My apologies, it was embedded in a *Caffeine Alert* section – Gulp . . . I trust you didn't skip it! If you did, no matter, read on.) There are two main take-aways here: (1) For an n-type semiconductor, $E_{F,n}$ is closer to E_C than to E_V, whereas for a p-type semiconductor, $E_{F,p}$ is closer to E_V than to E_C (take a peek at the figure). (2) One of the golden rules of semiconductors, meaning it is formally provable from very general principles, is that when Mr. Semiconductor is in equilibrium, E_F MUST remain constant in energy. No ifs, ands, or buts. E_F must be flat through the device in question while in equilibrium. Remember this.

- If the semiconductor becomes more n-type or p-type, E_F will necessarily change it's energy location within the bandgap. For example, let's say we make the semiconductor more n-type (i.e., more donor impurities, hence more majority electrons). Then $E_{F,n}$ moves closer to E_C. Hence the location of E_F is a good indicator of the majority carrier concentration in the material. There is no magic here; refer to Eqs. (5.9) and (5.10) in Chap. 5 – n and p depend exponentially on E_F. Change E_F and you change n or p, and vice versa. If the material is VERY heavily doped (say 1×10^{20} cm^{-3} in silicon), E_F will then be close to the band edge (E_C if n-type or E_V if p-type).

Great. (Re)armed with this information, let's look back at the energy band diagrams of an isolated n-type semiconductor and an isolated p-type semiconductor (our pn junction building blocks), which, because they are isolated, have no ability to exchange charge (Fig. 8.5 – don't worry about $\chi_{n,p}$, $\gamma_{n,p}$, and $\Phi_{n,p}$, they are just reference energies and fixed in value; material parameters). Observe that there is a difference in energy between $E_{F,n}$ and $E_{F,p}$ of $q\phi_{bi}$ (bi stands for "built-in"; more on this in a second). Logical conclusion? If we imagine bringing the n-type and p-type semiconductors into "intimate electrical contact" (don't sweat, this book isn't R-rated!), meaning that charge can be freely exchanged from n to p and p to n, it is demonstrably impossible for that system, now in intimate contact, to be in equilibrium. Picture Mother Nature's big fat pout! How do I know that it's not in equilibrium? E_F is not constant from side to side (refer back to Fig. 8.5). QED.

I have already waxed poetic on the fact that nature ALWAYS prefers the lowest-energy configuration for a given system in question, and it will thus try desperately to establish equilibrium by whatever means it has at its disposal. How will nature act to establish equilibrium in our pn junction? Only one option: Exchange of charge. Figure 8.6 shows the final equilibrium band diagram for

pn Junction Energy Band Diagram

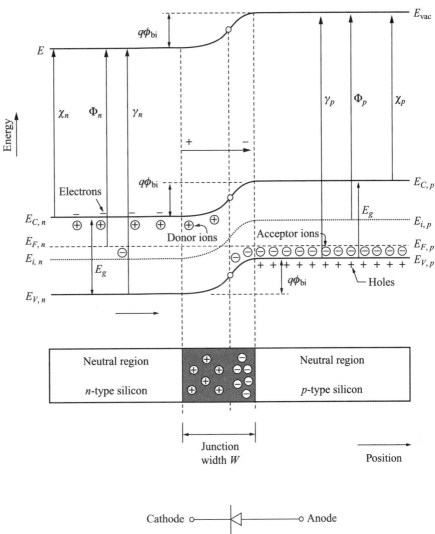

Figure 8.6 Equilibrium energy band diagram of a *pn* junction, showing charge neutral *n* and *p* regions, separated by the space-charge region that is depleted of free carriers. Also shown is the circuit symbol of a *pn* junction, with "anode" and "cathode" indicated.

Mr. *pn* Junction after this exchange occurs. Sure, it looks like a mess, but it's not that bad. Observations: E_F is now constant (flat); hence we are guaranteed that the system is now in equilibrium. This means there is no NET current flow in the junction. To achieve this constant E_F, there is induced "band-bending" present from side to side in the junction, and the total potential (voltage) drop side to side is ϕ_{bi}.

The question of the hour? What actually happened inside Mr. *pn* Junction to establish equilibrium? Consider: When brought into contact, the *n*-type side of the junction is electron rich, whereas the *p*-type side is electron poor. Read: We have a large driving force for electrons to diffuse from the *n* region to the *p* region. Oh yeah? Recall: There are two ways to move charge in a semiconductors: (1) drift, whose driving force is the electric field (applied voltage), and (2) diffusion, whose driving force is the carrier density gradient. The latter is exactly what we have here. Hence, once in contact, an electron moves from the *n* side to the *p* side, leaving behind a positively charged donor impurity (N_D). Careful: Far away from the junction, for each charged donor impurity there is a matching donated electron; hence the semiconductor is charge neutral. Once Mr. Electron leaves the *n* side, though, there is no balancing charge, and a region of "space charge" (literally, charge in space) results. Same thing happens on the *p* side. Hole moves from *p* to *n*, leaving behind an uncompensated acceptor impurity (N_A). Think way back to high school physics. What do you call positive charge separated from negative charge? Give up? A "dipole." Classic EM idea, but the result is simple: There is a field present between the positive and negative charges of a dipole, and this field points from + to − (to the right in this case). Question: How does that field affect the diffusion initiated side-to-side transfer of charge I just described? Answer: It opposes the diffusive motion of both electron and holes! Coulomb's law. Moral: When we bring Mr. *pn* Junction into electrical contact, the diffusion gradient moves electrons from *n* to *p* and holes from *p* to *n*; but as this happens a dipole of space charge is created between the uncompensated N_D and N_A, and an induced electric field (think drift!) opposes the further diffusion of charge. When does equilibrium in the *pn* junction result? When the diffusion and the drift processes are perfectly balanced. Want a formal statement of this? Sure! Think back to the drift–diffusion equations from Chap. 5. Now we have drift balancing diffusion such that the net current flow is zero:

$$\vec{J}_{n,\text{total}} = \vec{J}_{n,\text{drift}} + \vec{J}_{n,\text{diffusion}} = q\,\mu_n\,n\,\vec{\mathcal{E}} + q\,D_n\,\nabla n = 0 \qquad (8.2)$$

and

$$\vec{J}_{p,\text{total}} = \vec{J}_{p,\text{drift}} + \vec{J}_{p,\text{diffusion}} = q\,\mu_p\,p\,\vec{\mathcal{E}} - q\,D_p\,\nabla p = 0, \qquad (8.3)$$

so that

$$\vec{J}_{\text{total}} = \vec{J}_{n,\text{total}} + \vec{J}_{p,\text{total}} = 0. \qquad (8.4)$$

The Fermi level E_F is now flat through the device, and there is band-bending from side to side, yielding a "built-in" electric field and hence potential drop inside the junction of ϕ_{bi}. Make sense? Good. Be careful, though. I have been deliberate in saying the NET current flow is identically zero in equilibrium. This does NOT mean no carrier action is occurring in equilibrium. As illustrated in Fig. 8.7, generation and recombination of electrons and holes can and is still taking place, but the NET current flow side to side remains zero. Hook an ammeter to it and you measure nothing. So how fast does all this establish-equilibrium-thing happen? REALLY fast (see sidebar discussion on "Time scales in *pn* Junctions").

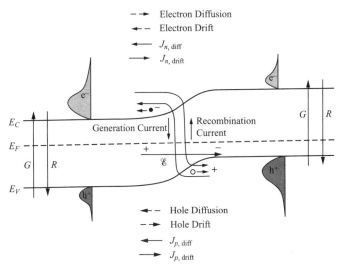

Figure 8.7 Carrier action inside the *pn* junction at equilibrium. The *net* current flow is zero.

Tick-Tock: Time Scales in *pn* Junctions

Question: You might logically wonder how fast the charge transfer between *n* and *p* regions required for establishing equilibrium actually happens? VERY fast. VERY, VERY fast. In semiconductors, the speed with which an electron (or hole) can respond to an applied electric field is governed by the so-called "dielectric relaxation time" in solids. How fast is that? Well, for *n*-type GaAs doped at 1×10^{17} cm^{-3}, the dielectric relaxation time is about 1.8 fs [5]. That's 1.8×10^{-15} s! That's 0.000,000,000,000,001,8 s! A trillion times faster than an eyeblink. Yep, that's fast. The diffusion process is a bit slower, but still blazing. Think 10s to 100s of femtoseconds for equilibrium to be reestablished in semiconductors when they are disturbed from their slumber by clever humans. Because, at a deep level, you will never build a practical semiconductor device that can switch faster than the electrons and holes can respond to an applied field in the semiconductor itself, the dielectric relaxation can be viewed as a fundamental speed limit of sorts for all micro/nanoelectronic devices. In practice, of course, parasitic resistance and capacitance, and other so-called "charge storage" mechanisms, will come into play long before we reach this speed limit. But as you saw earlier, modern devices can easily switch at picosecond (10^{-12}) speeds in real circuits. This is fundamentally why.

The so-called "electrostatics" of the junction are shown in Fig. 8.8 and represents a useful summary of all that I have described for the *pn* junction in equilibrium. Major points:

- The *pn* junction in equilibrium consists of a neutral *n* region and a neutral *p* region, separated by a space-charge region of width W. Observe that this forms a capacitor (conductor–insulator–conductor), and *pn* junctions have built-in capacitance that will partially dictate their switching speed.

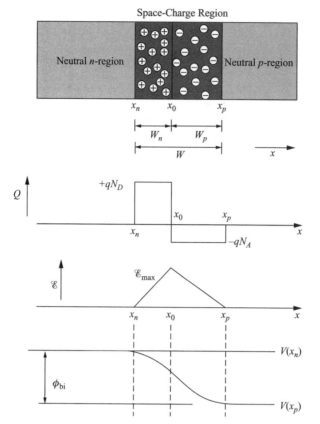

Figure 8.8 The electrostatics of the *pn* junction at equilibrium, showing charge, electric field, and potential distributions.

- The electric field in the space-charge region (for a step junction) is characteristically triangular shaped, with some peak value of electric field present (see sidebar discussion on the fields in junctions).
- There is a built-in voltage drop across the junction (ϕ_{bi}).
- As can be seen in the energy band diagram, there is a potential barrier for any further movement of electrons and holes from side to side. This barrier to carrier transport keeps the net current zero, the Fermi level flat, and the junction in equilibrium.

Shock Therapy: Built-in Fields in *pn* Junctions

At equilibrium, we have a side-to-side built-in voltage drop (ϕ_{bi}) required for flattening the Fermi level. How big is that voltage drop in practice? Well, we can easily show that

$$\phi_{bi} = \frac{kT}{q} \ln\left(\frac{N_A N_D}{n_i^2}\right), \tag{8.5}$$

and hence ϕ_{bi} depends on three things: (1) The semiconductor used (i.e., E_g), (2) the doping levels, and (3) the temperature. Consider a typical $p^+ - n^-$ silicon junction sitting at 300 K, and doped with 1×10^{19} cm^{-3} of boron and 1×10^{15} cm^{-3} of arsenic. In this case, $\phi_{bi} = 0.84$ V. Sounds small, right? After all, a tiny little AAA battery delivers 1.5 V. Question: At equilibrium, how big is the built-in electric field associated with this voltage drop inside Mr. Junction? Well, field is just voltage divided by distance. From the electrostatics solution, we can calculate the width of the space-charge region where band-bending occurs according to

$$W = \sqrt{\frac{2\kappa_{semi} \, \epsilon_0 \, \phi_{bi}}{q N_D}}, \qquad (8.6)$$

where κ_{semi} is the relative dielectric constant of the semiconductor (11.9 for silicon) and ϵ_0 is the permittivity of free space (8.854×10^{-14} F/cm). For our junction, we find that the width of the space-charge region over which ϕ_{bi} is dropped is about 1.05 μm at equilibrium. The built-in field is thus $\mathcal{E} = \phi_{bi}/W = 8,000$ V/cm. Big number or small number? Recall my example of standing in a pool of water and being asked to move your finger to within a centimeter of an electrode labeled "DANGER: 8,000 V! Would you do it? Only once! In human terms, this is a gigantic electric field. (As you will see in a moment, though, this field is well below the critical breakdown strength, the upper bound for applied fields in semiconductors, about 500,000 V/cm for silicon.) So, then... if the junction has a massive built-in electric field associated with it, why don't we get electrocuted when we pick it up? Good question! Look back to Fig. 8.6. Remember, outside of the junction the energy bands are flat (no field is present) and charge neutral. This lightening-bolt-waiting-to-happen is effectively buried inside two neutral conductors. Safe and sound – go ahead, pick it up! One last interesting question to ponder. What is the maximum allowable built-in voltage for an arbitrarily doped junction? Stare for a moment at Fig. 8.6 and see if you see it. Give up? E_g! Do you see it now? If I doped the junction as heavily as I can, E_F will to first order be aligned to the conduction and valence band edges on either side of the junction, and thus the total side-to-side voltage drop will be $q E_g$. Good rule of thumb to file away.

Fine. Here's where it gets fun. Let's now suppose you wanted to get current flowing again in the junction. How would you do it? Informed answer: We must unbalance the drift and diffusion mechanisms by lowering the potential barrier the electrons and holes "see" (forgive me my anthropomorphism!). How do you lower that barrier? Well, simply apply an external voltage to the n and p regions such that the p region (anode) is more positively biased than the n region (cathode). As shown in Fig. 8.9, this effectively lowers the side-to-side barrier by an amount $\phi_{bi} - V$, drift no longer balances diffusion, and the carriers will once again start diffusing from side to side, generating useful current flow. This is called "forward

pn Junction under Bias

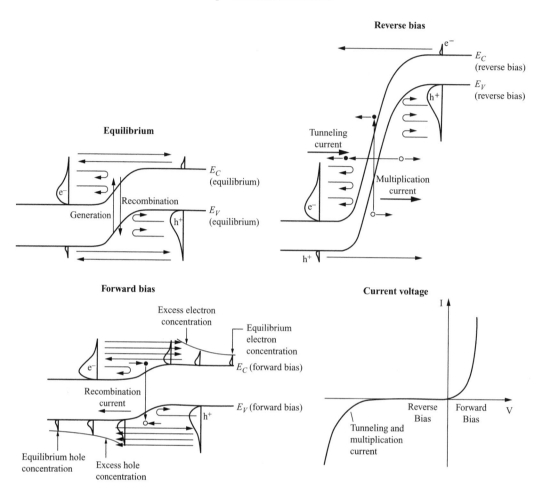

Figure 8.9   Carrier action in the *pn* junction at equilibrium and under both forward and reverse biases. The respective current–voltage characteristics are also indicated (inspired by [3]).

bias." What happens if we apply a voltage to the junction of opposite sign (*p* region more negatively biased than the *n* region)? Well, the barrier the carriers see grows by $\phi_{bi} + V$, effectively preventing any current flow (Fig. 8.9).

Congratulations! You have just created a solid-state switch (aka a "diode")! Consider: Apply a voltage of one polarity across Mr. Junction and lots of current flows. Apply a voltage of the other polarity across Mr. Junction and no current flows. On–off; on–off. Think light switch. Spend a moment with Fig. 8.9 and solidify this connection among applied bias, current flow, and band diagram in your mind's eye. Good. Let's move on.

Shockley shared the Nobel Prize with Bardeen and Brattain largely for explaining this phenomenon and of course by wrapping predictive theory around it, which led to the BJT. The result of that particularly elegant derivation (when I teach it, I follow his original derivation path) is the celebrated "Shockley equation," which governs the current flow in a *pn* junction:

$$I = qA \left(\frac{D_n n_i^2}{L_n N_A} + \frac{D_p n_i^2}{L_p N_D} \right) (e^{qV/kT} - 1) = I_S (e^{qV/kT} - 1), \quad (8.7)$$

where A is the junction area, V is the applied voltage, $D_{n,p}$ is the electron–hole diffusivity ($D_{n,p} = \mu_{n,p} kT$), $L_{n,p}$ is the electron–hole diffusion length ($L_{n,p} = \sqrt{D_{n,p}\tau_{n,p}}$), and I_S is the junction "saturation current" that collapses all of the mess into a single (measurable) parameter. Remember: Mobility (hence diffusivity) is simply a measure of how fast a carrier will move in response to an applied electric field, and diffusion length is how far a minority carrier will go before it recombines. Both are intuitively appealing definitions, and, importantly, independently measurable. One additional observation. Note that in term (1) of Eq. (8.7), D_n and L_n are electron parameters associated with the *p*-type region (N_A); that is, they are minority electron parameters in the *p*-type region. Same for term (2) – minority hole parameters in the *n*-type region. Hence the *pn* junction is often called a "minority carrier" device. All of the current contributors are minority carriers. If we build our junction with $N_A = N_D$, then the relative contributions of the electron and hole minority carrier currents to the total current flowing will be comparable (to first order).

So, what is really happening? Refer to Cressler's 5th law. Hold on tight. Under forward bias, electrons diffuse from the *n*-side to the *p*-side, where they become minority carriers. Those minority electrons are now free to recombine and will do so, on a length scale determined by L_n. So as we move from the center of the junction out into the neutral *p*-region, the minority electron population decreases because of recombination, inducing a concentration gradient (sound familiar?) as we move into the *p*-side, which drives a minority electron diffusion current. The same thing is happening with holes on the other side, and these two minority carrier diffusion currents add to produce the total forward-bias current flow. Question: What is the actual driving force behind the forward-bias current in a *pn* junction? Answer: Recombination in the neutral regions, because recombination induces the minority diffusion currents. There; not so bad.

Alas, simple theory and reality are never coincident. Sound like a law? Sure! *Cressler's 6th Law: Simple theory and reality are never coincident in semiconductor devices.* Figure 8.10 shows what you would actually measure it you took a real *pn* junction diode into the lab and hooked it up to a voltage source and an ammeter. I am plotting this on semilog scales (log of the absolute value of J vs. linear V), because the current depends exponentially on applied voltage, and thus we then obtain a straight line (think Moore's law). We won't worry much about the operative mechanisms, but let me summarize what you would actually see Mr. Junction do:

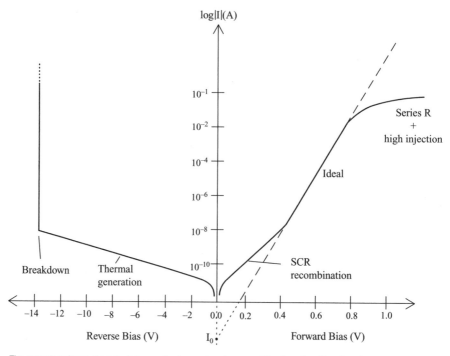

Figure 8.10 The current–voltage characteristics under forward and reverse bias for a "real" *pn* junction, but plotted on a logarithmic scale.

- Under forward bias, the ideal current–voltage behavior addressed by the Shockley equation occurs over only a fairly small range of applied voltages; say 0.4–0.7 V in a silicon junction.
- Below the ideal range, carrier recombination inside the space-charge region (SCR) is dominant, with a different (weaker) exponential voltage dependence.
- Above the ideal range, parasitic series resistance associated with the doped regions, together with other nonidealities known as "high-injection" effects become dominant, causing the current to "roll over" at high forward voltage, and we lose the nice exponential shape of the Shockley equation. Simply stated, high injection occurs when the injected minority carrier density exceeds the doping density in the region in question (e.g., electrons on the p-side). Because generally $J = q\,n\,v$, if the current density gets high enough, n will eventually surpass the doping level, triggering high injection (VERY tough to analytically treat). Practical result? There is only a finite current density that I can push through the junction, which is set by the doping levels used. Thus, if you need to scale up the current flow, the only alternative is to make the area of the junction larger, because $I = JA$. Interestingly enough, many, many types of semiconductor devices necessarily operate in high injection to achieve optimal performance, so its complexities are a fact of life.
- Under reverse bias, we indeed get small current flow, since a low-defect-density semiconductor like silicon will be dominated by thermal generation of carriers, which has a very weak bias dependence (ultimately, as the reverse bias grows,

the SCR spreads, presenting more volume for thermal generation, and hence slightly higher current).

- Rather alarmingly, at sufficiently high reverse bias, the current suddenly takes off due north (to infinity). This doesn't feel like a good idea, and it isn't (see sidebar on "smoking a device"). This reverse-bias "breakdown" voltage represents the maximum useful voltage that can be applied to the device and is obviously important to know a little about.

Smoking a Device

One of my favorite childhood books was Hans and Margret Rey's 1941 *Curious George*, about a cute little monkey named George who always seemed to get into trouble just because he was overly curious about things. Well, my name's not George, but on the first week on my new coop job at IBM in Research Triangle Park, NC, I pulled a George. Yep, a mighty-green, still-wet-behind-the-ears, second-year undergraduate coop student from Georgia Tech, having just been trained to use a tungsten needle probe to contact and measure his first semiconductor device (a MOSFET), decided it would be cool to see just how much voltage his tiny little device could actually handle. In those dark ages we used something called a "curve tracer" for such measurements; basically just a fancy variable voltage–current source and meter. On the front of the curve tracer was a switch that would allow one to remove the current compliance limit on such a measurement (okay, okay – it did indeed have a protective plastic cover that said "DANGER, DON'T REMOVE" on it). Just for fun, I intentionally defeated the compliance protection and proceeded to ramp said voltage on my device. I crossed the suspected breakdown voltage and just kept on going (it might be helpful to look back to Fig. 8.10 and notice just how steep that curve is in breakdown – a little voltage = a ton of current). Imagine my shock when I smelled something funny, glanced over at my probe station, and saw a small but clearly visible mushroom-shaped cloud of smoke rising from my device. Aghast, I raced to look into my microscope at the carnage, and to my horror, all I saw were peeled back, melted tungsten probes (melting point = 3,695 K = 3,422°C = 6,192°F), and underneath them, an ugly crater in the surface of my silicon wafer (melting point = 1,687 K = 1,414°C = 2,577°F), which said MOSFET used to call home. Alas, Mr. MOSFET was no more. SMOKED! Moral for George: In the absence of some mechanism to limit the current flow, breakdown in semiconductors will try VERY hard to reach infinite current. The IR drop associated with this now-very-large I will produce a massive temperature rise that will quickly grow to surface-of-the-sun-like temperatures! Not a good thing. pppsssttt. PPPPPSSSSSTTTTT. To all you budding device engineers – you haven't lived until you have smoked your first device! Give it a try!

So what actually happens in breakdown? Literally, all hell breaks loose! Look back to Fig. 8.8. Under reverse bias, the triangular built-in electric field grows; the higher the applied reverse bias, the higher the peak field. When a minority electron

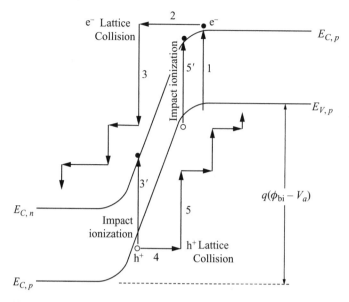

Figure 8.11   Illustration of impact ionization and the consequent avalanche multiplication process initiating junction break-down at high reverse bias.

enters the reverse-biased junction it quickly gains energy from this field. This is illustrated in Fig. 8.11. At some value of critical field, let's call it the "breakdown strength" (about 5×10^5 V/cm for silicon), energy becomes so large and the incident carrier is moving so fast, that the carrier can actually dislodge ("impact ionize") a valence electron from the host crystal lattice (remember silicon has eight valence electrons). The impact-ionized electron is boosted from the valence band to the conduction band, leaving behind a hole; and now both this "secondary electron" and this secondary hole (the electron–hole pair is created together) can themselves then be accelerated in this high electric field. When the secondaries gain enough energy, they can each create their own electron–hole pairs by means of impact ionization, and the numbers rapidly grow (unbelievably rapidly in fact). Clearly we have a positive-feedback mechanism in place. This entire process is called "avalanche multiplication" (imagine a snow avalanche thundering down the mountainside at 60 miles an hour, and you'll have the right visual image – yep, tough to stop once started). We can easily show that for a *pn* junction the break-down voltage depends reciprocally on the doping level (Fig. 8.12), such that $V_{\text{br}} \propto 1/N$. No mystery here. The higher the doping level, the more compressed the SCR width W is and hence the higher the junction field is for a given applied reverse-bias voltage. In a *pn* homojunction there is thus a fundamental trade-off between maximum blocking voltage in reverse bias and the forward-bias "on-resistance" of the junction (the junction's resistance with forward-bias current flowing), because increasing the doping to obtain a lower resistivity and hence lower on-resistance inevitably leads to lower breakdown voltage by means of the higher field. Curious about the mathematics of breakdown? Get a cup of coffee, kick your feet back, and look at the sidebar discussion (yep, time for a caffeine alert!).

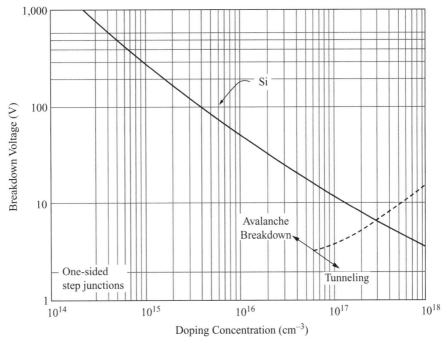

Breakdown voltage of a *pn* junction as a function of doping level.

Caffeine Alert #1 (Warning – Read at Your Own Risk!): The Mathematics of Breakdown

Impact ionization, like all things semiconductor, is a probablistic process. Let P be the probability that the incident electron creates an electron–hole pair by impact ionization. If n_{in} is number of incident electrons, Pn_{in} ionizing collisions will generate $n_{in}(1 + P)$ electrons (the 1 is for the initiating electrons). But Pn_{in} holes are also generated (electrons and holes are created as pairs, and assume for simplicity that they have equal creation probabilities), which will generate $P(Pn_{in}) = P^2 n_{in}$ electron–hole pairs. Next round gives me $P(P^2 n_{in}) = P^3 n_{in}$, etc. Hence the total number of total carriers (electrons + holes) crossing the junction (the current flow you would measure) during avalanche is given by

$$N = n_{in} (1 + P + P^2 + P^3 + P^4 + \cdots + P^{n-1} + \cdots +), \qquad (8.8)$$

which is otherwise known in mathematics as a "geometric series" or "geometric progression." What do geometric progressions do? They get very very large, very very quickly. As I like to tell my classes, avalanche multiplication represents one of the few true infinities in real life! If you know a little calculus (recall infinite series?) you can easily show that

$$N \cong n_{in} \frac{1}{(1 - P)}. \qquad (8.9)$$

(continued)

> **Caffeine Alert #1 (Warning – Read at Your Own Risk!): The Mathematics of Breakdown** (*continued*)
>
> so that the "multiplication factor" (M), the number of impact ionized carriers generated per incident carrier during the avalanche breakdown process, is just [3]
>
> $$M = \frac{1}{(1 - P)}. \tag{8.10}$$
>
> Moral: If $P = 1$ (the electric field is above the breakdown strength and thus the incident carrier has a unity probability of impact ionization), M tends to infinity, and breakdown occurs. Kaboom! Neat huh?!

8.3.3 What does it do for you?

So what makes Mr. Junction so compellingly useful? Well, as stated, he makes a nice on–off switch, with low loss when forward biased. Careful; he does NOT possess gain. And he can provide very nice electrical isolation when reverse biased (in power electronics circles he would be said to provide a "blocking" voltage, not allowing current flow in reverse bias up to some finite, and often huge, applied reverse voltage: 100s to even 1000s of volts). Very useful. He can also function as a wonderful solid-state "rectifier." Rectifiers are ubiquitous in power generation, conversion, and transmission, (e.g., to turn ac voltage into dc voltage). As we will see later, he can also emit and detect light! REALLY useful.

But from a transistor perspective (remember the hunt is on for a gain mechanism), Mr. Junction can be used to make a tunable minority carrier injector. Come again?! Okay. As you now know, forward-bias Mr. Junction and I get lots of electrons moving from the n-side injected into the p-side, hence becoming minority electrons; and vice versa for the minority holes on the p-side injected into the n-side. Importantly, I can trivially skew the relative magnitudes of the minority carrier injection from side to side. Careful – this is subtle but really important for the production of gain. As I said, make $N_A = N_D$, and the electron and hole contributions to the total current are of similar magnitude. BUT, if we intentionally make one side of the junction much more heavily doped than the other, then this current distribution changes profoundly. Consider an $n^{++}-p^-$ junction. One side is doped very heavily, and one side is doped lightly (say, for silicon, arsenic $= 1 \times 10^{20}$ cm^{-3} and boron $= 1 \times 10^{15}$ cm^{-3}). Fittingly, this is referred to as a "one-sided" junction. Now make your choice: "Alex, I'll take *pn* junctions for $100." "Answer:" Of electrons or holes, this carrier makes up most of the total current flow in forward bias in an $n^{++}-p^-$ junction." (Feel free to hum the Jeopardy theme song here.) Question: "What are electrons?" Correct! Look back to Eq. (8.7) and you will see this explicitly. The respective doping levels sit in the denominator of term (1) and term (2). Make N_D very large with respect to N_A, and the second term is now very small compared with the first. AND, the total current flow is primarily set by the doping concentration on the lightly doped side of the junction! So, imagine this. Say I want to use a *pn* junction under forward bias to enhance the "forward-injection" of electrons into the *p*-region, and suppress the "back-injection" of holes into the *n*-region. How would I do this? Use an n^{++}–p^- junction! Trivial to make, too. Just implant a high concentration of arsenic into the surface of a lightly boron-doped wafer and anneal it.

Careful folks, a gain mechanism is now officially lurking, and just begging for a creative mind! Nobel Prize, here we come! Read on.

8.4 The BJT

Okay. Mr. *pn* Junction can be made to serve as an efficient (and tunable!) minority carrier injector. BUT, Mr. Junction does NOT possess inherent gain. This is the fundamental reason why we do not build microprocessors from diode-resistor logic (one can in fact trivially built such a binary logic). Diodes make excellent binary switches, but without a gain mechanism to overcome nature's preference for attenuation, complex functions are not going to be achievable in practice. Sound like a law? Yep. *Cressler's 7th Law: "It is impossible to achieve voltage–current gain in a semiconductor device with only two terminals."* Bummer. We clearly need gain.

8.4.1 What is it?

What if, however, we imagined adding an additional *third* terminal to the device that somehow controlled the current flow between the original two terminals. Let terminal 1 = the input "control" terminal, and terminals 2 and 3 have high current flow between them when biased appropriately by the control terminal. Then, under the right bias conditions, with large current flow between 2 and 3, if we could somehow manage to *suppress* the current flow to and from 1, we'd be in business. That is, small input current (1) generates large output current (from 2 to 3), and hence we have gain! Make sense?[1]

How do we do this in practice? Get ready for the lightbulb! Let's use *two pn* junctions, placed back to back, such that the control terminal (our 1; which we will call the "Base" terminal – *B*) is in the central *p* region, and the two high-current-flow path output terminals (our 2 and 3; which we will call the "Emitter" and "Collector" terminals – *E* and *C*) are the two outside *n* regions[2]

---

[1] Historical aside: Prior to transistors, there was much precedence for using a multiterminal approach to achieve gain in vacuum tubes – the key here is to do it in semiconductors.

[2] The names emitter and collector make sense intuitively, but why "base"? In the original point-contact transistor, the physical base of the contraption that held up the assembly served as this control terminal – hence the name "base."

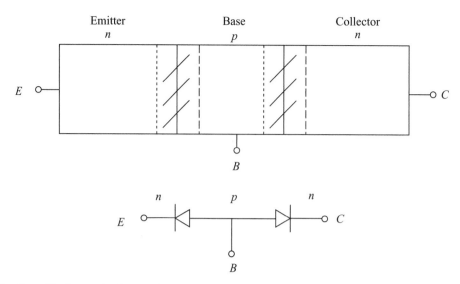

Figure 8.13   The emitter, base, and collector regions of an *npn* BJT, showing the two back-to-back *pn* junctions involved.

(see Fig. 8.13). Because the two central *p* regions are shared by both diodes, those can be coincident. That is, an *n* region separated from another *n* region by an intermediate *p* region actually contains two *pn* junctions. So far so good. Let's imagine forward biasing the emitter–base junction, and reverse biasing the collector–base junction, and then adding two more puzzle pieces: (1) We must dope the emitter very heavily with respect to the base, such that when we forward bias the emitter–base junction we have large electron flow from *E* to *B*; and simultaneously suppress the hole flow from *B* to *E* (our tunable minority carrier injector in action!). (2) We must make the central base region VERY, VERY thin. Why? Well, if we don't, then the electrons injected from *E* to *B* will simply recombine in the base before they can reach the collector (to be collected and generate our required large output current flow from *E* to *C*). Recall that the rough distance a minority carrier can travel before it recombines is given by the diffusion length ($L_{n,p}$). Clearly we need the width of the *p*-type base region to be much, much less than this number; in practice, a few hundred nanometers is required for a modern BJT. Final result? Mr. *npn* BJT is born! Cock your ear to the wind and faintly discern – "Rah-Rah! Nobel Prize!" . . . "Rah-Rah! Nobel Prize!" And did I tell you that that comes with a $1,000,000 check with your name on it?!

One could of course swap the doping polarities of *n* to *p* and *p* to *n* and achieve the same result – presto, instant *pnp* BJT. We thus have two flavors of BJT, and this is often VERY handy in electronic circuit design. Both devices, and their circuit symbols with applied driving voltages and currents indicated, are shown in Fig. 8.14. It is easy to remember which is which: The arrow is always drawn on the emitter terminal and always points in the direction of the emitter–base *pn* junction. In the *npn* BJT, the base is *p*-type and the emitter is *n*-type, and

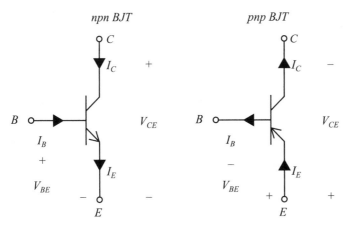

npn BJT   pnp BJT

Figure 8.14 Circuit symbols and bias orientations for both *npn* and *pnp* BJTs.

hence the base–emitter *pn* junction arrow points from base to emitter. For the *npn* BJT, the input current is I_B (*B* for base) and the output current is I_C (*C* for collector), or the input voltage is V_{BE} (*BE* for base–emitter), and the output voltage is V_{CE} (*CE* for collector–emitter). "Forward-active" bias in a BJT is defined in the scenario in which the *EB* junction is forward biased and the *CB* junction is reverse biased. This is the preferred mode of operation for amplifiers (aka miniature gain generators!).

8.4.2 How does it work?

Consider now how Mr. BJT actually works. (1) The reverse-biased *CB* junction has negligible current flow. (2) The forward-biased *EB* junction injects (emits) lots of electrons from *E* to *B* that diffuse across the base without recombining (because it is thin) and are collected at *C*, generating large electron flow from *E* to *C* (think current). BUT, because of the doping asymmetry in the *EB* junction (n^+–p), although lots of electrons get injected from *E* to *B*, very few holes flow from *B* to *E*. Forward electron current is large, but reverse hole current is small. Moral: Small input base current; large output collector. GAIN! Otherwise known to electronics geeks as "current gain" or β (ole timers often use h_{FE}). Formally,

$$\beta = \frac{I_C}{I_B}. \tag{8.11}$$

We're in business!

How do we make the BJT? Well, as might be imagined, it is more complex than a *pn* junction, but even so, the effort is worth it. Figure 8.15 shows the simplest possible variant. Start with a *p*-type wafer; ion implant a "buried" n^+ "subcollector"; grow a lightly doped n^- epitaxy on this; ion implant an n^+ region to contact the subcollector; ion implant a p^+ ring around the BJT to electrically isolate it; ion implant a thin *p*-type base together with p^+ contact to that base; ion

Top-Down and Cross-Sectional View of a BJT

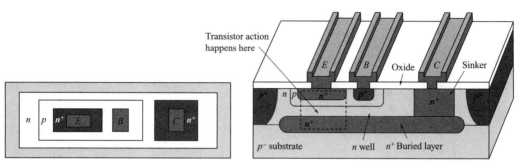

Figure 8.15   Simplest conceptual view of a BJT.

implant an n^+ emitter; apply metalization to contact the structure. Done! If you are feeling cocky, refer back to Chap. 7 for a modern interpretation of a real BJT. MUCH more complicated than this. Looking vertically down into the guts of the BJT directly under the emitter, we have an $n^+-p-n^--n^+$ doping structure – our BJT. (I'll explain the need for the n^--n^+ collector structure in a few moments.) Figure 8.16 also superposes both the equilibrium and forward-active bias energy-band diagrams, with the carrier minority and majority carrier distributions, to help you connect the pn junction physics to the BJT operation.

Within the band diagram context, here is intuitively how the BJT works. In equilibrium, there is a large barrier for injecting electrons from the emitter into the base. Forward bias the EB junction and reverse bias the CB junction, and now the EB barrier is lowered, and large numbers of electrons are injected from E to B. Because B is very thin, and the CB junction is reverse biased, these injected electrons will diffuse across the base, slide down the potential hill of the CB junction, and be collected at C, where they generate a large electron current flow from E to C. Meanwhile, because of the doping asymmetry of the EB junction, only a small density of holes is injected from B to E to support the forward-bias EB junction current flow. Hence I_C is large, and I_B is small. Gain. A different visualization of the magnitudes of the various current contributions in a well-made, high-gain, BJT is illustrated in Fig. 8.17. Remember, the current flow direction is opposite to the electron flow direction (Ben Franklin's questionable legacy for missing the call on the sign of the electron charge).

Shockley's theory to obtain an expression for β is fairly straightforward from basic pn junction physics (although you have two different ones to contend with obviously), provided you make some reasonable assumptions on the thickness of the base (base width $W_b \ll L_{nb}$). For the output and input currents under forward-active (amplifier) bias, we obtain

$$I_C \cong q A \left(\frac{D_{nb}\, n_i^2}{W_b\, N_{Ab}} \right) e^{q V_{BE}/kT} = I_{CS}\, e^{q V_{BE}/kT}, \qquad (8.12)$$

Operation of the BJT

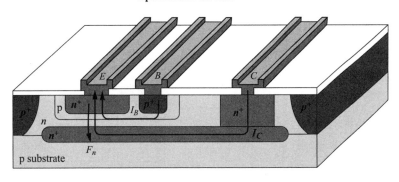

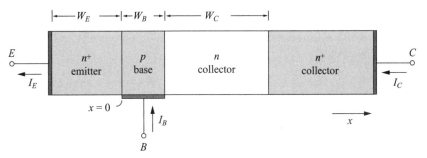

Equilibrium

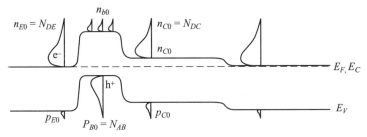

Forward Bias

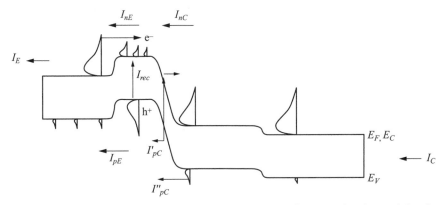

Figure 8.16

Current flow in an *npn* BJT under bias, together with the energy band diagrams and carrier populations for both equilibrium and forward-active bias (inspired by [3]).

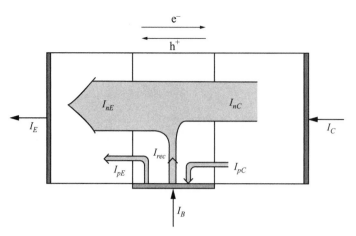

Figure 8.17   Relative magnitudes (not to scale) of the various current components in a "well-made" BJT exhibiting useful current gain.

$$I_B \cong qA \left(\frac{D_{pe}\, n_i^2}{L_{pe}\, N_{De}} \right) e^{q V_{BE}/kT} = I_{BS}\, e^{q V_{BE}/kT}, \tag{8.13}$$

where the b and e subscripts stand for base and emitter, respectively. Interestingly, the current gain does not to first order depend on bias voltage, the size of the junction, or even the bandgap! We finally obtain

$$\beta \cong \frac{I_C}{I_B} = \frac{I_{CS}}{I_{BS}} \cong \left(\frac{D_{nb}\, L_{pe}\, N_{De}}{D_{pe}\, W_b\, N_{Ab}} \right). \tag{8.14}$$

Just as with pn junctions, simple theory and reality are never coincident (you know, Cressler's 6th law). Figure 8.18 shows what you would actually measure if

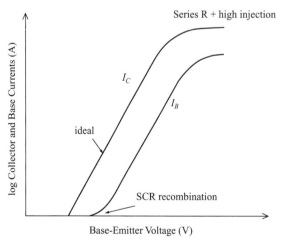

Figure 8.18   Current–voltage (Gummel) characteristics for npn BJTs, plotted on a semilog plot. Shown are the output current I_C and the input current I_B as functions of input voltage V_{BE} for fixed output voltage V_{CE}.

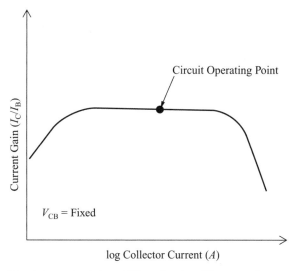

Circuit Operating Point

Current Gain (I_C/I_B)

V_{CB} = Fixed

log Collector Current (A)

Figure 8.19 Extracted current gain ($\beta = I_C/I_B$) as a function of bias current for an *npn* BJT.

you took a real *npn* BJT into the lab and hooked it up to some voltage sources and ammeters. As with the *pn* junction, I am plotting both I_C and I_B on semilog scales (this is called by all in-the-know a "Gummel plot" – see Geek Trivia) because they depend exponentially on V_{BE} (let's assume V_{CE} = fixed, say at 3.0 V, so that the device is under forward-active bias). We again won't worry much about the operative mechanisms, but let me summarize what you would actually see Mr. BJT do:

- Just as for a *pn* junction, under forward-active bias, the ideal current–voltage behavior of I_C and I_B occurs over a fairly small range of applied voltages; say 0.4–0.7 V in a silicon BJT.
- Below the ideal bias range, carrier recombination inside the *EB* SCR is dominant, with a different (weaker) exponential voltage dependence, producing a nonideal I_B. This does not affect I_C.
- Above the ideal bias range, parasitic series resistance associated with both the doped *E* and *B* regions, together with high-injection effects, becomes dominant, causing both currents to "roll over" at high V_{BE}.

Because $\beta = I_C/I_B$, we can directly extract the current gain from a Gummel plot, and traditionally this is plotted as linear β as a function of log I_C (Fig. 8.19). Observe that the base current nonideality at low bias directly reduces (degrades) the gain at low currents, and high-injection effects (not series resistance) degrade the gain at high bias. Still, there is a fairly wide region of bias conditions where the current gain is reasonably constant. A bias-independent current gain is a VERY nice feature for building many types of electronic circuits.

Geek Trivia: The Gummel Plot

Plotting the current–voltage behavior of a BJT on semilog scales (log I_C and I_B vs. V_{BE}, for fixed V_{CE}) produces what are known as the "Gummel characteristics" of the transistor, or simply a "Gummel plot." Why? Well, the first person to do it this way was Herman K. Gummel at Bell Labs in 1961 [6]. It might strike you as odd that simply a plotting data a particular way could make one famous, but actually Gummel used this as a clever way to indirectly measure the number of base dopants (formally, the integrated base charge; aka the "Gummel number"). The creative plotting of experimental data to extract maximal information is a true art form. Yep, you're on my wavelength! *Cressler's 8th Law: "Creative plotting of experimental data to extract maximal information is a true art form."* MUCH can be rapidly gleaned from a Gummel plot. Indispensable tool of the trade. In fact, VERY expensive instrumentation systems for measuring transistors (e.g., the Agilent 4155 Semiconductor Parameter Analyzer) pretty much exist solely to automatically measure the Gummel characteristics. No self-respecting device engineer would leave home without one!

Another common way to plot the BJT current–voltage characteristics is shown in Figure 8.20, where linear I_C is plotted vs. linear V_{CE} as a further function of I_B. Because I_C is larger than I_B, the gain is implicit here. This plot is known as the output "family" or "output characteristics." We use the output family to define the three regions of operation of Mr. BJT: (1) "forward active" (*EB* junction forward biased; *CB* junction reverse biased); (2) "saturation" (both *EB* and *CB* junctions forward biased), and (3) "cutoff" (both *EB* and *CB* junctions reverse biased – the $I_{B,0}$ line, so no output current flow). As indicated, forward-active bias is amplifier central, and as we will see, switching between cutoff and saturation will make an excellent regenerative digital switch!

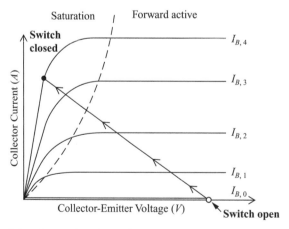

Figure 8.20   Output characteristics for *npn* BJTs, plotted on linear scales. Shown is the output current I_c as a function of output voltage V_{CE} for various input currents I_B. Also shown is a (resistive) "load line," indicating how the BJT would be used as a switch, as well as the demarcation between saturation and forward-active bias, the latter used for building amplifiers.

How about some simple take-aways from Eq. (8.14) for making a BJT with high current gain:

- Make the base width small (reason: to minimize minority carrier recombination in the base region).
- Dope the emitter much more heavily than the base (reason: to suppress base current).

Clearly, the current gain is a tunable parameter, giving us great flexibility. Interestingly, however, if you sampled all of the BJTs in the world, you would find that β is inevitably only about 100. Why? Good question. It has to do with junction breakdown. Unlike for pn junctions, the BJT, because it has two junctions, has an additional feedback mechanism that actually lowers the sustainable voltage that can be placed across the CE terminals of the device (BV_{CEO}). Recall that, in a pn junction, breakdown happens when $M - 1$ tends to infinity. In a BJT, however, because of a positive-feedback path initiated by the intrinsic gain of the transistor, breakdown will occur when $\beta(M - 1) = 1$. That is, M gets multiplied by β! Moral? As good as gain is, it will degrade the maximum voltage you can put across the device. Result: In practice one uses a value of β large enough to build good switches and amplifiers (100-ish) but not so large that the breakdown voltage of the BJT suffers too much.

Need to keep the current gain at about 100 or so? No problem. Not surprisingly, if you look at the doping profile of a real BJT (refer back to Chap. 7), the emitter is typically doped about 100 times higher than the base (say 1×10^{20} cm^{-3} for the emitter vs. 1×10^{18} cm^{-3} for the base). No coincidence!

So just how fast are these buggers? Pretty darn fast! The current speed record for a bipolar transistor digital switch is less than 10 ps (0.000,000,000,010 s – 10 trillionths of a second!). What limits that speed? Intuitively, the time it takes the electrons to be injected from the emitter, transit (diffuse across) the base, and then be collected by the collector. In other words, a transistor can't be faster than the time it takes the charge to move through it. In most transistors, step two is the limiting one, and the so-called "base transit time" (τ_b) sets the fundamental speed limit on how fast Mr. BJT can be. A first-order base transit time expression can be easily derived, with the result

$$\tau_b \cong \frac{W_b^2}{2D_{nb}}. \tag{8.15}$$

Hence, the smaller τ_b is, the faster Mr. BJT can switch. Clearly, making W_b as small as possible gives us a double benefit. It helps increase the current gain, yes, but even more important, it makes the transistor faster – quadratically faster! That is, decrease the base width by a factor of $2\times$ (say, from 200 nm to 100 nm), and you get a $4\times$ speed-up of the device (say, from 16 ps to 4 ps). Good bang-for-the-buck!

8.4.3 What does it do for you?

Okay. Now you have the scoop on BJTs. Let me restate some points for clarity. This beautiful three-terminal semiconductor device, if constructed correctly, will exhibit a (tunable) gain. Gain is the key to success in building any electronic system; hence the deserved fame of the BJT. This intrinsic gain will allow us to create a wide variety of amplifiers for use in a myriad of electronics applications. Amplifiers take (1) a small input current and turn it into a large output current (aka a "current amplifier"), (2) a small input voltage and turn it into a large output voltage (aka a "voltage amplifier"), (3) a small input current and turn it into a large output voltage (aka a "transconductance amplifier"), and (4) a small input voltage and turn it into a large output current (aka a "transimpedance amplifier"). Don't worry about the big names for the last two. Transconductance (g_m) in the electronics world just means the incremental change in current divided by the incremental change in voltage (see sidebar discussion for a bit more information on transconductance; it is important enough to know a little about). As a real-world example of amplifiers-in-action, at the input of your cell phone you have a handcrafted voltage amplifier that takes the tiny little RF signals and boosts them to a level sufficient to manipulate and decode them. In a receiver for a fiber-optic link, you have a handcrafted transimpedance amplifier that interfaces with the input photodetector to change the incoming photonic signals into electronic signals for processing.

In addition to building amplifiers, gain also allows us to construct nice regenerative binary switches. As can be seen in Fig. 8.20, if the input base current I_B (or input voltage V_{BE}) is zero, the output current I_C is zero, the on–off switch is now open and the output voltage V_{CE} is thus high. Let's call that state a logical "1." Conversely, if the input current I_B (or input voltage V_{BE}) is large enough to turn on the transistor, the output current I_C is large, output voltage V_{CE} drops to a low value, and the on–off switch is now closed. Let's call that state a logical "0." Open, close; open, close. Bamm! A regenerative binary switch to push back on nature's penchant for attenuating our signals! We're in business for digital logic.

What is Transconductance?

Here's a bit more on transconductance and the important role it plays in electronic systems. Formally, transconductance is defined as a mathematical partial derivative (think calculus, sorry!)

$$g_m = \frac{\partial I}{\partial V}. \tag{8.16}$$

For a BJT, for instance, we define

$$g_m(BJT) = \frac{\partial I_C}{\partial V_{BE}} = \frac{q}{kT} I_{CS} e^{q V_{BE}/kT} \tag{8.17}$$

so that transconductance is just the instantaneous rate of change in output current divided by the instantaneous rate of change in input voltage. Clearly, because I_C in a BJT depends exponentially on V_{BE}, and the derivative of an exponential is itself an exponential, the BJT will produce very high g_m, because exponential functions are the strongest nature gives us. This is a good thing. All this really says is that a small voltage wiggle on the input produces a large current wiggle on the output. This has many implications in practical circuits. For a voltage amplifier, for instance, the achievable amplifier voltage gain is directly proportional to g_m. Big g_m yields big voltage gain. In the digital world, g_m is also extremely important, because the ability of a logic gate to quickly charge a capacitively loaded line of magnitude C_W connecting two logic gates depends on the transconductance of the driving gate. That is, for a fixed logic swing (V_L – think of V_L as the difference in voltage between a logical "1" and a logical "0"), the logic gate's current drive I_{gate} charges that interconnect line capacitance in a time $\tau_{delay} \cong (C_W \, \Delta V_L)/\Delta I_{gate}$. $\Delta I_{gate}/\Delta V_L$ is basically just the transconductance of the logic gate [call it g_m(gate)]! Moral: The larger the g_m(gate), the faster the interconnect is charged up, the faster the logic gate can switch from a "1" to a "0," and the faster Mr. Microprocessor will be. Read: This is a big deal. Simply put, there is no such thing as having too much transconductance. Any self-respecting circuit designer, whether digital or analog or RF, will always welcome more g_m. The more the merrier.

One last point on the many merits of Mr. BJT. BJTs are capable of delivering VERY large amounts of current, very quickly, when called on to do so, for use in analog or digital or RF circuits. Consider: A SINGLE modern BJT can be easily and reliably biased to output 10 mA/μm^2 of current density when it is turned on. Is this a big number or a small number? Well, a quick dimensional change shows (try it) that this is equivalent to 1,000,000 A/cm^2 (yep, 1-million amps per square centimeter!). Said another way, if I had a wire with a 1-cm^2 diameter, Mr. Transistor (obviously suitably scaled up – yes, yes, I'm making a number of assumptions, see if you can tell me which ones) could push 1,000,000 A through this centimeter-sized wire! That's a lot of current! For reference, a run-of-the-mill summer-afternoon lightning bolt delivers around 40,000 A of current when it strikes the ground [4]. Let's assume that the diameter of said lightning bolt is about 10 cm (form a circle by connecting your two hands and you'll get the idea). The 40,000 A running through 10 cm^2 corresponds to a current density of about 127 A/cm^2 for our lightning bolt. Said another way, Mr. Transistor can source a current density equivalent to about 7,874 average lightning bolts! Moral to the story: 10 mA/μm^2 is a very, very big current density. Of course, the micro/nanolectronics world guarantees that even though the current densities sourced by transistors can be huge, the total current flow stays relatively small, because the transistor size is tiny ($I = J$ A). Still, be careful – there can be a gazillion of these transistors. Tiny current per transistor can still translate into large total chip current. A centimeter-sized modern microprocessor can easily sink 10s of amps of current. Think lightning bolts on a chip!

Enough said about BJTs. Let's move on to the second main class of transistors – MOSFETs. Very different animal! Read on.

8.5 The MOSFET

So . . . why would we need a second type of transistor if we already have BJTs? Remember Cressler's 1st law? "Transistor form and function are inseparable." The point is that MOSFETs, by virtue of the way they designed, inevitably have attributes that the BJTs do not possess. Important attributes. This clearly must be the case because MOSFETs dominate the world's transistor population by a wide margin, and one does not see, for instance, BJTs in modern microprocessors. Why? Good question. Read on.

8.5.1 What is it?

I would love to be able to tell you that the MOSFET works just like the BJT; but alas, it is fundamentally different. Sorry! There are, however, interesting similarities between the two. For one, we follow the same conceptual path to creating gain by introducing a third control terminal to modulate the current flow between the other two terminals. The MOSFET can be used as either an amplifier or a regenerative switch, the same as for a BJT. And the MOSFET also contains two *pn* junctions. But, importantly, the third control terminal in a MOSFET is not a *pn* junction and is instead electrically insulated from the semiconductor (the fundamental difference between FETs and PETs).

MOSFET: Metal-Oxide Semiconductor Field-Effect Transistor. It is just what it says; sort of. Consider the left-hand drawing in Fig. 8.21, which shows an "*n*-channel" MOSFET (aka the *n*FET). The "substrate" is just the semiconductor (silicon) platform we build the device on [more commonly called the "body" (*B*) of the transistor], and two *n*-type regions are introduced [the "Source" (*S*) and the "Drain" (*D*)] into the body to form two *pn* junctions. The source and

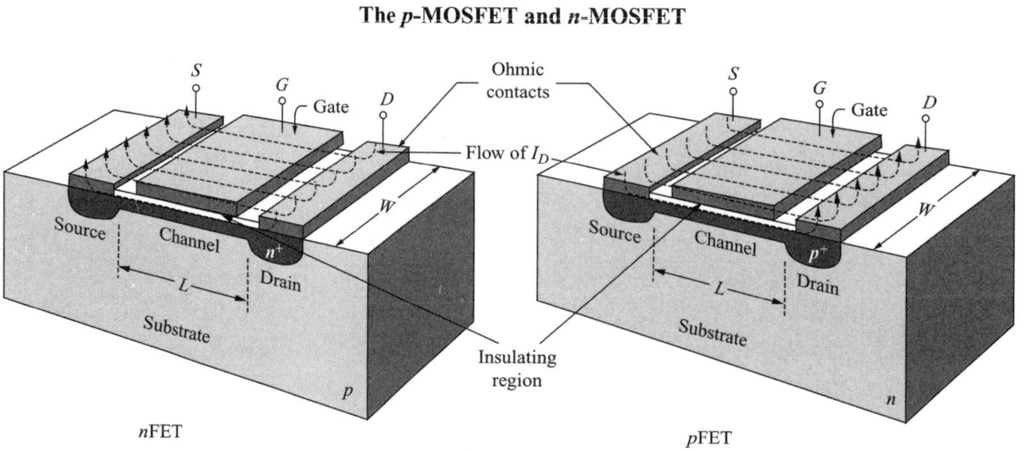

The *p*-MOSFET and *n*-MOSFET

Figure 8.21   Conceptual view of both *n*-channel and *p*-channel MOSFETs (*n*FET and *p*FET).

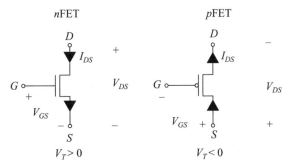

Figure 8.22   Circuit symbols and bias orientations for both *n*FETs and *p*FETs.

the drain are spaced a short distance L away from one another, and this distance between source and drain is called the "gate length." (Recall that the gate length of a MOSFET defines the "CMOS technology node" in semiconductor IC fab; e.g., a 90-nm fab.) As we will see, the gate length is everything in the MOSFET world, and to make a good transistor it needs to be a very, very small number. The "gate width" (W) is just the width of the S/D regions in the direction opposite to the gate length. The "channel" region lies inside the semiconductor and between the S/D junctions, and this is where all of the transistor action occurs. On top of the surface of the channel we grow a thin oxide layer (SiO_2), the "gate oxide," and then our conductive third control terminal, very heavily doped n-type polysilicon, in modern MOSFETs known as the "gate" (G) terminal (it "gates" the current flow from source to drain).[3] Note that since the gate oxide is insulating, negligible current flows into the gate. Instead, it influences the channel region capacitively by indirectly applying an electric field; this is the fundamental difference between FETs (e.g., MOSFETs) and PETs (e.g., BJTs). Just as with the BJT, we can construct the same type of device by simply swapping the doping polarities of all of the doped regions (Fig. 8.21), yielding a p-channel MOSFET (aka the pFET). One can easily tell which is which: The nFET has n-type source–drain–gate regions, and the pFET has p-type source–drain–gate regions. The circuit symbols, and the voltage–current polarities for the nFET and the pFET, are shown in Fig. 8.22. Again, it is easy to tell one from the other: We simply place a circle on the gate of a pFET, just to easily tell them apart. Strictly speaking, the MOSFET is a four-terminal device (body = 4), but in common practice, the body is usually electrically shorted to the source, resulting in only three active terminals. If the body terminal is shorted and hence electrically out of the picture, it is typically omitted from the circuit

---

[3] Historically, MOSFETs had metal gates, typically aluminum, hence the name. Interestingly, however, today MOSFET gates are made from heavily doped polysilicon (still a good conductor). In an ironic twist of fate, future MOSFETs will again have metal gates, albeit significantly more sophisticated than in the old days.

symbol, and the circle on the *p*FET gate then becomes necessary for identification (Fig. 8.22).

8.5.2 How does it work?

So far so good. Now how does the beast actually work? Ready for another law? Sure. *Cressler's 9th Law: "An energy-band diagram is the simplest path to understanding how a given semiconductor device works."* I could of course have given this to you under *pn* junctions and BJTs, but what the heck, better late than never! The guts of MOSFET physics are contained in the "MOS" (metal-oxide semiconductor) capacitor formed by the gate–gate oxide–semiconductor (conductor–insulator–conductor = capacitor) composite. Careful: This is not your run-of-the-mill electrolytic capacitor! Care to the see the band diagram of the *n*MOS capacitor as a 1D cutdown through the center on Mr. *n*FET? Go on, be bold, and check out Fig. 8.23. Focus first on the upper-left-hand image, which shows the equilibrium energy band diagram of the *n*MOS capacitor. Here, both the gate and the body are electrically grounded. We assume for simplicity that there is no charge present in the gate oxide (this will never be strictly true) and that the Fermi level in the heavily doped polysilicon gate is coincident with the conduction band edge. Okay – here's your big chance! Given the opportunity to pin a name to this equilibrium situation, what would you call it? "Cressler Condition"? Alas, no. Give up? How about the "flat-band" condition? Look back at the figure and the reason will be obvious. Now, because we are dealing with a capacitor, it is useful to talk about the charges present for a particular bias condition. Clearly, in equilibrium, in flat-band, with no voltage difference between the gate and the body, no charge is present. Fine.

Now let's imagine changing the bias on the gate (keep the body grounded). What happens? Because this is an MOS capacitor, no dc current can flow side to side no matter how I bias it; hence the Fermi level MUST remain constant inside the semiconductor. Always. The "splitting" between the gate and body Fermi levels is just equal to the applied voltage, and because this is a plot of energy, we must multiply V by q to make it dimensionally correct ($V = -E/q$). In addition, positive applied voltage on the gate pushes its Fermi level *downward* on an energy band diagram, because energy is the negative of potential (voltage). Because the Fermi level must remain constant, band-bending is induced in both the oxide and the semiconductor, completely analogous to what happens in a *pn* junction. Remember, if band-bending is present and electric field is present, charge must also be present.

So . . . let's first bias Mr. *n*MOS capacitor with a negative voltage on the gate. A negative applied gate voltage means I have negative charge on the gate, and, because of the principle of "charge balance," I need a balancing charge to maintain charge neutrality (recall from Chap. 5 that this is a golden rule

nFET Energy Band Diagram with Bias

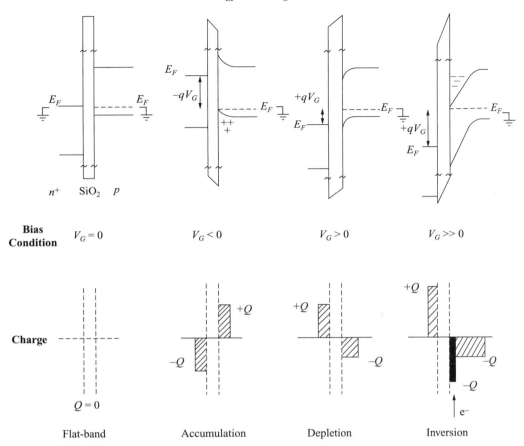

| Bias Condition | $V_G = 0$ | $V_G < 0$ | $V_G > 0$ | $V_G \gg 0$ |

| | Flat-band | Accumulation | Depletion | Inversion |

Figure 8.23 The energy band diagram of the core MOS capacitor that defines an *n*FET. Shown are the four important regions of operation in a MOSFET: (1) equilibrium, (2) accumulation, (3) depletion, and (4) inversion. The latter is used for circuits.

in semiconductors), which must obviously be positive. Where do we get positive charge in a *p*-type semiconductor? Yep, holes! Under negative gate bias, the energy bands in the semiconductor bend upward, inducing holes to collect (accumulate) at the surface. This is called the "accumulation" region (go figure). How can you tell from the energy band diagram that holes are actually present at the surface? Well, recall that the hole density depends exponentially on the energy difference between the valence band edge and the Fermi level. Because E_F gets close to E_V in accumulation, holes must be present. Turns out that, for most practical uses of MOSFETs, accumulation is not so interesting. Sigh.

How about we instead bias Mr. *n*MOS capacitor with a positive gate voltage with respect to the body? With the gate voltage slightly positive, I now have positive charge on the gate and must find some negative charge to balance things out. Where? Think space charge. In this case, the Fermi level in the gate pushes down, bending the bands downward. This band-bending is completely analogous to what happens in the SCR of a *pn* junction. Space charge in a *p*-type semiconductor gives me negatively ionized acceptors (N_A^-), aka the needed depletion charge to balance the gate charge. Name of this *n*MOS operating regime? The "depletion" region. Fine.

Hold on tight, the moment of truth is at hand. Thought experiment. What happens if we continue to increase the positive voltage on the gate? Logical inference? Mr. *n*MOS capacitor MUST come up with additional negative charge to balance this growing positive gate charge; some way, somehow. Alas, only so much negative charge can be garnered from the SCR by means of band-bending, and we soon run out of that. Solution? We bend the bands so far downward that electrons are actually induced to collect at the surface. Huh? Look back at the right-most energy band diagram in Fig. 8.23. Observe that, at the surface, the Fermi level is coincident with the conduction band edge. Implication? Because the electron density depends exponentially on the difference in energy between E_F and E_C, the surface is now electron rich! In this scenario, the doping polarity of the semiconductor is actually inverted from *p*-type to *n*-type, and we thus call this region of operation "inversion."

Now THIS is useful. Consider. Look back to Fig. 8.21 and imagine morphing this *n*MOS capacitor into the guts of an *n*MOSFET by simply adding n^+ source and drain regions on either end of the channel. Ground the body and ground the source, and then put a positive voltage on the drain. Good. Now, if the gate voltage stays grounded, absolutely nothing happens. Open circuit between source and drain. BUT, if we now bias the gate to a positive voltage large enough to invert the surface of the *n*MOS capacitor, we now have an n^+ source, an n^+ drain, and, importantly, an induced *n*-type inversion region in the channel. What do you call an object built from an n^+(source)–n(channel)–n^+(drain)? A fancy resistor! What happens when you apply a voltage across a resistor? Electron current flows from source to drain (I_{DS})! Said another way, I have made a wonderful binary switch. If the gate voltage is at ground, I have an open circuit, but if the gate voltage is large enough to invert the surface, I have a closed circuit. On–off; on–off. A switch! Excellent. Not only this, but because there is no gate current flow ($I_G =$ tiny), the current gain (I_D/I_G) associated with Mr. *n*FET is HUGE! Moral: We don't just have any ole switch, we have a regenerative switch! Microprocessors here we come!

One could of course build the identical object by using a *p*MOS capacitor and a p^+ source and drain. The voltage polarities change and current flow direction changes, but it is exactly the same idea. We can thus construct Mr. *n*FET and Mr. *p*FET. As we will see in a moment, having both transistor topologies at our fingertips is pivotal (think CMOS).

Some practical details. It would obviously be very helpful to know exactly how much voltage I have to apply to the gate to invert the surface, and importantly how that value depends on the details of how I build Mr. MOSFET. Read: So I can design a needed value. The gate voltage value in question is logically named the "threshold voltage" (V_T), and to first order is given by

$$V_T = 2\phi_F + t_{ox} \frac{\kappa_{semi}}{\kappa_{ox}} + \sqrt{\frac{4q\ N_A\ \phi_F}{\kappa_{semi}\ \epsilon_0}}, \qquad (8.18)$$

where N_A is the body (substrate) doping level, $\phi_F = kT/q\ ln\ (N_A/n_i)$ is called the "Fermi potential," t_{ox} is the oxide thickness, κ_{semi} you know already, κ_{ox} is the relative dielectric constant of the gate oxide (3.9 for SiO_2), and ϵ_0 you know already. Clearly the body doping level plays a strong role in setting V_T. For practical silicon, MOSFET's V_T is typically only a fraction of a volt and is a function of the technology node (gate length L). For instance, for a 90-nm node, V_T might be ± 0.4 V for the nFET and the pFET, respectively, and often multiple V_T's are included to enable circuit design flexibility.

Okay, so how much current actually flows when $V_{GS} > V_T$ and the switch is closed? Depends (smile). Alas, the simplest theoretical path to I_{DS} involves making a not-so-accurate assumption called the "long-channel approximation" (in a word, L is large enough to enable a number of simplifications) that does not hold well in modern 100-nm-sized MOSFETs. Still, the result provides useful insight, and we find

$$I_{DS} = \left(\frac{W}{L}\right) C_{ox}\ \mu_{eff} \left[(V_{GS} - V_T)\ V_{DS} - \frac{V_{DS}^2}{2}\right], V_{DS} \le V_{DS,sat}, \qquad (8.19)$$

$$I_{DS} = \left(\frac{W}{L}\right) \frac{C_{ox}\ \mu_{eff}}{2} \left[(V_{GS} - V_T)^2\right], V_{DS} \ge V_{DS,sat}, \qquad (8.20)$$

where $V_{DS,sat} = V_{GS} - V_T, C_{ox} = (\kappa_{ox}\epsilon_0)/t_{ox}$, and μ_{eff} is the "effective mobility" (a scalar multiple of the bulk mobility μ – see sidebar discussion on inversion in MOSFETs). Interestingly, the current flow for Mr. MOSFET is driven by majority carrier (electron) drift transport, unlike for the BJT, which you will recall occurred by means of minority carrier diffusion transport. Observe that, once the channel dimensions, oxide thickness, body-doping level, semiconductor, and temperature are specified for Mr. MOSFET, all is known, and we have $I_{DS} = I_{DS}(V_{GS}, V_{DS})$. Give me the applied voltage, and I'll tell you exactly how much current flows. Clearly, though, I_{DS} depends on V_{GS}, and for large enough V_{DS}, I_{DS} becomes independent of V_{DS}. The plots of I_{DS} vs. V_{DS} for both an nFET and a pFET are shown in Fig. 8.24. As with BJTs, a plot of linear output current on linear output voltage is called the transistor output characteristics (or output family). Note that, in the case of MOSFETs, however, the input drive is a voltage (V_{GS}) not a current, a fundamental difference between BJTs and MOSFETs. Consider the left-most curve in Fig. 8.24; our nFET. If $V_{GS} > V_T$ and $V_{DS} < V_{DS,sat}$, we are biased in the sublinear region (sometimes called the "triode" region, which harkens back to vacuum tube days). Note that at a low value of V_{DS} the $I - V$ characteristics are quite linear; meaning that Mr. MOSFET indeed acts like a resistor ($V = IR$),

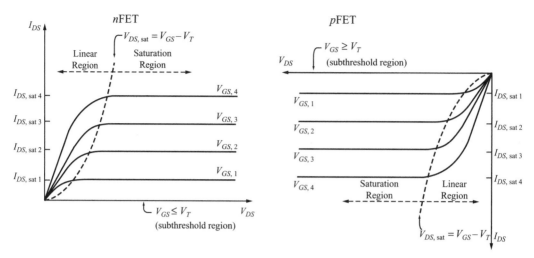

Output characteristics for an *n*FET and a *p*FET, plotted on linear scales. Shown is the output current I_D as a function of output voltage V_{DS}, for various input voltages V_{GS}. The demarkation between linear and saturation regions, as well as the subthreshold region, is also shown.

as previously argued. If, however, $V_{DS > V_{DS},\text{sat}}$, the current saturates at a constant value (conveniently called the "saturation" region).[4]

Inversion in MOSFETs

Let's look a bit more carefully at the induced inversion layer in the MOSFET channel when $V_{GS} > V_T$. Some relevant questions to ponder: How thick is the inversion layer when it forms? Answer: In the range of <10 nm. Read: VERY thin. Quantum well thin. You might logically wonder if this is thin enough to induce significant quantum-mechanical effects in the electron inversion layer, and the answer would be yes, most definitely. Viewing the MOSFET from the top, we can see that the inversion layer is actually a 2D system; to those in-the-know, a 2DEG (2D electron gas). More than one Nobel Prize was won by understanding the physics of such systems (sounds like a trivia question – go for it!). Quantum mechanics in action. Mr. MOSFET thus brings into each and every computer owner's hands millions of intensely quantum-mechanical objects to play with. To my mind, kind of neat.

Next. How much electron charge can we place in this ultra-thin inversion layer. Answer: It depends, but think of the order of 10^{12} electrons/cm^2. Why cm^{-2}? It is a 2D electron layer; a sheet charge density. More electron sheet charge translates directly into more current flow, so the more the merrier.

Finally, what is the carrier mobility in this electron inversion layer? Answer: NOT the standard carrier mobility you learned about in Chap. 5. Why? Consider what we do to the MOS capacitor to invert the surface – we apply a very large vertical electric field. When the electrons in the inversion layer are created, they thus hug tightly (okay, I'm again getting a little anthropomorphic

---

[4] Do note some pathological transistor name games. Saturation in a MOSFET and saturation in a BJT mean exactly opposite things! Compare Figs. 8.20 and 8.24.

here, but what the heck!) to the interface between the SiO_2 and Si, and now when we apply a V_{DS} to transport them from source to drain to generate I_{DS}, the electrons strongly "feel" (smile!) the atomic-scale roughness of the silicon surface as they move (Fig. 8.25). Said another way, the inversion-layer electrons directly feel the local variations in the actual surface atom arrangements. Recall that anything in a semiconductor that impedes carrier transport (scattering) affects its velocity and hence its low-field carrier mobility. Our case? This "surface roughness scattering" inevitably degrades the electron mobility. This inside-the-MOSFET mobility is called the "surface mobility," or the "field-effect mobility," or more commonly, the "effective mobility" (μ_{eff}). A good rule of thumb for the surface-scattering-induced mobility degradation? Sure. About $3\times$. For example, in silicon, the bulk electron mobility at a doping level of 1×10^{17} cm^{-3} is about 800 cm^2/V s. We would thus expect an effective mobility of about 267 cm^2/V s (800/3) for Mr. MOSFET with a 1×10^{17} cm^{-3} body doping.

As can be gleaned from Eq. (8.20), I_{DS} is directly proportional to μ_{eff}, and hence the surface scattering directly translates to reduced current drive, never a good thing. In addition, given the fundamental disparity between the bulk electron and hole effective mobilities (in silicon, at 1×10^{17} cm^{-3}, $\mu_n = 800$ cm^2/V s but $\mu_p = 330$ cm^2/V s), if we want to build a pFET to deliver an identical amount of current as an nFET, all else being equal, we have to increase the width W of the pFET by about $2.4\times$ (267/110). As we will shortly see, this will play an important role in the design of CMOS circuits.

Question. Why on earth does the transistor output family look the way it does: linear, then sublinear, then constant? Good question! Consider a water-flow analogy, as depicted in Fig. 8.26 [3]. Let the source and drain regions be lakes S and D, respectively, connected by a canal of some fixed depth (our channel), through which water can flow unimpeded. The total depths of lakes S and D are

Surface Roughness Scattering

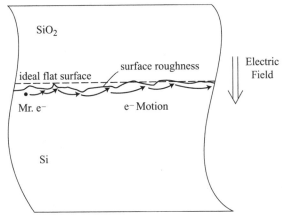

Figure 8.25   Schematic atomic-scale magnification of electron transport along the SiO_2 to Si interface, with consequent surface roughness scattering.

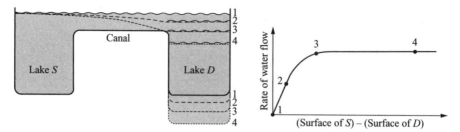

Figure 8.26 **Water-based analogy between the draining of Lake *S* into Lake *D* via a canal, and the output current of a MOSFET (inspired by [3]).**

fixed. The bottom of lakes S and D have an analog in the potential energy for electrons as function of position in our MOSFET (think energy band diagram). Now imagine varying the depth of lake D (i.e., varying V_{DS}). Let's then imagine plotting the rate of water flow from S to D (I_{DS}) as a function of the difference in surface position between S and D (V_{DS}). In case 1, the surface of lake D is at the same height as the surface of lake S, and there is no water flow from S to D (Mr. MOSFET is off). In case 2, however, the surface of D is lowered with respect to S, and water begins to flow from S and D (I_{DS} grows). Fine; makes sense. For case 3, in which the surface of lake D is at the same depth as the bottom of the channel, the slope of the water across the channel increases, and hence so does the rate of water flow from S to D. If, however, we continue to lower the surface position of D to below that of the bottom on the channel, the slope (rate of water flow) has reached its maximum and hence cannot continue to rise. The water flow (I_{DS}) saturates, no matter how far down lake D is pushed. Make sense? Good.

If we plot the log of I_{DS} as a function of linear V_{GS}, for fixed V_{DS}, we will obtain Fig. 8.27. This is called the "subthreshold" characteristics of the MOSFET, because it shows the current flow below V_T (clearly Mr. MOSFET is not entirely "off" when $V_{GS} < V_T$). There is something particularly intriguing about this plot. Do you see it? I am plotting log output current on linear input voltage and getting a straight-line behavior at low values of V_{GS}. Conclusion: I_{DS} depends *exponentially* on V_{GS} in this subthreshold regime. Careful: This result has MAJOR implications for CMOS scaling limits (more on this in a later chapter). Here is a head-scratcher for your consideration. WHY does output current depend exponentially on input voltage in subthreshold in a MOSFET? Give up? First cry "UNCLE"! Fine. Look back to Fig. 8.21, which shows the cross section of Mr. MOSFET. Imagine that we are biased at $V_{GS} < V_T$, and hence no channel inversion region is formed. What do you have between source–body–drain regions? Drumroll please! An *npn* BJT! How does current depend on voltage in an *npn* BJT? Exponentially! Moral: When Mr. MOSFET is biased in subthreshold, diffusion transport in the underlying BJT takes over its current–voltage characteristics. Mr. MOSFET morphs into a BJT! Cool.

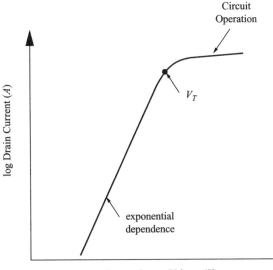

MOSFET Subthreshold Characteristics

Figure 8.27 The subthreshold characteristics of an *n*FET, plotted on semilog scales. Shown is the output current I_D as a function of input voltage V_{GS} for fixed output voltage V_{DS}.

Are there any nonidealities lurking in MOSFETs, where simple theory fails to capture reality, or is that just a BJT thing? You bet there are! That's why Cressler's 6th law is a law! For fun, let's list a few of the more important failures of simple theory:

- Clearly, because I_{DS} depends directly on V_T, any unexpected change to V_T is important. Alas, there are lots of problems associated with V_T. The threshold voltage given by Eq. (8.18) is in fact only approximate, and in practice one also has to deal with nonidealities like parasitic residual charge in the gate oxide as well as interface traps between the SiO_2 and Si, both of which can never be entirely eliminated and can lead to unanticipated changes in V_T.
- It turns out (rather annoyingly) that V_T can be a strong function of the gate length L, and this behavior is known as the infamous "short-channel effect" in MOSFETs. In short, with decreased L in an *n*FET, V_T collapses. For example, a 250-nm *n*FET might have a V_T of 0.4 V, but if L is now shrunk down to 90 nm, V_T decreases to 0.1 V. Not good. It is easy to understand why this is so. In essence, when the source and drain regions, which have finite depth into the semiconductor, become close enough to electrically "feel" each other's presence (formally, the electric-field lines begin to overlap), they interact in a manner to decrease V_T.
- In addition, V_T also depends on the body potential, and thus if it is not grounded we will have $V_T = V_T(V_B)$, which must be accounted for.
- What is the maximum voltage that can be applied to Mr. MOSFET? Well, it is the maximum voltage we can put on the drain to first order on the body-to-drain

pn junction breakdown voltage (there is no feedback mechanism tied to the gain in this case). The maximum voltage we put on the gate (V_{GS}) depends on the field strength the gate oxide can support before it gets smoked! As stated several times, SiO_2 is a wonderful insulator, with a breakdown strength in excess of 10 MV/cm; but in modern MOSFETs that gate oxide might be only 1–2 nm thick. If we imagine putting 1.5 V across 1 nm, that corresponds to an electric field of 15 MV/cm. Read: "Houston, we have a problem" (see Geek Trivia on electrostatic discharge).

- In practice, saturation in a MOSFET is really a misnomer – the current doesn't actually saturate! That is, for fixed applied $V_{GS} > V_T$, I_{DS} continues to rise as V_{DS} increases. In a well-made MOSFET this I_{DS} rise will be modest, but it's always there and can cause serious difficulties in certain types of circuits. The physics is related to the V_{DS}-imposed modulation (shortening) of the channel. Observe from Eq. (8.20) that $I_{DS} \propto 1/L$, so if $L = L(V_{DS})$, decreasing with increasing V_{DS}, I_{DS} will rise with increasing V_{DS}. This effect is logically called "channel length modulation."

- Remember velocity saturation in semiconductors? At a sufficiently high electric field, the maximum velocity of an electron is fixed (about 1×10^7 cm/s in Si at 300 K), representing the effective speed limit of the material. Well, in a MOSFET, V_{DS} provides a lateral field to move electrons from source to drain. If that field becomes too large, then we reach velocity saturation and our simple equations for current and speed change (in unfavorable ways). It is easier than you think to velocity saturate a MOSFET. Apply a V_{DS} of 1.2 V to a 130-nm MOSFET. The lateral field is thus 92.3 kV/cm (try it!), well into velocity saturation.

Geek Trivia: ESD

Ever wonder why that extra memory chip, or game card, or new microprocessor, arrives in that thin, black "antistatic" wrap with all the yellow warning labels on them? Electrostatic discharge – ESD. What is ESD? Well, recall what happens when you walk across a carpet on a clear winter day and then touch the doorknob. Yep, that annoying zap of an electrical spark! An electrical spark is triggered when the electric field exceeds approximately 30 kV/cm (the dielectric breakdown strength of air). One of the causes of ESD events is static electricity [4]. Static electricity is often generated through what is called "tribocharging," the separation of electric charges that occurs when two materials are brought into contact and then quickly separated. Examples of tribocharging would include our carpet-walking feat or removing some types of plastic packaging. In these cases, the friction between two materials results in tribocharging, thus creating a difference in potential that can lead to a spark – aka an ESD event. Another cause of ESD damage is by electrostatic induction. This occurs when an electrically charged object is placed near a conductor that is isolated from ground. The presence of the charged object creates an electrostatic field that causes charges on the surface of the other object to redistribute. Even though the net electrostatic charge of the object has not changed, it now has regions of excess positive and negative charge. An ESD event may occur when the object comes into contact with a conductive path to ground. For example, charged regions on the surfaces of styrofoam cups or plastic bags can induce potential on nearby ESD-sensitive components by electrostatic induction, and an ESD event may occur if the component is then touched with a metallic tool.

So what's the big deal with ESD? Well, the fact that a spark results should tell you that the voltages built up can be large; hundreds to even thousands of volts, depending on the conditions. So why don't you get electrocuted when you walk across the carpet? The current flow is tiny. But then so are our transistors. Very tiny. Refer back the maximum voltage a MOSFET gate oxide can withstand: 1.5 V on a 1-nm oxide puts me above breakdown strength. Now we are talking 100s of volts. Yikes! Question: What is by far the easiest and quickest way to kill an IC? Yep, lack of adequate ESD protection. Hence the black antistatic bags; the black antistatic foam; the grounding straps for anyone handling the IC; the antistatic floors in the IC fab; and the air deionizers (to increase the breakdown strength of the air). At the IC level, we also ESD protect all of the I/O pins with special ESD diodes that shunt charge away from the input transistors if they get inadvertently exposed to excessive voltages. All of this hassle in the name of keeping ESD off the precious gate oxide in Mr. MOSFET. Yet, even with this multitiered ESD protection, tons of ESD-triggered IC deaths occur every day.

For testing the susceptibility of ICs to ESD from human contact, a simple test circuit called the "Human Body Model" (HBM) is typically used (to mimic a human touching an IC). This circuit consists of a capacitor in series with a resistor. Experimentally, the capacitor is charged to a high voltage from an external source and is then discharged through the resistor directly into the input terminal of the electronic device (our IC). Question of the day? Does the part still speak to you post-ESD event? A common HBM definition is defined in the U.S. military standard, MIL-STD-883, Method 3015.7, Electrostatic Discharge Sensitivity Classification, which establishes the equivalent circuit to use (e.g., a 100-pF capacitor and a 1.5-kΩ resistor), as well as the necessary test procedures required for mimicking a realistic human-induced HBM ESD event.

So just how fast is Mr. MOSFET? Well, intuitively it can't be any faster than the time it takes the carriers to travel from source to drain. What does that depend on? Well, we can easily show that (assume that we are NOT velocity saturated)

$$\tau_{\text{FET}} \cong \frac{L^2}{\mu_{\text{eff}}\, V_{DS}}.$$

(8.21)

Hence the speed of Mr. MOSFET depends quadratically on the distance the carriers travel from source to drain, and reciprocally on the channel mobility and drain-to-source voltage. Moral: It is very easy to understand why so much attention is paid to gate length scaling in CMOS, because a $2\times$ decrease in L gives me a $4\times$ speed improvement, just as with the BJT. In addition, you ALWAYS want as large a mobility as you can get, to improve both current drive and speed, and MOSFETs are just another case-in-point. Sound like a law? Yep. *Cressler's 10th Law: "There is no such thing as too high a mobility in semiconductor devices."* We also clearly desire higher operating voltages for improved speed, but alas, there are interesting complexities associated with MOSFET reliability that will unfortunately get in the path of using higher V_{DS} with smaller L. We'll talk about this in the chapter on ultimate scaling limits.

Care for a trivia question with a hidden agenda? Sure! Which is faster, a BJT or a MOSFET? Ahhhh, good question. Answer: Depends! If we consider only the raw transistor speed (its "intrinsic" speed) of Mr. BJT and Mr. MOSFET, and plug in some reasonable values, we will find that the race is a wash. That is, the very best BJTs and FETs, when configured as simple binary switches, can both achieve sub-10-ps types of speed numbers. Blazing. BUT, the answer changes if that BJT or FET binary switch now has to drive a large capacitive load, as might be encountered, for instance, in charging a long interconnect wire in the guts of the complex logic of a microprocessor. In this case the BJT has a significant advantage, and it will inevitably win the sprint. Why? It's all about the transconductance (g_m). Remember, the logic gate with the largest transconductance (I_{gate}/V_L) can charge a capacitively loaded line the fastest; and although it might not be immediately obvious, the transistor with the largest g_m will enable the gate with the largest g_m(gate). Consider the transconductance of the nFET in saturation [it's not hard – just take the derivative of Eq. (8.22)]:

$$g_m(\text{FET}) \;=\; \frac{\partial I_{DS}}{\partial V_{GS}} \;=\; \left(\frac{W}{L}\right) C_{\text{ox}}\, \mu_{\text{eff}}\, (V_{GS} - V_T). \qquad (8.22)$$

Observe that, although g_m depends exponentially on input voltage for a BJT [Eq. (8.17)], it depends only linearly on input voltage in a MOSFET. Exponential vs. linear? No contest.[5] Under large loading, the BJT wins that footrace every time. Here's the kicker, though. MOSFETs dominate microprocessors, for which the rule of the day is a binary switch driving a capacitive load, so there must be more going on that meets the eye. Yep. Read on.

8.5.3 What does it do for you?

Okay, first things first. Clearly Mr. MOSFET is a fundamentally different kind of transistor than Mr. BJT, with different underlying physics, different operational principles, and different resultant electrical characteristics. Still, it can do the same types of things that all transistors can do: (1) Act as a voltage/current amplifier in analog–RF circuits, and (2) serve as a regenerative binary switch for digital circuits. That is not to say that Mr. MOSFET necessarily makes a BETTER amplifier than a BJT. Mr. MOSFET does have the virtue of a very high (near-infinite, actually) imput resistance (i.e., you see an open circuit looking into the gate because no dc current flows into the gate), which can prove advantageous in

---

[5] One does have the advantage in a MOSFET of having g_m proportional to the size of the transistor (W/L), and hence if you need more g_m you can just make a device with LOTS of W. Appreciate, however, that scaling up a transistor's size to improve g_m yields a large area penalty (and often degrades other performance metrics), because for equal g_m, Mr. MOSFET is going to have to be huge compared with Mr. BJT.

certain amplifier topologies, but its lower g_m compared with that of a BJT typically results in lower amplifier gain for, say, standard voltage amplifiers. Ultimately, whether you use a MOSFET or a BJT for an given amplifier depends on what you are trying to accomplish, and importantly the target circuit performance specifications that must be satisfied within the available cost structure for the circuit (and system). Economics most definitely plays a role here. You guessed it! *Cressler's 11th Law: "The choice of transistor type for a given circuit depends on the performance specifications that must be satisfied within the available cost structure."*

Mr. MOSFET's supreme virtue actually rests with his use as a regenerative switch, IF AND ONLY IF we add a clever twist. It should be intuitively obvious that an nFET can be configured as a switch. Apply $V_{GS} > V_T$, and the transistor turns on, closing the switch. Apply $V_{GS} < V_T$, and the transistor turns off, opening the switch. On–off. On–off. Binary switch. Fine. How about if we instead use not just Mr. nFET, but also Mr. pFET, in tandem, to form a switch, as shown in Fig. 8.28. In this case the gates of the nFET and the pFET are electrically tied together, as are their drains. The body terminals of each are shorted to their sources (hence, often not explicitly shown), which are connected to V_{DD} (>0 V) and ground; the voltage "rails" of the circuit. Let the commoned gates be the input (IN) of the circuit and the commoned drains be the output (OUT) of the circuit. What you have just constructed is the famous and now absolutely ubiquitous complementary metal-oxide semiconductor (CMOS) "inverter." Complementary just means that we use BOTH types of MOSFETs (an nFET and a pFET). Own a computer? Then you own a gazillion of these puppies!

An "inverter" is the simplest binary logic gate; it does just what it says – inverts the signal presented at the input. If IN $= V_{DD} =$ a logical "1," then OUT $=$ ground $= 0$ V $=$ a logical "0"; and vice versa. BUT, it does this with intrinsic gain, and hence is regenerative. Need a CMOS operational walk-through? Sure. If IN $= V_{DD}$ (say +1.2 V), then $V_{GS,n} = V_{DD} > V_{T,n}$, and the nFET is in inversion and hence on; switch closed. At the same time, $V_{GS,p} = 0$ V $> V_{T,p}$, and the pFET is turned off; switch open (careful: V_{DD} is on both the gate and the source of the pFET, so $V_{GS,p} = 0$ V; and $V_{T,p}$ is negative). If Mr. pFET is off, but Mr. nFET is on, then OUT is directly connected to ground, and hence OUT $= 0$ V (obviously if OUT is capacitively loaded, it may take some time to reach this final state). Moral: IN $=$ "1" produces OUT $=$ "0"; an inverter. Similarly, if IN $= 0$ V, then $V_{GS,n} = 0$ V $< V_{T,n}$ and the nFET is now off, switch open. At the same time, $V_{GS,p} = -V_{DD} < V_{T,p}$, and the pFET is in inversion, and hence on; switch closed. If Mr. pFET is on, but Mr. nFET is off, then OUT is directly connected to V_{DD}, and hence OUT $= V_{DD}$ (again, if OUT is capacitively loaded, it may take some time to reach this final state). Moral: IN $=$ "0" produces OUT $=$ "1"; an inverter. Excellent, we're in business.

The resultant V_{out} vs. V_{in} characteristics of the CMOS inverter, usually called the "voltage transfer characteristics" (VTC), are shown in Fig. 8.28. Now, hold on tight. Drumroll please . . . Here's the kicker. Question: How much current flows

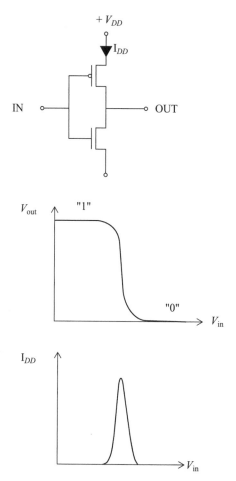

Figure 8.28   The simplest implementation of a CMOS logic gate – an inverter. Also shown are the voltage transfer characteristics of the gate (V_{out} vs. V_{in}), as well as the current flow during the logic transition.

from V_{DD} to ground during the operation of the inverter? Appreciate that this is a vital question because it translates directly into the power dissipated (heat) by the logic operation ($P = VI$). Said another way, how much power does it take me to turn a "1" into a "0" and vice versa? Well, look back at Fig. 8.28. Observe that when $V_{in} = 0$ V and $V_{out} = V_{DD}$, Mr. pFET is on (closed) but Mr. nFET is off (open), and thus there is no direct electrical connection between V_{DD} and ground. Read: No current flow! Same thing when $V_{in} = V_{DD}$ and $V_{out} = 0$ V. Mr. nFET is on (closed) but Mr. pFET is off (open). No direct electrical connection between V_{DD} and ground. No current flow! The only time we have current flow in a CMOS inverter is when we are *transitioning* between a "1" and a "0," and vice versa; i.e., only during dynamic switching. In this case both transistors are momentarily on, with one turning on and one turning off, and thus current flows from V_{DD} to ground; but only for an instant. As soon as the second transistor turns off, that power-to-ground path is broken and current flow halts. Result? The

CMOS inverter consumes no static power! If it just sits and holds its logic state, to first order it burns no energy doing this.[6] What you have with CMOS, folks, is a clever combination of two transistor types to produce a regenerative binary switch that is EXTREMELY power efficient. In a nutshell, this is why MOSFETs dominate the world transistor count (for an interesting digression, see the sidebar discussion on complementary BJT logic).

Complementary BJT Logic?

But wait a second (the clever among you say). You used pFETs $+$ nFETs to build a CMOS. Fine. But you also said that the g_m of MOSFETs is most definitely inferior to the g_m of BJTs at fixed geometry. Because you have npn BJTs and pnp BJTs available, couldn't you do the same thing as in a CMOS and simply connect the npn and pnp BJTs together to form a complementary BJT (C-BJT) logic that is not only power efficient but is also MUCH faster than CMOS under realistic loading? Answer? Yes and no. Yes, you could in fact do this, and yes it would in fact produce a power-efficient binary logic; but, alas, the resultant C-BJT logic would be slow as midnight, and hence a disaster for building a complex logic function. Why? Well, you have to look back at the physics of BJTs (go figure). If you talk your way through the operation of your intended C-BJT inverter, you will discover that, if IN $=$ "1," Mr. pnp is off and Mr. npn is on, and Mr. npn is actually biased in saturation (low V_{CE}). Why does this matter? Well, in saturation in a BJT, both the emitter–base junction AND the collector–base junction are forward biased, and this dramatically increases the base transit time of the BJT (in a word, there is additional charge storage in the base associated with the forward bias of the collector–base junction). Net result: Biasing a BJT into saturation slows it down. Low power $+$ slow speed is NEVER going to be a winning combination for binary logic. There are, of course, other secondary issues. The complexity and area (read: ultimate cost) required for building a CMOS inverter are far smaller than for a C-BJT inverter anyway. Moral: You won't find C-BJT logic in microprocessors. Careful, though; this does NOT mean that BJT logic is not alive and kicking. Classical npn BJT logic families such as "emitter-coupled logic" (ECL) still hold the speed records for fast binary logic under loading, albeit at substantially increased complexity and power compared with these of a CMOS. Thus, if pure speed is the ultimate driving force, BJT logic may well be the best solution at the end of the day.

CMOS. A brilliantly simple but incredibly powerful idea. AND, as I like to remind my graduate students when in need of inspiration, the brainchild of a Ph.D. student (refer to Chap. 4). The rest of the binary logic world can be built up from

[6] Clearly this can't be strictly true, because we know that pn junctions under reverse bias (our source–drain–body junctions) have tiny but measurable reverse current flow. Hence, we will in fact have a tiny, but finite, "leakage" current flowing under static conditions, and alas this leakage current has recently come back to haunt us in major ways. This topic will be revisited in the chapter on ultimate scaling limits.

our simple inverter concept while maintaining this inherent power efficiency – binary OR gates, NOR gates, AND gates, NAND gates, and XOR gates. Everything you need to build the complex binary logic of a computational engine like a microprocessor is now at your fingertips. Rejoice!

8.6 X-Men transistors

So, we now have Mr. BJT and Mr. MOSFET. These guys are alive and well and used in just about any electronic product you'd care to look inside. You can easily imagine, however, that the quest for faster–smaller–cheaper transistors drives a lot of creative folks around the world to scratch their heads and ponder such questions as "How can I use my cleverness to make Mr. BJT and Mr. MOSFET do their transistor thing just a little bit better?" Let's end this chapter by briefly examining what lies BEYOND BJTs and MOSFETs – the future of the transistor world. Here we will go back to our semiconductor toolkit and pull out a few tricks for reengineering BJTs and FETs. I call these new transistors "X-Men transistors"; you know, mutants. More accurately, they should be called "bandgap-engineering-enhanced" transistors, and this subject occupies a tremendous amount of present-day research (including my own). You owe it to yourself to at least know a little something about these mutant devices.

Let me first remind you what we mean by bandgap engineering. Instead of building Mr. BJT and Mr. MOSFET from a single semiconductor (say, silicon), what if we had the freedom to combine multiple semiconductors with differing bandgaps to engineer the properties of our devices? Presto: Bandgap engineering! Modern epitaxial growth techniques (Chap. 7), although not cheap, make this a snap, particularly if you are dealing with lattice-matched materials that are chemically compatible (e.g., AlGaAs and GaAs). Enter the world of heterostructures. Don't sweat; we're just going to skim the surface of this fascinating field.

8.6.1 HBTs

First stop: X-Men BJTs. Let's imagine how we would use a bandgap-engineered heterostructure to enhance the electrical performance characteristics of Mr. BJT. Not surprisingly, a good understanding of how BJTs work will point the way. Recall that for a fixed base profile (base width and base doping) the current gain in a BJT is limited by the hole current that flows from the base to emitter when the emitter–base junction is forward biased. To suppress this current, making I_B small and hence β large, we have to dope the emitter VERY heavily with respect to the base (say $100\times$ larger). This indeed produces a high gain, but the high emitter doping necessarily makes the emitter–base junction capacitance large. This is because the SCR inside the emitter–base pn junction is squeezed down by the high doping [refer to Eq. (8.6)], effectively decreasing the dielectric thickness of the junction capacitor. Implication? Well, large parasitic capacitance will slow down the transistor when it switches, limiting its dynamic response. Never good. This produces a fundamental gain–speed trade-off in BJTs. In addition, we cannot dope the base very heavily in a BJT, because it has to be $100\times$ or so lower than the emitter. For fixed base width, this means that the base resistance will inevitably be fairly large in a BJT. Again, not good from a dynamic perspective, because base resistance limits the maximum switching speed, the gain at high frequency while driving a load, and even key amplifier metrics such as noise performance. Again, this produces an unfavorable gain–speed trade-off in BJTs.

Solution? How about we make the emitter of the transistor out of a different semiconductor; one with a larger bandgap! Said another way, let's use an emitter–base "heterojunction" to reengineer Mr. BJT. Result? X-Men BJTs = Heterojunction Bipolar Transistors (HBTs), as depicted in Fig. 8.29. Appreciate the elegance of this solution [7,8].[7] With a difference in bandgap between the base and emitter, the induced "valence band offset" (ΔE_V) increases the potential barrier the holes "see" when looking into the emitter, suppressing the hole current injected from base to emitter in forward bias, decreasing I_B, *even if the emitter is more lightly doped than the base*. In addition, if the base bandgap has a conduction band offset (ΔE_C) that lowers the potential barrier seen by the electrons looking into the base, we will get an enhanced electron flow in forward bias, increasing I_C, *even if the base is more heavily doped than the emitter*. If I_B is suppressed and I_C is enhanced, we have current gain to burn! AND, the parasitic capacitance of the emitter–base junction will be strongly decreased, because we can use a more lightly doped emitter; and the base resistance will be strongly decreased, because we can now dope the base much more heavily. Net result? Well, we can easily make the HBT have tons of current gain, but that is typically not what we are after. Instead, at a fixed current gain (say, 100–200), the HBT will have much less emitter–base capacitance and much lower base resistance than a BJT, providing a substantial speed advantage over the BJT. In a word, the HBT removes the gain–speed trade-off faced in BJTs. We could label such a HBT an *Npn* HBT, because the *n*-type emitter has a larger bandgap than the base and collector, as indicated by the capital N.

As illustrated in Fig. 8.29, depending on the semiconductors used and the consequent "band alignments" (how E_C and E_V line up to one another when the materials are connected), we may inadvertently introduce band "spikes," but no worries, these can be easily smoothed out by using bandgap grading techniques, in which we change from semiconductor 1 to semiconductor 2 over some finite "grading" distance. A good example of such a HBT would be an AlGaAs/GaAs

---

[7] Amazingly, Shockley's original transistor patent, U.S. Patent 2,502,488, filed in June of 1948 and issued April 4, 1950, covered the basic idea behind the HBT!

HBTs

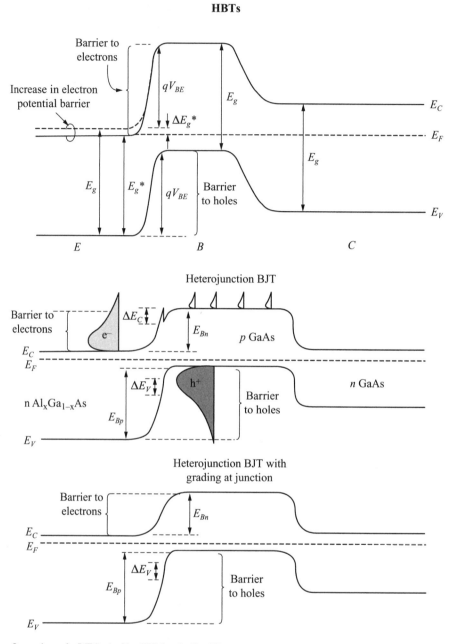

Figure 8.29 Comparison of a BJT (top) with a HBT (inspired by [3]).

HBT, often used as a power amplifier inside your cell phone! HBTs currently hold the speed records for all transistors, with usable gain at frequencies exceeding 700 GHz in the InGaAs/InP material system, with simulated projections to well above 1,000 GHz (1 *THz*). That's fast!

There are other clever tricks to play. How about we also use a large bandgap semiconductor in the collector; one that has a much higher breakdown field strength. This will boost the overall breakdown voltage of the transistor because the collector–base *pn* junction now has a larger breakdown voltage. Combining techniques 1 (emitter–base heterojunction) and 2 (collector–base heterojunction) into one transistor yields the so-called "double heterojunction bipolar transistor" (DHBT). We might label this transistor an *NpN* DHBT.

Care for one last clever HBT trick? Sure. Instead of using a large bandgap emitter and collector for all of the action, let's do our X-Men thing in the small bandgap base by compositionally grading the bandgap of the base to induce a position-dependent conduction band edge. As you will recall, a position-dependent band edge produces an electric field that accelerates the carriers as they transit the base. Because the base transit time typically sets the speed limit of BJTs, this bandgap engineering trick can dramatically improve the speed of the transistor, with only modest changes to the underlying doping profile. Even small bandgap grading over very tiny distances can induce huge accelerating electric fields, greatly speeding up the device (e.g., 100-meV grading over 100 nm = 10,000 V/cm field, enough to put an electron at scattering-limited velocity in silicon). This is exactly what is done in silicon germanium (SiGe) HBTs, my own field of research [9]. The SiGe HBT has an advantage over III-V HBTs, in that dramatic speeds can be realized while maintaining the economy of scale of conventional silicon manufacturing and compatibility with silicon CMOS. An IC technology that contains both BJTs–HBTs and CMOS is called BiCMOS (bipolar CMOS) and gives circuit designers added leverage for system optimization by using the transistor best suited for the task at hand (traditionally, the BJT–HBT for analog–RF circuits and CMOS for digital circuits and memory). Having a HBT that is silicon compatible is clearly a huge plus from a cost perspective. SiGe HBTs with usable gain at frequencies above 500 GHz have been demonstrated, and thus bandgap engineering with SiGe can clearly yield extremely impressive speeds while maintaining strict compatibility with conventional silicon (CMOS) fab. Win–win.

8.6.2 HFETs

Okay. Last stop: X-Men FETs. Let's imagine how we would use a bandgap-engineered heterostructure to enhance the electrical performance characteristics of Mr. FET. Recall first what fundamentally limits the speed of the FET: At fixed gate length and bias, μ_{eff} sets the speed, because mobility always measures the carrier velocity in response to an applied electric field. In a silicon MOSFET, not only are the bulk electron and hole mobilities quite modest by semiconductor standards, but because all of the transport action is localized to the Si/SiO$_2$ interface, surface roughness scattering limits μ_{eff} to roughly 1/3 of the achievable bulk mobility. Bummer. Solution? How about we (1) change to a faster semiconductor (for instance, at 1×10^{16} cm^{-3} the 300 K electron mobility in GaAs is about 7,000 cm^2/V s vs. only 1,250 cm^2/V s for silicon); and (2) get rid of the gate oxide between the gate and the body to remove surface roughness scattering that limits the mobility! Result? X-Men FETs = heterostructure FETs (HFETs), as depicted

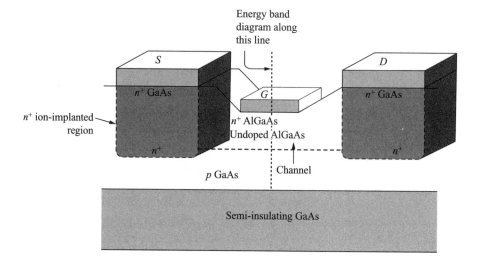

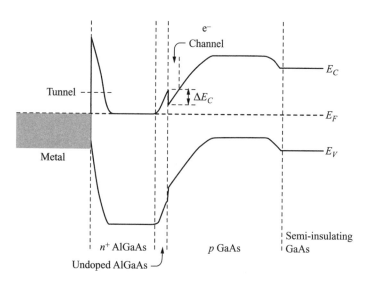

Figure 8.30   Cross-sectional view and resultant energy band diagram of a HFET.

in Fig. 8.30. Note that there is no gate oxide in this device, and thus you might naively think that this is not in fact a FET, but it is. The gate terminal is still capacitively coupled to the channel, but via a reverse-biased Schottky barrier (metal–semiconductor contact). The conducting channel between the source and drain occurs at the interface inside the AlGaAs/GaAs heterostructure, where a high density of electrons accumulate when the HFET is turned on. Viewed top-down, the electron channel is confined to a 2D layer (2DEG), as with a MOSFET, but there is no Si/SiO$_2$ interface to cause problems! This type of 2D carrier transport at a heterointerface was first demonstrated in 1979 [10], ultimately leading to a Nobel Prize, and shortly thereafter applied to FETs in 1980 [11]. HFETs come in many forms and flavors, and go by many different names, including [12] the high-electron-mobility transistor (HEMT) and the modulation-doped FET (MODFET), but think of them all as X-Men FETs. HFETs, like HBTs, are capable of delivering usable gain at frequencies above 500 GHz, and importantly, hold the transistor records for high-frequency noise performance, a key metric in high-sensitivity radio receivers and instrumentation.

There are games to be played. HFETs clearly have a fundamentally different structure from that of MOSFETs and can be built only from "fast" III-V semiconductors like GaAs. In addition, it is VERY difficult to build a true CMOS by using HFETs, because getting viable n-type HFETs and p-type HFETs together on the same wafer is very difficult, to say the least. A logical question presents itself. Is there a viable compromise between silicon MOSFETs and full-blooded HFETs that might bear fruit? Indeed there is! A strained-silicon CMOS is such beast, and should rightly be considered a close cousin of the HFET [13]. In strained-silicon CMOS, we keep the gate oxide, despite its limitations, and instead work to improve the transport properties of the underlying silicon channel. Sound impossible? It's not. Recall what you know about semiconductors. The band structure, effective masses, and ultimately the carrier mobility are strong functions of the lattice arrangement. Change the lattice structure and you can thereby change the mobility. At least in theory. How? In practice, we induce large amounts of stress and strain in the silicon channel (think pushing or pulling on the silicon to distort the lattice in controllable ways) by a variety of techniques, including putting a different semiconductor (e.g., a SiGe alloy) into the source and drain regions to squeeze on the channel; or we might wrap a thin-film dielectric (say, nitride) around the gate of Mr. MOSFET to tug on the channel. The result is the same. The energy bands are distorted and, if done correctly, the carrier mobility is thereby increased, often by as much as 50% or more over standard silicon, WITHOUT changing the basic MOSFET structure or moving away from conventional silicon fabrication. Presto! Strained-silicon CMOS. Looks, feels, and smells like an ordinary CMOS, only more oomph under the hood! Such MOSFETs are often called "transport-enhanced" devices, for obvious reasons. They are becoming increasingly common these days, for reasons we'll soon address.

Good as HBTs and HFETs are, why isn't the whole world now dominated by X-Men transistors? In a word – economics. Traditional bandgap engineering for HBTs and HFETs requires III-V materials like GaAs and InP, and ALL III-V materials necessarily represent smaller wafers, more difficult fabrication, poorer yield, lower integration densities, poorer reliability, etc., than . . . gulp . . . standard silicon. There is a law lurking. *Cressler's 12th (and final!) Law: "If any electronic function can be successfully implemented in silicon, it will be."* That is not to say that III-V HBTs and HFETs have no role to play in modern electronics. They do; BUT, it does say that that role is typically niche oriented, satisfying a special demand for an electronic circuit that needs something very special: say, ultra-high speed, or very large breakdown voltage, or photonic capability, etc. III-V HBTs and HFETS are not, and likely never will be (I'm being bold again), mainstream for things like computers and cell phones.

What we consequently see at present is instead the absorption of HBT and HFET design TECHNIQUES into the core silicon manufacturing world, with prevalent examples being SiGe HBTs and strained-silicon CMOS. Are they as good as their III-V brethren? Nope. But will economics keep them in the game as a logical follow-on to conventional silicon electronics? Yes, very likely. In fact, it is already happening. SiGe HBTs are already widely available commercially and in lots of gadgets, and virtually all CMOS technologies at the 90-nm node and below embody some form of strained-silicon CMOS.

Whew! I titled this chapter "Transistors – Lite!," but we sure have covered quite a bit of territory! Device menagerie, to *pn* junction, to BJT, to MOSFET, to HBT, to HFET! You now have some steady legs under you, should you care to engage in a meaningful discussion (who wouldn't!) of how semiconductors are actually used to architect the electronic components fueling our Communications Revolution. Truthfully, though, we have scratched only the surface of this fascinating field. No worries; we'll save "Transistors – Not-so-Lite!" for another day! Now on to bigger and better things.

References and notes

1. K. K. Ng, *Complete Guide to Semiconductor Devices,* 2nd ed., Wiley Interscience, New York, 2002.
2. S. M. Sze (Editor), *High-Speed Semiconductor Devices,* Wiley Interscience, New York, 1990.
3. B. L. Anderson and R. L. Anderson, *Fundamentals of Semiconductor Devices,* McGraw-Hill, New York, 2005.
4. See, for instance, Wikipedia: *http://en.wikipedia.org*.
5. S. Tiwari, *Compound Semiconductor Device Physics,* Academic, New York, 1992.
6. H. K. Gummel, "Measurement of the number of impurities in the base layer of a transistor," *Proceedings of the IRE*, Vol. 49, p. 834, 1961.
7. H. Kroemer, "Theory of wide-gap emitter for transistors," *Proceedings of the IRE,* Vol. 45, p. 1535, 1957.
8. H. Kroemer, "Heterostructure bipolar transistors and integrated circuits," *Proceedings of the IEEE,* Vol. 70, p. 13, 1982.
9. J. D. Cressler and G. Niu, *Silicon–Germanium Heterojunction Bipolar Transistors,* Artech House, Boston, 2003.
10. H. L. Stormer, R. Dingle, A. C. Gossard, W. Wiegmann, and M. D. Sturge, "Two-dimensional electron gas at a semiconductor-semiconductor interface," *Solid State Communications,* Vol. 29, p. 705, 1979.

11. T. Mimura, S. Hiyamizu, T. Fujii, and K. Nambu, "A new field effect transistor with selectively doped $GaAs/n-Al_x Ga_{1-x}As$ heterojunctions," *Japanese Journal of Applied Physics,* Vol. 19, p. L225, 1980.

12. M. Shur, *GaAs Devices and Circuits,* Plenum, New York, 1987.

13. J. D. Cressler (Editor), *Silicon Heterostructure Handbook: Materials, Fabrication, Devices, Circuits, and Applications of SiGe and Si Strained-Layer Epitaxy,* CRC Press, 2006.

9 Microtools and Toys: MEMS, NEMS, and BioMEMS

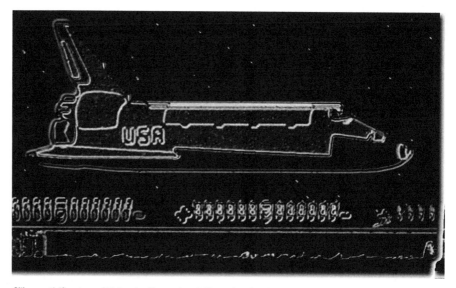

Silicon art! (Courtesy of Molecular Expressions.) (See color plate 26.)

The Real Voyage of Discovery
Consists Not in Seeking New Landscapes
But in Having New Eyes.

Marcel Proust

Two words: *Fantastic Voyage*. Think back. Think way back. I have very fond memories of this 1966 landmark science fiction movie starring Steven Boyd (as Grant) and of course Raquel *1,000,000 B.C.* Welch as Cora (hey, I was a highly impressionable young man when I first saw it). Based on a screenplay by Harry Kleiner and (reluctantly) novelized by none other than Isaac Asimov, *Fantastic Voyage* centered on a new top-secret technology that could be used to miniaturize objects (and people!) to submicron dimensions. The "spaceship" Proteus and its adventuresome crew are shrunk and injected into the bloodstream of Jan Benes (Jean Del Val) to laser-zap an inoperable blood clot in his brain and save his life. Memorable scenes include (if you haven't seen it, go rent it!) (1) the unanticipated trip through Benes' defibrillated heart, (2) the attack of the killer antibodies (poor

Cora had a close call), (3) Cora getting plucked from a tear duct at the last second before growing back to life size, and of course (4) the classic ingestion scene of the traitor Dr. Michaels (Donald Pleasance) by a giant marshmallow-like white blood cell (burp!). What a way to go![1] So what does *Fantastic Voyage* have to do with MEMS, NEMS, and BioMEMS? More than you might think. Read on.

MEMS is the real-life equivalent of *Fantastic Voyage*. Well, sort of! So what exactly are MEMSs? Well, simply put, micro–electro–mechanical systems (MEMSs). Said another way, MEMSs exploit microelectronic fabrication techniques to miniaturize commonplace macroscopic objects (say, a motor, or a gear drive, or a cantilever beam – or the Proteus!) to micron-sized dimensions. You might reasonably wonder WHY we would want to go to the considerable effort of doing this. Hold tight, I'll deal with that in a moment. These tiny MEMS microgadgets typically exploit a combination of electrical mechanical and/or properties (hence electromechanical) of the host building material (e.g., the semiconductor silicon) to do their thing. Presto – MEMS! Want to morph a MEMS gadget from micrometer- to nanometer-sized dimensions (1,000 times smaller than MEMS)? Sure! Then you would have NEMSs (Nano–Electrical–Mechanical Systems). Care to let your MEMS gadget do something biological; say, analyze the chemistry of a micromilliliter of blood? Sure! Then you would have BioMEMS (BIOlogy + MEMS). The list could go on, but these three, MEMS–NEMS–BioMEMS, with MEMS being historically first and still far-and-away the most important, are the lifeblood of an exciting and burgeoning new field of advanced technology that is a direct offshoot of micro/nanoelectronics technology. Definitely something to know a little about!

Geek Trivia: MEMS – A Rose By Any Other Name

1987 was a watershed year for silicon micromachining, because techniques for integrated fabrication of mechanical systems (e.g., rigid bodies connected by joints for transmitting, controlling, or constraining relative motion) were demonstrated for the first time. The term micro–electro–mechanical system (MEMS) was coined during a series of three technical workshops held on this new field of microdynamics in 1987. Although the acronym MEMS is reasonably ubiquitous today, equivalent terms in common usage include "microsystems" (often preferred in Europe), and "micromachines" (often preferred in Japan).

The term "micro/nanomachining" refers to the set of fabrication techniques needed for building the MEMS/NEMS/BioMEMS menagerie and is appealingly descriptive (refer to sidebar discussion on the naming of MEMS). We machine, literally carve out, the guts of the toy we want to build, using our standard microelectronics fabrication toolkit of lithography, etching, oxidation, thin-film

---

[1] Supposedly, Donald Pleasance's excessive screaming while being eaten alive by the white blood cell was due to rogue soapsuds used to make the white blood cell (hey, it was low budget) that had gotten into his eyes. You will recall that he was trapped in the command chair of the Proteus and was thus unable to receive any medical attention until the scene was safely "in the can" [1].

deposition, etc. (refer to Chap. 7). Increasingly today, however, this toolkit that was historically based on classical semiconductor fab has broadened considerably to include a host of new MEMS–NEMS–BioMEMS-specific processes, which might include micromolding, laser machining, ion milling, and wire-electrodischarge machining. New, more exotic, non-semiconductor-fab materials are also now being routinely engaged, especially to support BioMEMS; things like hydrogels, gas-permeable membranes, enzymes, antigens, and even antibodies.

A wide variety of MEMS–NEMS–BioMEMS gadgets are being commercialized, and most of you, in fact, currently own more than a few, whether you know this or not. But fair warning, MEMS–NEMS–BioMEMS is a rapidly moving target, supporting an ever-increasing set of remarkably clever ideas, from airbag sensors, to gyroscopes, to HDTV displays, to blood monitors, to DNA assays. In this chapter, we'll scratch the surface of this amazingly diverse field of microtools and toys, and see what all the fuss is about.

9.1 Micro-intuition and the science of miniaturization

So, let's begin at the beginning: Why miniaturize an object? Good question! Although we humans are reasonably adept at thinking in terms of macroscopic distances and sizes, we are often at a loss with respect to correctly guessing how large-scale systems will behave when dramatically downsized. Developing some level of "micro-intuition" [2] is thus important in the context of MEMS.

Nature can be a powerful teacher. First, some reminders on how nature does business. As an example, we'll focus on size, surface area, and volume in the animal kingdom (very interesting takes on the impact and implications of miniaturization on surface tension, diffusion, electrochemistry, optics, and strength-to-weight ratio can be found in [3]). As an animal shrinks in size, the ratio of surface area to volume increases, rendering surface forces more important [4]. Consider. There is no land animal currently larger than an African elephant (3.8 m tall). Why? Well, land animals, which inevitably have to work against the force of gravity (acting on their surfaces), can grow only so large before they become too slow and clumsy to remain evolutionarily viable (read: if too slow and too clumsy, they will quickly become someone else's meal!). Large animals also have far fewer eco-niches that can support their big size and requisite appetite, and hence there can't be many of them (Fig. 9.1). For sea creatures, however, this gravity constraint is effectively removed, enabling much larger body sizes. The blue whale is currently the largest animal on Earth and can grow to 25 m or more in length (stump-your-friends trivia challenge: The record length of a blue whale is 33 m), far larger than Mr. Elephant (okay, dinosaurs were larger still, but you won't find many brontosauruses around today). At the smaller end of the animal kingdom, birds and small mammals are intrinsically more fleet of foot (or wing) and hence more agile. Go even smaller and we find the nimble insects, far and away the largest population on the planet (no coincidence).

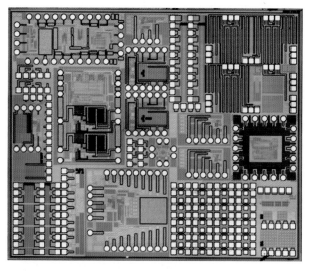

An integrated circuit designed at Georgia Tech and fabricated at IBM.

A replica of the first transistor, a germanium point-contact device, demonstrated on December 23, 1947. The transistor is roughly the size of your thumbnail (courtesy of Science and Society Picture Library).

A 300-mm, 90-nm CMOS wafer (courtesy of Intel Corporation).

A bird's-eye view of a Pentium 4 microprocessor fabricated in 90-nm CMOS technology. This integrated circuit contains over 42,000,000 transistors and operates at a greater than 2.0-GHz frequency. The integrated circuit is roughly 1.0 cm on a side in size (courtesy of Intel Corporation).

The "Deep Field" image of the universe from the Hubble Telescope (courtesy of NASA).

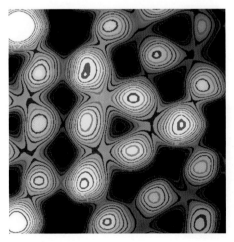

Image of the atomic surface of a silicon crystal from an atomic force microscope (courtesy of Professor Mannhart, University of Augsburg).

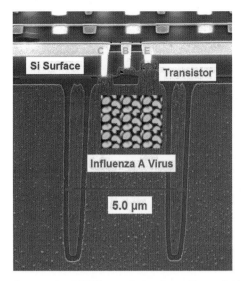

Plate 7 Cross-sectional SEM image of a transistor with a scaled Influenza A virus shown for comparison. (Courtesy of International Business Machines Corporation. Unauthorized use is not permitted.)

Plate 8 An example of cell tower "sculpture" forming the outside chapel at Saint Ignatius House Retreat Center, in Atlanta, GA.

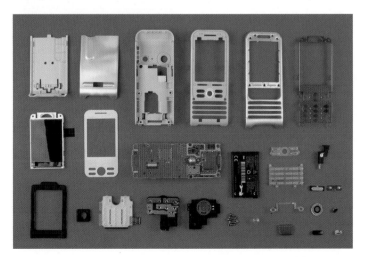

Plate 9 Stripped-down components of a generic 3G cell phone (courtesy of Portelligent, Inc.).

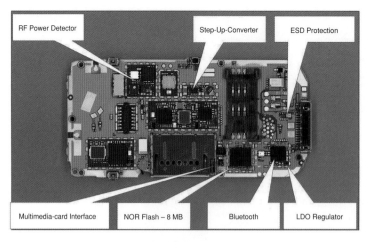

| RF Power Detector | Step-Up-Converter | ESD Protection |
| Multimedia-card Interface | NOR Flash – 8 MB | Bluetooth | LDO Regulator |

Plate 10 Rear view of a generic 3G cell phone with the covers removed, showing the various integrated circuit components (courtesy of Portelligent, Inc.).

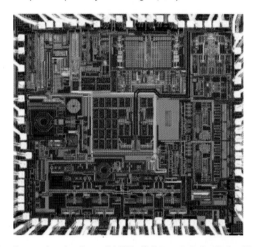

Plate 11 A zoom-in, showing a "delidded" integrated circuit, in this case one of the RF transceivers (courtesy of Portelligent, Inc.).

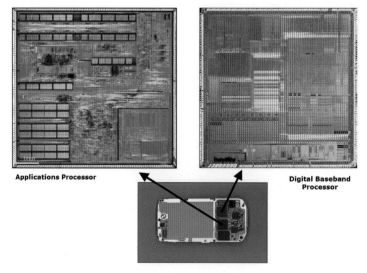

Applications Processor

Digital Baseband Processor

Plate 12 A zoom-in, showing "delidded" integrated circuits, in this case the application processor with embedded memory and the digital baseband processor (courtesy of Portelligent, Inc.).

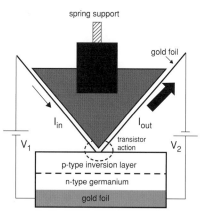

spring support

gold foil

I_{in} I_{out}

V_1 transistor action V_2

p-type inversion layer

n-type germanium

gold foil

The famous point-contact transistor, the first solid-state amplifying device, invented in 1947 by John Bardeen and Walter Brattain at Bell Laboratories in the United States. Bardeen and Brattain discovered that, by placing two gold contacts close together on the surface of a crystal of germanium through which an electric current was flowing, a device that acted as an electrical amplifier was produced (photograph courtesy of Alcatel – Lucent). A schematic diagram of the device is shown on the right.

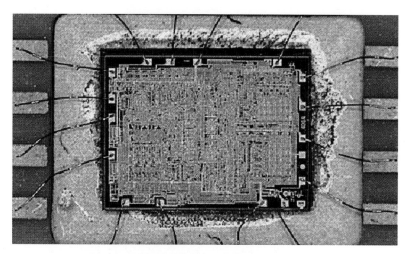

The first microprocessor, the Intel 4004 (courtesy of Intel Corporation).

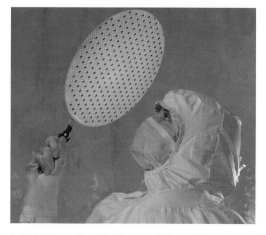

A fabrication engineer holding up a 300-mm silicon water. (Photograph courtesy of International Business Machines corporation. Unauthorized use is not permitted.)

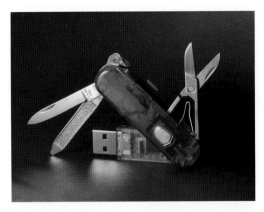

A thumb drive for the outdoor enthusiast (courtesy of CustomUSB).

Close-up of a read–write "head" hovering over a HDD "platter" (courtesy of International Business Machines Corporation; unauthorized use is not permitted).

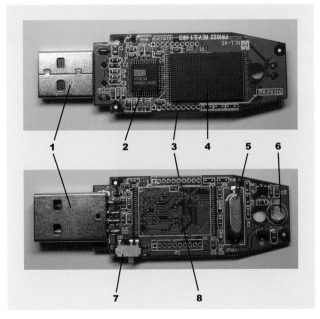

Front and back side views of a naked thumb drive, showing the building blocks. Components 1–8 are defined in the text (for reference, component 4 is the NAND flash IC). (Photograph and annotations by J. Fader, 2004.)

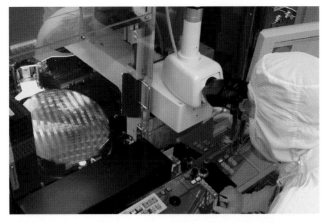

Photograph of IC fabrication equipment in use (courtesy of International Business Machines Corporation. Unauthorized use is not permitted).

Growth of a silicon crystal "boule" from a seed crystal lowered and rotated in a vat of highly purified molten silicon (courtesy of Texas Instruments, Inc.).

A 300-mm silicon crystal boule (Courtesy of Texas Instruments, Inc.).

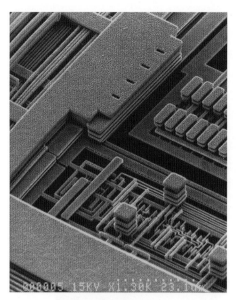

Plate 22 Top view of the copper metalization of a modern IC. (Courtesy of International Business Machines Corporation. Unauthorized use is not permitted.)

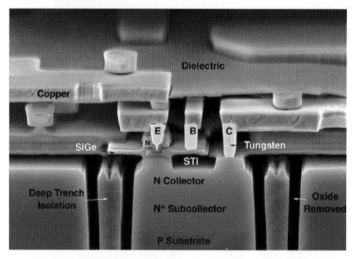

Plate 23 Decorated cross-sectional view of a modern bandgap-engineered SiGe bipolar transistor (Courtesy of International Business Machines Corporation. Unauthorized use is not permitted.)

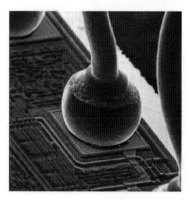

Plate 24 Decorated view of a single gold wirebond on the surface of an IC. (Courtesy of International Business Machines Corporation. Unauthorized use is not permitted.)

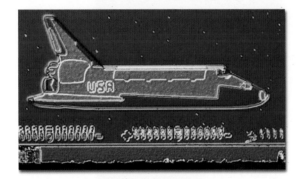

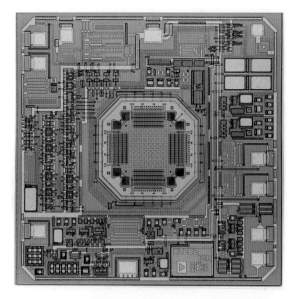

The Analog Devices ADXL 50 accelerometer. This two-axis MEMS accelerometer is shown at the center of the die, surrounded by support electronics. It can measure both static and dynamic acceleration. (Copyright Analog Devices, Inc., all rights reserved. Used with permission.)

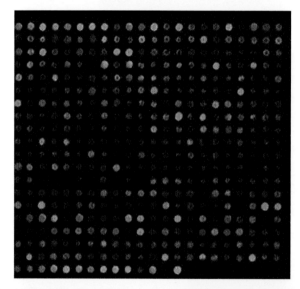

A typical DNA chip used in cancer gene analysis (courtesy of National Human Genome Institute).

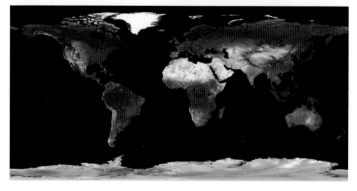

Our Earth (courtesy of NASA).

Arguably the most famous picture ever taken of the Earth (courtesy of NASA).

Artist's conception of a GPS satellite in Earth orbit (courtesy of NASA).

Exploded view of naked Garmin zūmo 550. (Image produced with the permission of Garmin. Copyright 2008, Garmin Ltd. or its subsidiaries, all rights reserved.)

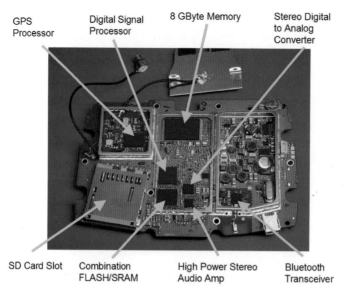

GPS Processor
Digital Signal Processor
8 GByte Memory
Stereo Digital to Analog Converter

SD Card Slot
Combination FLASH/SRAM
High Power Stereo Audio Amp
Bluetooth Transceiver

Plate 34 Back-side view of the IC board of a Garmin zūmo 550. (Image produced with the permission of Garmin. Copyright 2008, Garmin Ltd. or its subsidiaries, all rights reserved.)

Plate 35 Alfred Sisley's "The Effect of Snow at Argenteuil" (1874).

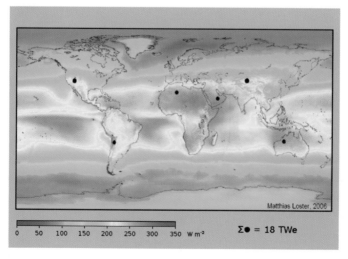

Matthias Loster, 2006

0   50   100   150   200   250   300   350   W m⁻² $\Sigma\bullet = 18$ TWe

Plate 36 The solar irradiance map of the Earth's surface, indicating solar "sweet spots" (courtesy of NASA).

A solar panel from the largest photovoltaic solar power plant in the United States (2007), located at Nellis Air Force Base. The solar arrays will produce 15 MW of power at full capacity. (Courtesy of the U.S. Air Force. Photo taken by Senior Airman Larry E. Reid, Jr.)

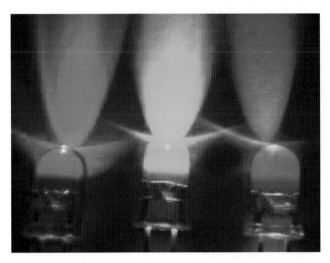

LEDs emitting primary colors.

A packaged GaAs laser diode (courtesy of NASA).

Plate 40 A plasma-enhanced chemical vapor deposition (PECVD) chamber modified to grow CNTs. The dc heating element glows red (hot!), and the nanotube-laden plasma is blue.

Plate 41 SEM image of aligned ZnO nanowire arrays synthesized by a new vapor–solid process (courtesy of Z. L. Wang, Georgia Tech).

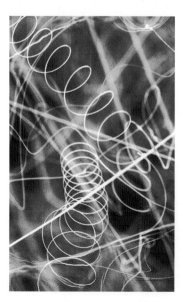

Plate 42 Nanoscale helixes of ZnO that exhibit superior piezoelectric properties, an important effect for converting a mechanical signal into an electrical signal, and vice versa (courtesy of Z. L. Wang, Georgia Tech).

Aligned propeller arrays of ZnO, which have applications in nanoscale sensors and transducers (courtesy of Z. L. Wang, Georgia Tech).

Hierarchical nanostructures of ZnO for chemical and biomedical sensor applications. These types of nanomaterials have the potential of measuring nanoscale fluid pressure and flow rate inside the human body (courtesy of Z. L. Wang, Georgia Tech).

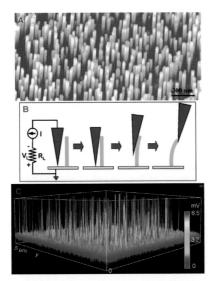

Piezoelectric nanogenerators based on aligned ZnO nanowires: **(A)** SEH images of as-grown ZnO nanowires on a sapphire substrate, **(B)** schematic experimental procedure for generating electricity from a nanowire by using a conductive AFM, **(C)** piezoelectric discharge voltage measured at an external resistor when the AFM tip is scanned across the nanowire arrays (courtesy of Z. L. Wang, Georgia Tech).

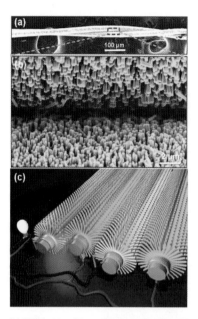

(a) SEM image of two entangled microfibers that were covered radially with piezoelectric ZnO nanowires, with one of them coated with gold. The relative scrubbing of the two "brushes" generates electricity. **(b)** A magnified SEM image at the area where the two "brushes" meet teeth-to-teeth, with the top one coated with gold and the bottom one with as-synthesized ZnO nanowires. **(c)** A schematic illustration of the microfiber–nanowire hybrid nanogenerator, which is the basis of using fabrics for generating electricity (courtesy of Z. L. Wang and X. D. Wang, Georgia Tech).

The integrated circuit meets Mother Nature. (Courtesy of International Business Machines Corporation. Unauthorized use is not permitted.)

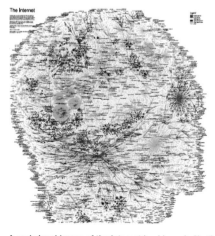

A cool visual Image of the Internet backbone in North America (courtesy of Lumeta Corporation).

Number of Animal/Species vs. Size

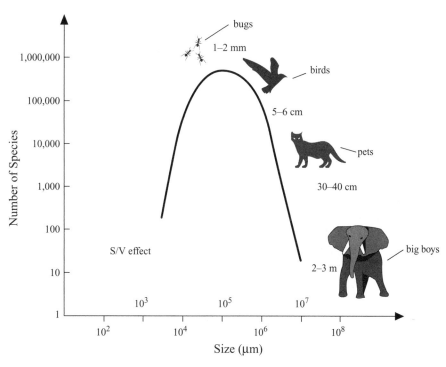

Figure 9.1   Number of species as a function of size for the animal kingdom (inspired by [3]).

But, alas, there can be serious problems with being TOO small. The pygmy shrew and the hummingbird, for instance, must eat almost continuously or freeze to death. Don't believe me? The heat loss from a living creature of length l is proportional to its surface area ($\propto l^2$), and the rate of compensating heat generation obtained through eating is proportional to its volume ($\propto l^3$). Thus, as an animal gets smaller, a greater percentage of its energy intake is required for overcoming this inexorable rising heat loss. Moral: A warm-blooded animal smaller than a shrew or hummingbird is simply impractical and hence evolutionarily nonviable – it simply can't eat enough to maintain its body temperature. Those pesky insects, on the other hand, get around this heat-loss constraint by being cold blooded and are amazingly prolific as a result (it is estimated that there are between 1×10^{16} and 1×10^{17} ants on the Earth at present – though transistors still have them beat in sheer numbers!). Even so, nature's scaling laws limit the sizes of animals (in a dry environment) to a minimum of about 25–30 μm in length. Any animal smaller than this simply cannot retain its vital fluids long enough to reproduce, because the larger surface-to-volume ratios lead to much faster evaporation of life juices (bummer!). No independent living organism (this does not include viruses) is smaller than the smallest bacterium, about 0.2 μm long. Water-based creatures can stretch these boundaries a bit, but not dramatically.

Fine. Now let's consider miniaturizing a complex electromechanical system (say, a motor). If the system is reduced in size isomorphically (that is, it is scaled

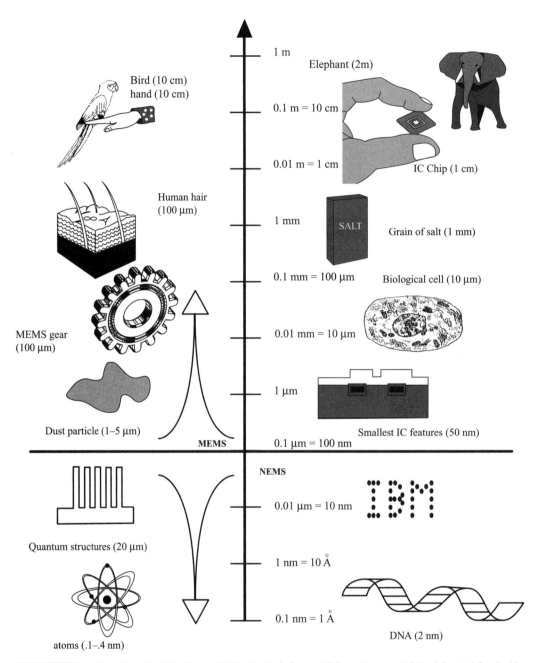

Figure 9.2   Linear size of earthly objects, with the lands of micromachining and nanomachining delineated (inspired by [3]).

down with all of the system dimensions decreased uniformly), the changes in length, surface area, and volume ratios will inevitably alter the relative influence of various physical effects that determine the overall system functionality; often in unusual and unanticipated ways. With MEMS, we zany engineers work hard to exploit these changes for very useful ends. It's what we do best. The land of MEMS (micromachining) and NEMS (nanomachining), within the context of macroscopic objects, is illustrated in Fig. 9.2.

Depending on the intended application, MEMS can have many advantages over conventional macroscopic solutions, including [3]: (1) much smaller energy is expended per function, (2) much less material is needed to implement a given function, (3) we have the ability to massively parallelize a complex system within a tiny form factor, (4) we can dramatically improve system sensitivity and dynamic range, (5) we can discover and exploit new physics and chemistry that are operative only at micron-sized dimensions, (6) we can exploit favorable scaling laws to enhance system performance (e.g., surface tension becomes more important than gravity), (7) we can readily integrate MEMS with conventional electronics, and (8) the net cost can be far lower per unit function. The list could go on. Read: MEMS is a good thing!

9.2 MEMS classifications

MEMSs can be used to do many clever functions, but they primarily serve as either miniaturized "actuators" or "sensors." We'll talk about some concrete examples in a moment. Actuators and sensors are simply objects that convert energy from one form into another. In the case of MEMS actuators, one harnesses the ensuing movement (action – hence the name) of the MEMS to perform some useful function; say, rotate a gear assembly in a MEMS motor. With MEMS sensors, we harness the information gained by what the MEMS tells us about its local environment (what it senses in its surroundings); say, the chemical composition of a gas ambient containing the MEMS sensor. Often a MEMS actuator is used AS a sensor. For example, the deflection of a miniature MEMS cantilever beam can be used to sense acceleration (a MEMS accelerometer). Why electromechanical? Well, both electrical and mechanical effects are commonly exploited in the same MEMS object. In a MEMS accelerator, for instance, the mechanical bending of a MEMS cantilever beam is used to generate a tiny electrical signal that can then be amplified (yep, using transistors!) and subsequently processed by a microprocessor to fire the airbag in your car in the event of a crash. MEMS as a subdiscipline of engineering thus beautifully marries BOTH the electrical engineering and the mechanical engineering fields (as well as chemistry, physics, biology, and increasingly today, medicine). MEMS is the quintessential interdisciplinary merger – an important take-away for budding engineers; no longer is it sufficient to only know your own narrow field of study.

MEMS can be roughly divided into four main categories, based on what moves or rubs and what doesn't [5]:

- Class I: MEMS with no moving parts. Examples include accelerometers, pressure sensors, ink-jet print heads, and strain gauges.
- Class II: MEMS with moving parts, but no rubbing or impacting surfaces. Examples include gyroscopes, comb drive gears, RF resonators, RF filters, and optical switches.

- Class III: MEMS with moving parts and impacting surfaces. Examples include deformable micromirrors, relays, valves, and pumps.
- Class IV: MEMS with moving parts and rubbing and impacting surfaces. Examples include optical switches, shutters, scanners, locks, and discriminators.

As you can easily imagine, the engineering problems faced in actually implementing such complex systems abound. For instance, with microsized surfaces rubbing or impacting each other at very high repetition, "stiction" (just what it sounds like – surfaces sticking together when we don't want them too) can become a nightmare because of the onset of newly magnified surface forces one never deals with in macroscopic systems, including (strongest to weakest) capillary forces, hydrogen bonding forces, electrostatic forces, and van der Waals forces. Ensuring adequate reliability and long life are the key challenges in MEMS technology. For virtually all MEMS, robust packaging also represents a significant challenge, because moving parts at micron dimensions are clearly a recipe for disaster if not properly protected. Complicating this picture is the fact that most MEMSs must operate in vacuum under pristine "hermetic" conditions to avoid frictional forces that are due to the impact of air molecules on their tiny little moving parts. Still, MEMS as a field has come a long way, very quickly, and MEMS as a commercially viable enterprise is well entrenched today, with over $10B (U.S.) in revenue in 2005 [6,7].

9.3 A grab bag of MEMS toys

Go-Go-Gadget-Go![2] Okay, so you now have a good idea that MEMSs are pretty darn tiny and pretty darn complex. But just how tiny, and just how complex? Let's go on a quick tour of some interesting MEMS gadgets to give you a feel for the amazing diversity found in the MEMS world. First, check out Figs. 9.3 and 9.4, which show some classic images of spider mites (refer to Geek Trivia sidebar discussion) crawling around on some MEMS gear trains with their furry little appendages (ugly little cusses: I suppose beauty is indeed in the eye of the beholder).

Geek Trivia: Spider Mites

Family, *Tetranychidae*; order, *Acari*. The unassuming spider mite, barely visible to the naked eye, is not actually an insect, but instead is closely related to spiders, harvestmen (aka daddy longlegs), and ticks. Unlike insects, which have six legs and three body parts, spider mites have eight legs and a one-part body. They also lack wings, antennae, and compound eyes. Individual spider mites, although almost microscopic, can in large numbers inflict serious damage on a wide variety of shade trees, shrubs, and herbaceous plants. Very photogenic too!

[2] Those of you that are Gen-Xer's will surely recognize the *Inspector Gadget* lingo from television. My kids were fans. Okay, me too!

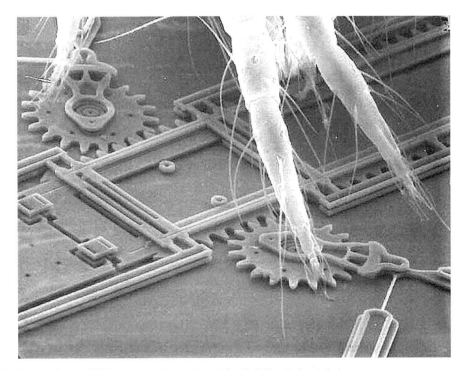

Figure 9.3 Spider mite on a MEMS gear assembly (courtesy of Sandia National Laboratories).

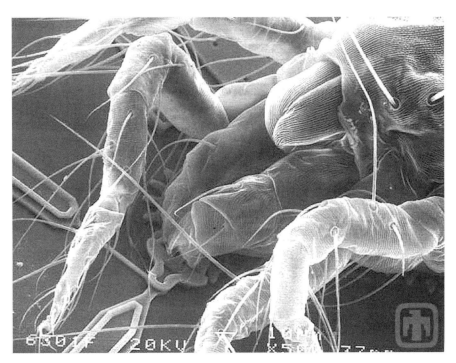

Figure 9.4 Spider mite trampling a MEMS gear assembly (courtesy of Sandia National Laboratories).

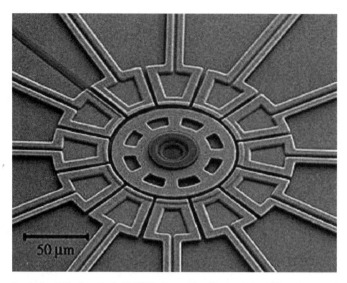

Figure 9.5 An electrostatically actuated MEMS micromotor. The central rotating element of the motor (e.g., the rotor) is the circular structure in the middle held to the substrate by the central bearing. Properly phased voltage potentials are placed on the motor stators (typically 120° advanced in phase sequentially around the stators), which are equally spaced around the perimeter of the rotor, and these applied voltages on the stators cause the central rotor to turn around the bearing at extremely high angular velocities. This MEMS device was made through the MEMS and Nanotechnology Exchange fabrication network (courtesy of the MEMS Exchange).

Our MEMS menagerie also includes [8] (for some very slick movies of some actual operating MEMS gadgets, see [9]): a MEMS motor (Fig. 9.5); a MEMS ratchet (Fig. 9.6); a MEMS thermal actuator (Fig. 9.7); a MEMS triple-piston steam engine (Fig. 9.8); a MEMS sixfold gear chain (Fig. 9.9); MEMS micromirror assemblies (Figs. 9.10 and 9.11); a MEMS indexing motor (Fig. 9.12); a MEMS torsional ratcheting actuator (TRA) (Fig. 9.13); a MEMS electrostatic comb drive (Fig. 9.14); a MEMS vibrational gyroscope (Fig. 9.15); an example of RF MEMS, in this case an indicator–capacitor (LC) bandpass filter (Fig. 9.16); and a looks-great-but-doesn't-do-squat monster MEMS "contraption," no doubt fabricated just for fun by an aspiring graduate student (Fig. 9.17). The list could on for several miles, but you get the idea: lots of cool-looking objects for doing things we normally associate with the macroscopic world (gears, motors, engines, etc.). So just how do we actually make these microtools and toys? Simple – micromachining! Read on.

9.4 Micromachining silicon

MEMS is a surprisingly recent endeavor. Think late 1980s. A little MEMS history? Sure! Interest in silicon micromachining first centered on the development of microsensors (i.e., the application of microfabrication techniques to building sensors) [6]. The first demonstrated microsensor was a pressure sensor. Why? Well, in 1954 it was discovered that the "piezoresistive effect" (squeeze on a

Figure 9.6 A MEMS ratchet assembly (courtesy of Sandia National Laboratories).

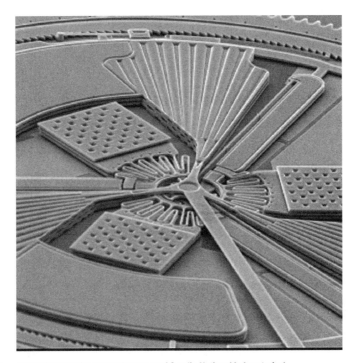

Figure 9.7 A MEMS thermal actuator (courtesy of Sandia National Laboratories).

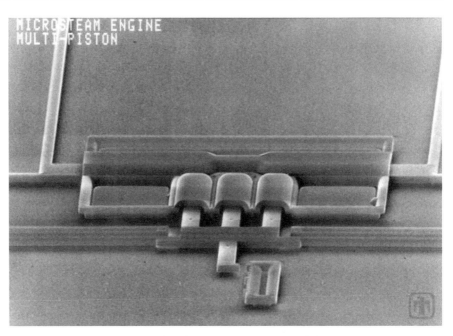

Figure 9.8 Triple-piston microsteam engine. The water inside of three compression cylinders is heated by electric current and vaporizes, pushing the piston out. Capillary forces then retract the piston once the current is removed (courtesy of Sandia National Laboratories).

Figure 9.9 Six-gear MEMS chain (courtesy of Sandia National Laboratories).

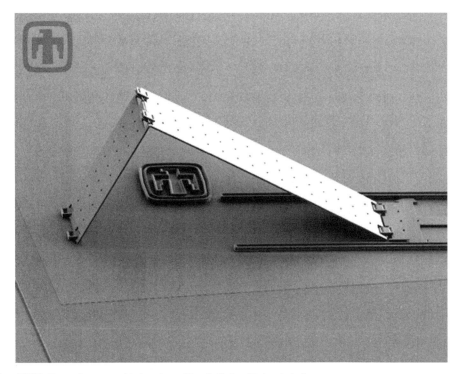

Figure 9.10   MEMS silicon mirror assembly (courtesy of Sandia National Laboratories).

sample and its resistance changes – key obviously to electrically sensing pressure variations) in silicon had the potential to produce silicon strain gauges (a fancy name for pressure sensors) with a gauge factor (the key figure-of-merit for strain-gauge sensitivity) 10 to 20 times larger than those based on *de facto* standard metal films then in use. Silicon strain gauges thus began to be developed commercially as early as 1958, and the first high-volume pressure sensor was marketed by

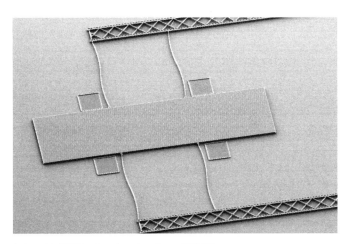

Figure 9.11   A steerable MEMS micromirror (courtesy of MEMX).

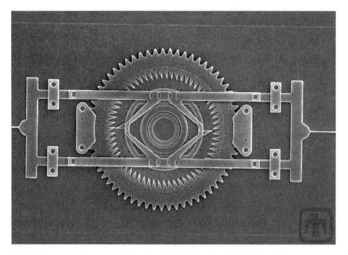

Figure 9.12   MEMS indexing motor (courtesy of Sandia National Laboratories).

National Semiconductor in 1974. MEMS-based silicon pressure sensors are now a billion-dollar industry.

In 1982, the term "micromachining" came into use to designate the fabrication of micromechanical parts (such as pressure-sensor diaphragms or accelerometer suspension beams) for use in silicon microsensors. The micromechanical parts were fabricated by selectively etching areas of the silicon substrate away to leave

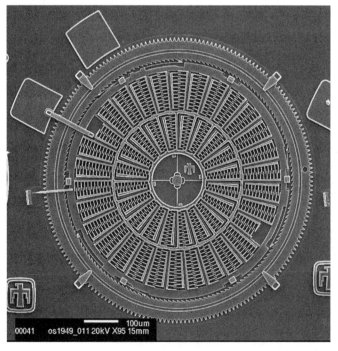

Figure 9.13   MEMS TRA. The TRA uses a rotationally vibrating (oscillating) inner frame to ratchet its surrounding ring gear. Charging and discharging the inner interdigitated comb fingers causes this vibration (courtesy of Sandia National Laboratories).

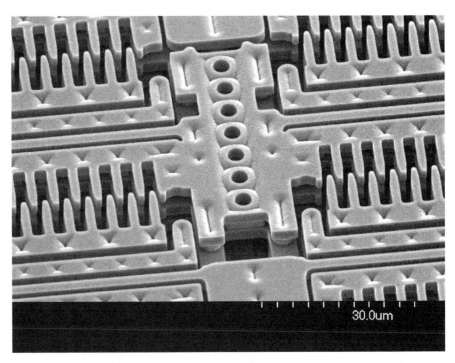

Figure 9.14   A large-force electrostatic MEMS comb drive (courtesy of MEMX).

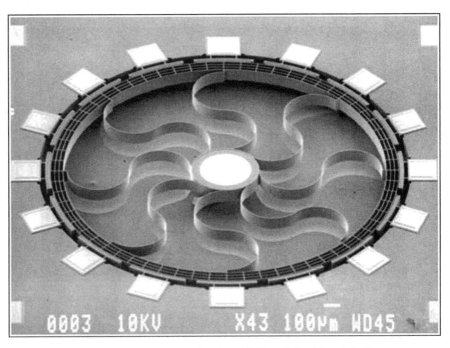

Figure 9.15   A MEMS shell-type vibrational microgyroscope (courtesy of Dr. Farrokh Ayazi, Georgia Tech).

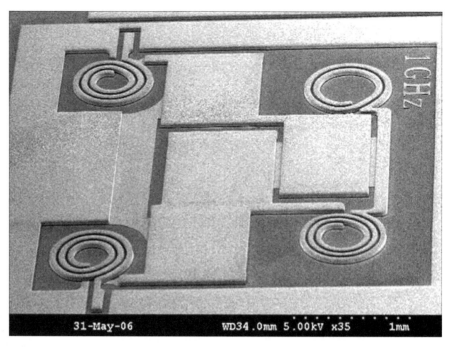

Figure 9.16   A MEMS LC bandpass filter for cognitive radio applications (courtesy of Dr. Farrokh Ayazi, Georgia Tech).

behind the desired geometries. Isotropic wet etching of silicon was developed in the early 1960s for doing this, followed by anisotropic wet etching about 1967. These microelectronics etch techniques form the basis of "bulk micromachining" in MEMS technology. Various etch-stop techniques were subsequently

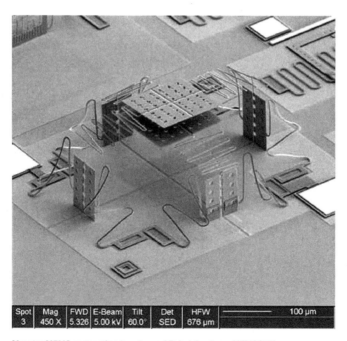

Figure 9.17   Monster MEMS contraption (courtesy of Rob Johnstone, MEMSCAP).

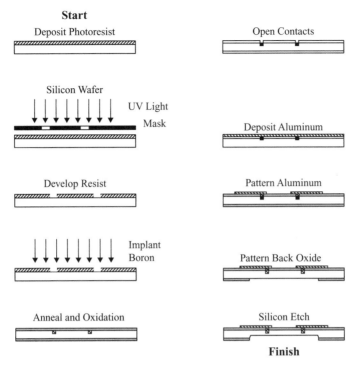

Start

Deposit Photoresist

Silicon Wafer

UV Light

Mask

Develop Resist

Implant Boron

Anneal and Oxidation

Open Contacts

Deposit Aluminum

Pattern Aluminum

Pattern Back Oxide

Silicon Etch

Finish

Figure 9.18   Process flow for bulk micromachining of silicon.

developed to provide process flexibility for use in micromachining. Among these was the pivotal "sacrificial-layer" technique, first demonstrated in 1965, in which a layer of material (say oxide) is deposited between structural layers (say silicon and polysilicon) for mechanical separation and isolation. This layer is removed during the "release" etch to free the structural layers and to allow mechanical appendages (say a polysilicon beam) to move relative to the substrate material (e.g., silicon). The application of the sacrificial-layer technique to micromachining in 1985 gave rise to "surface micromachining," in which the silicon substrate is primarily used as a mechanical support upon which the micromechanical elements are fabricated. Prior to 1987, these micromechanical structures were fairly limited in motion. During 1987–1988, however, a turning point was reached when the first integrated electromechanical mechanisms were demonstrated. MEMS as a field of microelectronics was born, and the rest, as they say, is history!

9.4.1 Bulk micromachining

Bulk micromachining refers to selective removal by etching of the bulk of the silicon substrate, leaving behind the desired MEMS elements. The process flow for bulk micromachining is illustrated in Fig. 9.18, resulting in a suspended silicon "cantilever beam." If that beam is made thin enough, it can flex in response to applied external forces (e.g., acceleration or pressure), serving as a nice little

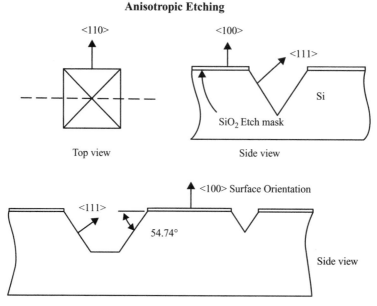

Anisotropic Etching

Figure 9.19 Etching grooves and notches into silicon by use of selective etchants.

MEMS sensor. The defining feature in bulk micromachining is deep etching by use of various etchants, and, in this context, the unique etching properties of silicon are cleverly exploited. Silicon is a 3D crystal, and its etch properties vary widely, depending on the crystal plane one exposes the etchant to. For instance, a KOH (potassium hydroxide) wet etch on the <111> surface of silicon (one the angled crystal planes) can be used to produce self-limiting (i.e., independent of etch time, an excellent feature for manufacturing control) notches, grooves, and other assorted 3D shapes onto the silicon surface (Figs. 9.19 and 9.20). Bulk micromachining was the workhorse for early MEMS technology, but increasingly it has been supplemented by "surface micromachining techniques," which offers a number of advantages.

9.4.2 Surface micromachining

In contrast to bulk micromachining, in which we subtract material to form MEMS gadgets, in surface micromachining, the MEMS features are added to the surface of the silicon crystal, and the wafer serves simply as a "host" or "handle" for the built-up MEMS. In this case, dry etching (say, RIE) is used to define surface features, and then wet etching is used to undercut and "release" the MEMS element. Figure 9.21 shows a polysilicon pressure sensor made by surface micromachining, and a process flow for surface micromachining is depicted in Fig. 9.22. Surface micromachined MEMSs have a number of advantages over their bulk micromachined cousins. For instance, a deposited polysilicon beam serves as the active MEMS element in the surface micromachined MEMS, and we have

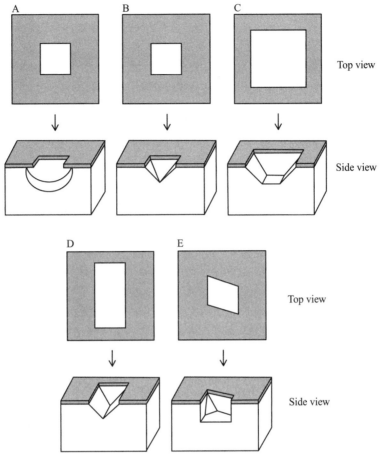

Figure 9.20 A collection of bulk micromachined shapes in silicon.

excellent thickness control in that CVD process, down to submicron thicknesses. This level of thickness control is tough to implement with bulk micromachining, and translates directly into improved sensor response, because the thinner the beam is, the more responsive it is to external forces. One can also easily build up multiple MEMS features, one MEMS layer at a time, using surface micromachining techniques, and thus most of the more sophisticated MEMS gadgets are constructed this way (most of the MEMS gadgets previously shown were fabricated with surface micromachining). A compelling feature of MEMS technology may have escaped you: Because MEMS and microelectronics are made with the same toolkit of fabrication techniques, we can very easily combine our MEMS gadget with transistor-based electronics for on-die signal detection, amplification, and manipulation, the bread-and-butter of conventional silicon-based microelectronics. Hence, MEMS + electronics represents a powerful merger of two compelling ideas: MEMS for sensors–actuators, married to microelectronics, sometimes referred to as "mechatronics" (mechanics + electronics). An example of this mechatronics merger is illustrated in Fig. 9.23 and is typical of virtually

Surface Micromachining

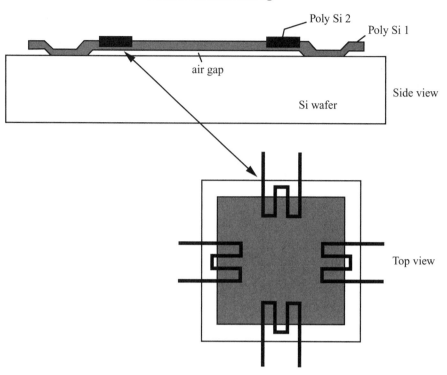

Figure 9.21 An example of surface micromachining (inspired by [3]).

MEMS Suspended Diaphragm

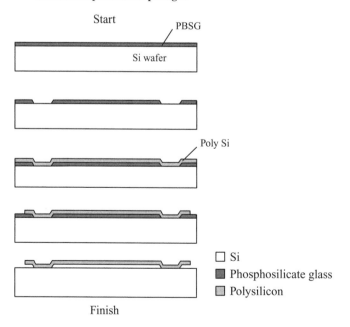

Figure 9.22 Process flow for surface micromachining of silicon (inspired by [3]).

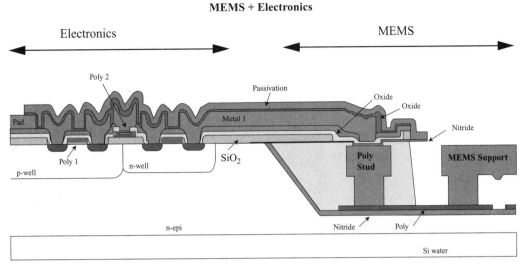

MEMS + Electronics

Monolithic integration of CMOS electronics with MEMS devices (inspired by [3]).

all commercial MEMS-based widgets one can buy today. Now on to some cool, real-world examples [3,10].

9.5 Cool app #1 – MEMS accelerometers

Whether you are aware of it or not, MEMS gadgets surround you and are a part of your daily life. A prevalent example resides in the car you drive each day. The modern automobile is actually chock full of MEMS, as are almost all modern vehicles (planes, trains, buses, spacecraft, you name it). As shown in Fig. 9.24, MEMS sensors abound in cars and are used in many and varied ways, including torque sensors, position sensors, accelerometers, temperature sensors, oxygen sensors, humidity sensors, light sensors, pressure sensors, load–force sensors, air-flow sensors, etc. [3]. As a concrete example, let's focus on the ubiquitous Mr. Accelerometer. An accelerometer is just what its name suggests; a MEMS gadget for sensing acceleration. In the automotive context, there are many occasions during which we would like to detect, in real time, the acceleration of our car. One potentially life-saving example is in the airbag system, those modern safeguards for reckless driving and unexpected accidents.

Know how airbags work? Quite simple, actually. When you crash into a brick wall (sorry!), the objects inside the car (you!) are still moving at high speed, and contact with your now-abruptly-stopped steering wheel is likely to do you some serious harm. The MEMS accelerometer in the front end of the car and attached to the airbag system detects the sudden deceleration when you hit the wall, amplifies and massages the electrical signal and sends it rapidly (remember transistors are FAST) to an explosive firing mechanism that uses sodium azide (NaN_3) to rapidly inflate the airbag and insert it between you and the steering wheel before you

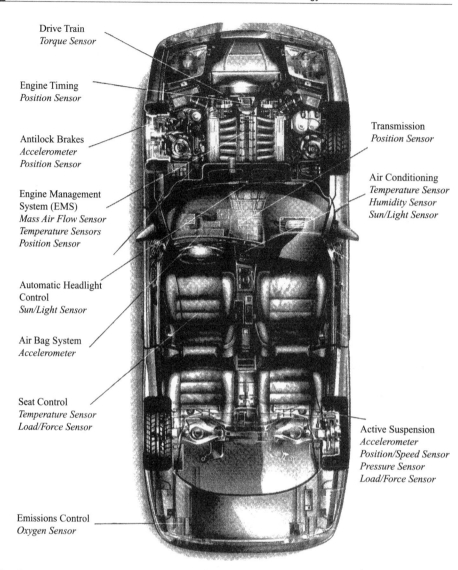

Drive Train
Torque Sensor

Engine Timing
Position Sensor

Antilock Brakes
Accelerometer
Position Sensor

Engine Management
System (EMS)
Mass Air Flow Sensor
Temperature Sensors
Position Sensor

Automatic Headlight
Control
Sun/Light Sensor

Air Bag System
Accelerometer

Seat Control
Temperature Sensor
Load/Force Sensor

Emissions Control
Oxygen Sensor

Transmission
Position Sensor

Air Conditioning
Temperature Sensor
Humidity Sensor
Sun/Light Sensor

Active Suspension
Accelerometer
Position/Speed Sensor
Pressure Sensor
Load/Force Sensor

Figure 9.24   The menagerie of automotive sensors (inspired by [3]).

reach it. How fast must this happen? Depends on your impact speed obviously, but current-generation airbags can deploy within 30–50 ms! Plenty fast enough for any survivable crash. Pretty clever, huh? For a brief history of the airbag, refer to the Geek Trivia sidebar.

Beginning in 1998, all new cars have been required to have airbags on both driver and passenger sides. Statistics show that airbags reduce the risk of dying in a direct frontal crash by about 30%, no small improvement, and are credited with saving at least 3,100 lives in the United States alone, according to estimates from the National Highway Traffic Safety Administration (NHTSA). You can (and should) thank your resident MEMS accelerometer for his or her always watchful guardian angel protection!

Geek Trivia: A Little Airbag History

The airbag was invented by John W. Hetrick of Newport, PA, in 1951 (patented in 1952) [1]. Hetrick, an ex-naval engineer, came up with the idea to help protect his own family (teenage drivers?). Building upon similar devices used in airplanes as early as the 1940s, his early airbags were large and bulky air-filled bladders. Settling on a lightning-quick impact sensor proved to be difficult, however. Inventor Allen K. Breed developed a crude but effective ball-in-tube sensor for crash detection and marketed this innovation to Chrysler in 1967. The automobile airbag was soon born. Ford built an experimental fleet of cars with airbags in 1971, and General Motors followed with a fleet of 1,000 experimental vehicles in 1973, and these Chevrolet cars equipped with dual airbags were sold to the public through GM dealers two years later. In 1980, Mercedes-Benz first introduced the airbag in Germany as an option on its high-end S-Class car. In the Mercedes airbag system, the impact sensors would first tighten the seat belts and then deploy the airbag during a crash (pretty clever). In 1987 the Porsche 944 turbo (don't you wish!) became the first car in the world to have both driver and passenger airbags as standard equipment. The first airbag on a Japanese car, the Honda (Acura) Legend, also came out in 1987. The rest is history.

So how, exactly, does Mr. Accelerometer work? Figure 9.25 shows a commercial two-axis (meaning, it senses acceleration in the 2D x–y plane) MEMS accelerometer. As with most microelectronics widgets, once it is packaged it looks deceptively simple and you have no clue what actually lurks inside. If you pry off the covers, however, you would see the integrated circuit shown in Fig. 9.26, in this case the Analog Devices ADXL 50. This is a nice example of the merger of a MEMS actuator–sensor and electronics (mechatronics). The MEMS accelerometer is the circular centerpiece, and everything else is support electronics. It is a beautifully sophisticated little subsystem, unfortunately hidden under the cover of its ugly opaque package. If we zero in on the MEMS, it would look like that shown in Fig. 9.27, a surface micromachined comb-like plate, with fixed polysilicon fingers attached to the surface of the silicon wafer, alternated with a released (hence movable) polysilicon inertial mass anchored on four corners by a flexible polysilicon spring. Presto! A capacitive MEMS accelerometer. Why capacitive? Look again. There is a tiny capacitor formed between each of the 2-μm-thick conductive polysilicon inertial mass fingers and the conductive stationary polysilicon fingers (the dielectric is just the vacuum gap in between). Because the inertial mass is free to move with the applied acceleration, the distance between the capacitor fingers (d) will change, increasing or decreasing the capacitance ($C = \epsilon A/d$). Set up some circuitry to dynamically sense and amplify that capacitance change, and we are in business! In the ADXL 50, rated to $+/-$ 5 g, the equilibrium comb capacitance is tiny, only 100 fF, with an inertial mass of 0.3 μg (!), and a capacitance change under 5-g acceleration of only 0.1 fF, with a 1-MHz (1,000,000 times a second) sensing frequency. Sounds like an awfully

Figure 9.25 Commercial two-axis MEMS accelerometers. (Copyright Analog Devices, Inc., all rights reserved. Used with permission.) (See color plate 27.)

tiny capacitance change, but it gets the job done! This is a nice testament to the utility of microminiaturization. Tiny MEMS sensors can sense big effects with very small changes. Care for another cool example? Sure!

9.6 Cool app #2 – MEMS micromirror displays

If you are richer than me, you may well already have a large-screen high-definition TV (HDTV) display (okay, I'm jealous!). Depending on your needs (and $!), you can buy either a LCD, a plasma display, or something called a DLP™ (Digital Light Projection) HDTV display (Fig. 9.28). Interestingly enough, this latter type of HDTV display is based on a sophisticated MEMS chip *tour de force* (Fig. 9.29). Read on. In 1987, from seemingly out of nowhere, Larry J. Hornbeck of Texas Instruments introduced the Digital Micromirror Device (DMD™), patented on October 7, 1986 (U.S. Patent 4,615,595) [3]. A DMD is a MEMS gadget consisting of a 2D array of steerable MEMS optical-mirror pixels fabricated on a silicon substrate by surface micromachining. Each MEMS pixel is made up of a reflective aluminum micromirror supported on a central post, and this post is mounted on a lower aluminum platform, the mirror "yoke" (Figs. 9.30 and 9.31). The yoke is then suspended above the silicon substrate by compliant L-shaped hinges anchored to the substrate by two stationary posts. As you might imagine, control electronics is then embedded in the silicon wafer under each

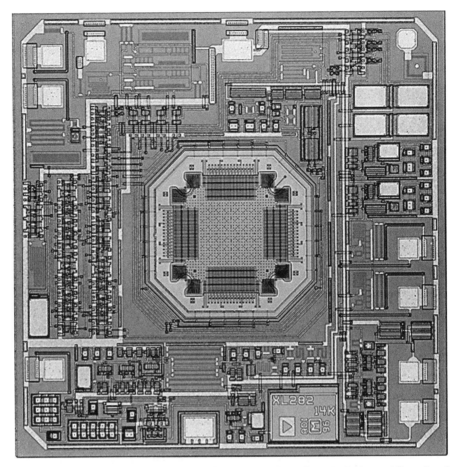

Figure 9.26 The Analog Devices ADXL 50 accelerometer. This two-axis MEMS accelerometer is shown at the center of the die, surrounded by support electronics. It can measure both static and dynamic acceleration. (Copyright Analog Devices, Inc., all rights reserved. Used with permission.) (See color plate 28.)

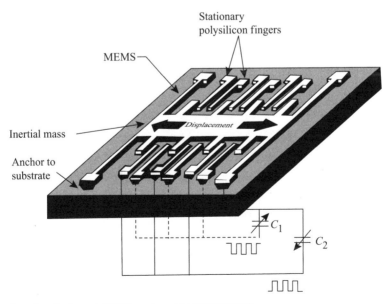

Figure 9.27 Operation of a two-axis MEMS accelerometer (inspired by [3]).

Figure 9.28   An example of a HDTV utilizing DLP MEMS technology.

Figure 9.29   Texas Instruments' DLP chip and package (courtesy of Texas Instruments, Inc.).

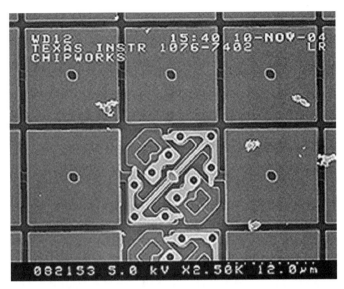

Figure 9.30   Texas Instruments' DLP MEMS micromirror assembly, with two mirrors removed to show the underlying actuation assemblies (courtesy of Texas Instruments, Inc.).

pixel. The two bias electrodes can tilt the micromirror either $+10°$ or $-10°$ with the application of 24 V between one of the electrodes and the yoke (Figs. 9.32 and 9.33). In other words, we can controllably steer the micromirror!

How do we use it to make a display? Well, off-axis illumination of the micromirror reflects into the projection lens only when the mirror is in its $+10°$ state, producing a very bright appearance (the pixel is "ON"). In the flat position (no bias applied) or the $-10°$ state the pixel does not reflect the light source and remains dark ("OFF"). Cool, huh? The DMD MEMS IC contains...hold

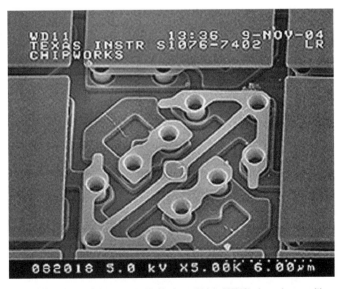

Figure 9.31   A blowup of Texas Instruments' DLP display utilizing MEMS micromirrors, with one mirror removed to show the underlying actuation assembly. For a cool movie showing how it actually works, visit *http://www.dlp.com/tech/what.aspx.* (Courtesy of Texas Instruments, Inc.)

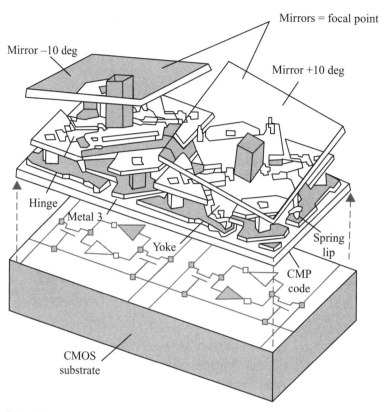

Figure 9.32   Schematic cross section of the DLP MEMS micromirror assembly (inspired by [3]).

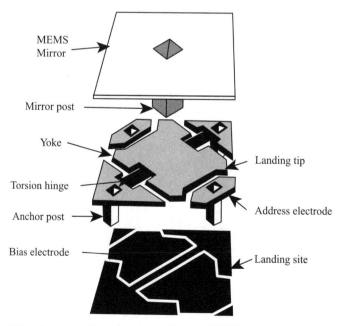

Figure 9.33   Schematic cross-section actuator assembly of the DLP MEMS micromirror (inspired by [3]).

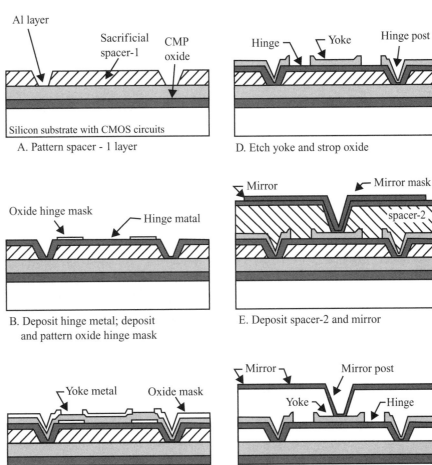

A. Pattern spacer - 1 layer

B. Deposit hinge metal; deposit
   and pattern oxide hinge mask

C. Deposit yoke and pattern
   yoke oxide mask

D. Etch yoke and strop oxide

E. Deposit spacer-2 and mirror

F. Pattern mirror and
   etch sacrificial spacers

Figure 9.34   Fabrication flow for making the DLP MEMS micromirror (inspired by [3]).

your breath . . . 442,000 of these MEMS micromirror pixels, each of which can be toggled on or off at 100,000 times per second! Gray-scale images can be achieved either with appropriate toggling of multiple adjacent mirrors, or even with rapid toggling of a single pixel between its on and off states. Color can be generated by shining the incident light through a rapidly rotating disk divided into red, yellow, and blue (primary) color segments. A separate image is generated for each color, timed to appear when the appropriate color segment is in front of the light source. Pretty amazing!

Figure 9.34 shows an abbreviated process flow for making the DMD MEMS micromirror. Does the whole assembly of 420,000 MEMS micromirrors seem impossibly fragile? It's not. This is another case in which miniaturization works on our side. Because the mass of the micromirror is tiny, it can withstand up to a 1,500-g mechanical shock, and the mean time between failures (MTBF) is more than 100,000 h (that's a lot of football games!). Not only is MEMS-based

DMD technology helping to bring affordable HDTV projectors into the home, but they are also the centerpiece of next-generation digital cinema for large-screen theaters (as you may recall, first introduced in the *Star Wars* saga by Lucasfilm). As a fun stump-your-friends trivia footnote, Hornbeck, obviously an engineer by trade and not an actor, was actually awarded an Emmy for his MEMS invention! Moral to the story: Next time you walk by those cool HDTV displays in the store – think MEMS!

9.7 Cool app #3 – BioMEMS

In 2003, per-capita spending on health care in the United States was $5,635, and the current global health market is estimated to be well over $2 trillion, including over $100 billion in medical instruments and over $300 billion in pharmaceuticals. No surprise then that there are some real financial incentives for figuring a good way to introduce MEMS into the biomedical market. Presto! BioMEMS! BioMEMS is very rapidly evolving application sphere, but I'd like to at least give you a feel for some of the excitement currently being generated from the application of MEMS technology to the biomedical domain. Read: VERY hot topic!

Because most traditional MEMS devices are made from hard materials (think silicon), they are not especially conducive for use in implantable biomedical devices, which often require compliant materials like polymers for improved compatibility with biological tissues. In addition, if we want to use MEMS for diagnostic sensing the components of blood, say, we need to bring on board a host of new biological materials (enzymes, proteins, etc.), and thus the development of BioMEMS technology centers on extending traditional MEMS fabrication techniques to include these new types of biocompatible materials.

BioMEMS applications currently include pacemakers, angioplasty catheters, blood analyzers, genetic (DNA) testing, electrophoresis (an electrical technique used in molecular biology and medicine for separating and characterizing proteins, nucleic acids, and even sub-cellular-sized particles like viruses and small organelles), liquid-handling systems, and even surgical cutting tools. BioMEMS can be roughly categorized into three groups: (1) diagnostic BioMEMS; (2) therapeutic BioMEMS, and (3) surgical BioMEMS [5,11].

Diagnostic BioMEMS is far and away the most mature market sector for BioMEMS technology and is typically composed of a microfluidic MEMS assembly (a way to pump tiny amounts of fluids from point A to point B by using a MEMS actuator) and some sort of MEMS sensor. As an an example, consider a diagnostic BioMEMS microfluidic chamber used in cytometry (blood analysis), shown in Fig. 9.35. This type of (disposable) BioMEMS assembly can then be embedded in a handheld clinical analyzer, like the i-STAT portable clinical analyzer (PCA) (Fig. 9.36) for whole blood analysis [3] (feels kind of like something Bones McCoy would carry on "Star Trek"!). These types of sophisticated

MEMS Fluid Chamber for Cytometry

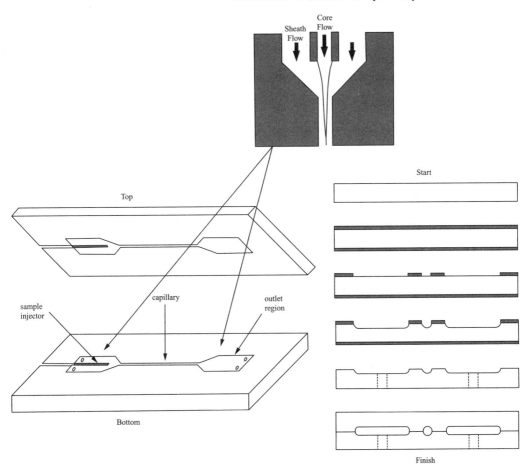

Figure 9.35   A MEMS-based microfluidic chamber for cytometry (inspired by [3]).

BioMEMS diagnostic tools are often referred to as "labs-on-a-chip." Another rapidly emerging area for diagnostic BioMEMS is in genetic testing. "DNA chips" can be used to look at one or many of our 24,000 different genes. In a DNA chip, "reference fragments" of DNA (for example, gene fragments known to be altered by a particular cancer) are placed on a thumbprint-sized BioMEMS DNA chip. Which versions of the genes are present in a sample of DNA from a patient can then be quickly analyzed, simply by applying the patient's DNA sample to the DNA chip and seeing if they chemically react with the reference fragments. The test sample of DNA is first "unzipped" (i.e., the DNA double helix is unwound and pulled apart) by heating, then "labeled" with a chemical dye, and literally washed over the DNA chip. The DNA chip relies on the natural ability of a DNA strand to stick to a "complementary copy" of itself [e.g., adenine (A) always pairs up with thymine (T) to form a base pair, cytosine (C) always pairs up with guanine (G) to form a base pair, etc.]. Any gene in the sample that matches a fragment on the chip will stick to it, showing up as a brightly colored dot on

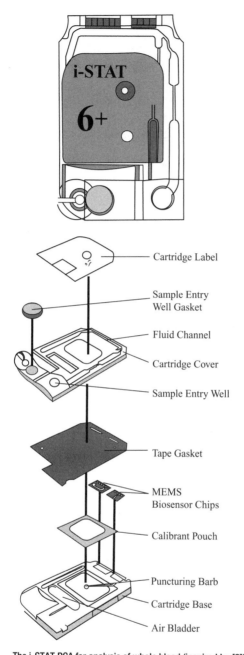

- Cartridge Label
- Sample Entry Well Gasket
- Fluid Channel
- Cartridge Cover
- Sample Entry Well
- Tape Gasket
- MEMS Biosensor Chips
- Calibrant Pouch
- Puncturing Barb
- Cartridge Base
- Air Bladder

Figure 9.36   The i-STAT PCA for analysis of whole blood (inspired by [3]).

the DNA chip (Fig. 9.37). Near-instant DNA assaying with a cheap, disposable BioMEMS! Pretty cool.

Most therapeutic BioMEMS research centers on developing novel implantable or transdermal (through the skin) drug-delivery systems for patients suffering from chronic diseases. In this case, a BioMEMS micropump, with on-board electronic control, can precisely deliver tiny amounts of a given drug to a particular location. A logical target is an implantable insulin delivery system for diabetics.

Finally, BioMEMS can also be attractive for a variety of ultra-delicate surgical instruments. An "inchworm actuator" [5] uses a piezoelectric BioMEMS for

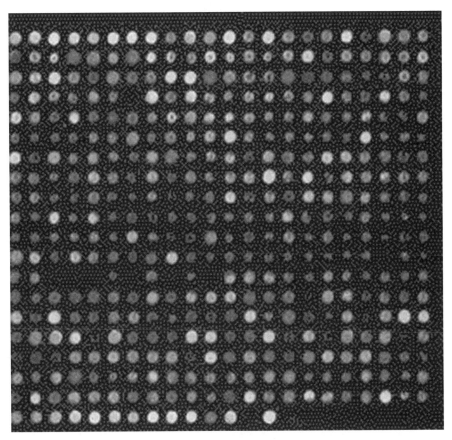

Figure 9.37 A typical DNA chip used in cancer gene analysis (courtesy of National Human Genome Institute). (See color plate 29.)

precision eye surgery. Ultra-sharp 3D BioMEMS brain probe arrays have also garnered interest for precision electrical stimulation during brain surgery.

This brief introduction to MEMS gadgetry has scratched only the surface of this fascinating field. Key take-away? Sure. MEMS technology is a direct benefactor and unanticipated descendent of the techniques developed for fabricating ICs, and serves as a nice example of powerful manner in which an advanced technology can morph from its intended application sphere to launch other, often nonobvious, fields. The driving force? Good ole human ingenuity. Although MEMS research has been a mainstay for some time now, MEMS as a commercial enterprise is literally exploding before our eyes, and you ain't seen nothin' yet! *Fantastic Voyage* suddenly seems much less fantastic. Stay tuned.

References and notes

1. See, for instance, Wikipedia: *http://en.wikipedia.org*.
2. T. Trimmer, "Micromechanical systems," in *Proceedings, Integrated Micro-Motion Systems: Micromachining, Control, and Applications (3rd Toyota Conference)*, October 1990, Aichi, Japan, pp. 1–15.

3. M. J. Madou, *Fundamentals of Microfabrication: The Science of Miniaturization,* 2nd ed, CRC Press, Boca Raton, FL, 2002.

4. D. W. Thompson, *On Growth and Form,* Cambridge University Press, New York, 1992.

5. M. Taya, *Electronic Composites: Modeling, Characterization, Processing, and MEMS Applications,* Cambridge University Press, New York, 2005.

6. S. A. Vittorio, "MicroElectroMechanical systems (MEMS)," *http://www.csa.com/ discoveryguides/mems/overview.php.*

7. The MEMS Exchange, a clearinghouse for companies that provide various MEMS services: *https://www.mems-exchange.org/.*

8. See, for instance, Sandia National Laboratories: *http://www.mems.sandia.gov.*

9. For a cool set of MEMS movies (really!), showing MEMS-in-motion action shots, visit *http://www.memx.com/products.htm* and *http://www.memx.com/ movie_gallery.htm.*

10. M. Gad-el-Hak (Editor), *The MEMS Handbook,* CRC Press, Boca Raton, FL, 2002.

11. D. L. Polla, A. G. Erdman, and W. P. Robbins, *Annals of Biomedical Engineering,* Vol. 2, pp. 551–576, 2000.

10 Widget Deconstruction #3: GPS

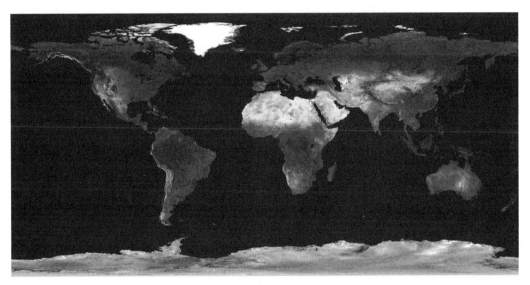

Our Earth (courtesy of NASA). (See color plate 30.)

If a Man Wishes To Be Sure
Of The Road He Treads On,
He Must Close His Eyes
And Walk In the Dark.

Saint John of the Cross

Location. Location. Location. It is quintessentially human to wonder where the heck we are in the grand scheme of things. Our location defines where we live; how much sunshine we get in October; when we plant our tomatoes; how we interact with our neighbors; how history ebbs and flows and intersects our lives; ultimately, how we know and understand ourselves. We humans are natural-born explorers. Adventurers. Wayfarers. We're awfully curious creatures. We wonder if the Earth is truly the center of the universe; or perhaps it is instead the Sun? We wonder which is the best path for traveling across a barren wilderness. We wonder if we'll indeed sail right off the Earth if we get too close to the ocean's edge. And, not surprisingly, we draw detailed maps to guide us on our many and

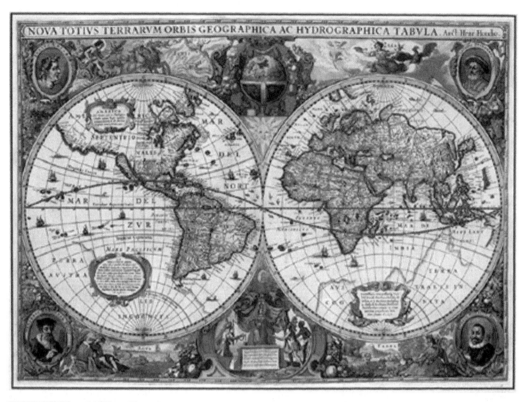

Figure 10.1 A 1629 world map by cartographers Henricus Hondius and Jan Jansson. This map was a revision of the famous Mercator–Hondius world map created 35 years earlier. Portraits of Julius Caesar and Claudius Ptolemy are featured in the upper corners, with portraits of the father of Henricus Hondius (Jodocus, the Elder) and Gerard Mercator in the lower corners (courtesy of U.S. Library of congress).

varied (con)quests. We create great works of art (Fig. 10.1), we pull out the Rand McNally atlas, and today, of course, we go to Map Quest and print out directions!

Homo audens. Man the bold adventurer. Man the daredevil. Figure 10.2 shows arguably the pinnacle picture of humankind's travel journal to date. To look back and see our own world from the confines of another. Majestic conquest and technological triumph. Stunning in its implications! It is patently ASSUMED, today, that we trivially possess precise knowledge of our own location on the surface of planet Earth. And we do. But it was not always so. Nope, it was otn so until very, very recently in fact.

Location. Location. Location. One hundred years ago, if you asked a geographer where I am sitting and writing this book, the answer would have been 33° N latitude, 84° W longitude; for the curious among you, in the Technology Square Research Building, on the campus of Georgia Tech, at 85 5th Street, N.W., in Atlanta, Georgia, USA. (For a quick reminder on latitude and longitude, see the sidebar discussion.) Ask that same question today and you would get a very different answer. I am actually located at 33° 46′ 39.05″ N, 84° 23′ 25.02″ W Some improvement in accuracy, huh?! This coordinate is precise enough that you can hop onto Google Earth and zoom in to see my building; heck, even the

Figure 10.2 Arguably the most famous picture ever taken of the Earth. Here's the scoop: In December of 1968, the Apollo 8 crew flew from the Earth to the Moon and back again. Frank Borman, James Lovell, and William Anders were launched atop a Saturn V rocket on December 21, circled the Moon 10 times in their command module, and returned to Earth on December 27. The Apollo 8 mission's impressive list of firsts includes the first humans to journey to the Moon, the first manned flight using the Saturn V rocket, and the first to photograph the Earth from deep space. As the Apollo 8 command module rounded the far side of the Moon, the crew could look toward the lunar horizon and see the Earth appear to rise, because of their spacecraft's orbital motion. A now-famous picture resulted of a distant blue Earth rising above the lunar surface (courtesy of NASA). (See color plate 31.)

corner of the 5th floor where my office is located! Please call before dropping by! How is it that we have suddenly gotten so amazingly accurate with our locating? Simple answer – GPS. The Global Positioning System [2–4]. A very slick, world-flattening piece of technology enabled by Mr. Transistor. Curious how it does its magic? You should be. Read on.

Geek Trivia: Latitude and Longitude

Latitude gives the north–south geographic coordinate of a particular spot on the Earth (Fig. 10.3). Lines of latitude are the horizontal lines shown running east to west on maps. Classically, latitude is an angular measurement expressed in degrees, minutes (60 minutes per arc degree), seconds (60 seconds per arc minute), with units (°, ′, ″), ranging from 0° at the Equator (low latitude) to 90° at the poles (90° N for the North Pole or 90° S for the South Pole; high latitude). Some geographic trivia? Sure! Some important (east–west) circles of latitude include the following:

Arctic Circle, 66° 33′ 39″ N
Tropic of Cancer, 23° 26′ 21″ N
Tropic of Capricorn, 23° 26′ 21″ S
Antarctic Circle, 66° 33″ 39″ S

(continued)

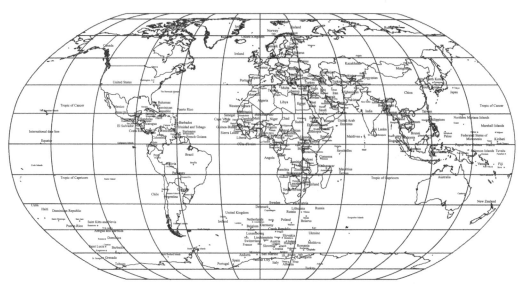

Figure 10.3 A map of the world, showing lines of latitude and longitude.

Geek Trivia: Latitude and Longitude (*continued*)

Longitude, on the other hand, gives the east–west (side-to-side) geographic coordinate of a particular spot on the Earth. Unlike latitude, which has the Equator as a natural reference circle, there is no natural starting position for measuring longitude. Therefore a reference meridian had to be chosen. Although British cartographers had long used the meridian (north–south circles connecting pole to pole) passing through the Royal Observatory in Greenwich, England (near London), other references were used elsewhere throughout history (for often-obvious historical reasons – you can almost guess the period), including Rome, Jerusalem, Saint Petersburg, Pisa, Paris, Copenhagen, Philadelphia, and Washington. In 1884, however, the International Meridian Conference, recognizing the need for a universal standard (think of all those maps that would have to be redrawn!), adopted the Greenwich meridian as the "prime" meridian, or the zero point of longitude. It stuck. Just as for latitude, longitude is classically given as an angular measurement ($^\circ$, $'$, $''$) ranging from 0° at the Greenwich (prime) meridian to $+180^\circ$ eastward and -180° westward. Cartography in action!

10.1 With a broad brush

GPS is the only fully functional global navigation system in town and utilizes a "constellation" of 24 medium Earth orbit (MEO – see sidebar discussion on space weather) communications satellites that broadcast precise RF signals back to the Earth. The GPS satellite constellation enables a GPS receiver anywhere on the surface of the Earth to determine its precise location, speed, direction, and time. Slick! Similar systems currently under development in other countries include the Russian GLONASS system; the European Union's GALILEO system[1]; the proposed COMPASS system of China; and the IRNSS system of India. Originally envisioned and developed by the U.S. Department of Defense (DoD), GPS is officially named NAVSTAR GPS (contrary to popular belief, NAVSTAR is not an acronym, but rather simply a name given by Mr. John Walsh, a key budget planner for the DoD GPS program). The GPS satellite constellation is managed by the U.S. Air Force 50th Space Wing, and it isn't cheap to run. GPS currently costs approximately $750 million per year, including operations, the replacement of aging satellites, and research and development for new GPS systems. Following the very sad shootdown of Korean Air Lines (KAL) flight 007 by the Russians in 1983, U.S. President Ronald Reagan issued a bold directive making the GPS system available for civilian use (for free!), and since that time GPS has become THE ubiquitous navigational aid worldwide (see sidebar discussion on the history of GPS).

So what exactly is involved? Well, the GPS satellites are continuously transmitting coded position and timing information at (several) RFs to all locations on the Earth. The GPS receiver in your hand must pick up (receive, lock onto) those transmitted RF signals, decode them, and then use the information within them to calculate its position on the Earth. How? Well, the GPS receiver first cleverly calculates the time it takes for the RF signals transmitted by each GPS satellite it can "see" to reach the receiver (clearly the satellite itself must know precisely its own location and transmit that as a part of the signal). Multiplying the time it takes the signal to get from satellite to receiver by the speed the RF signal is traveling (roughly the speed of light), the receiver can then determine precisely how far the receiver is from EACH of the transmitting satellites. With these accurate distance data, the receiver can then mathematically "triangulate" [e.g., the intersection of three ("tri" – think triangle) satellite transmission spheres, say, as depicted in Fig. 10.4] to calculate the receiver's precise location on the Earth. Now, mathematically translate that location to an appropriate coordinate system, and then spit out the location coordinates (or better yet, show it visually on a map). Oh, and also use lots of HIGHLY complex and very subtle error correction schemes along the way to improve the accuracy (read: heavy math). Done! You now know where you are. For me, 33° 46' 39.05" N, 84° 23' 25.02" W, in downtown Atlanta.

In only a few short years, GPS has become much more than a simple navigational tool and is now firmly embedded in the technology fabric of planet Earth. Applications are growing by the day [1]. Although GPS was originally conceived and funded as a military project, it is now officially considered a "dual-use" (military–civilian) technology. A few military examples? Sure! Handheld GPS units allow soldiers and commanders to find objectives in the dark or in unfamiliar territory, and to accurately coordinate the movement of troops and supplies, as well as create maps for reconnaissance. These military GPS receivers are called the "Soldier's Digital Assistant" (really!). Many military weapons systems use

[1] For an "interesting" (read: not very flattering) perspective on GALILEO's development, see [5].

GPS Locating

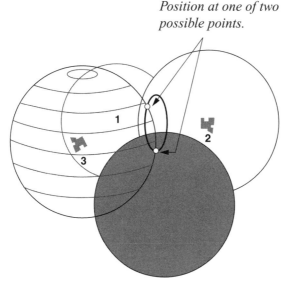

Figure 10.4 Illustration of two possible intersecting points from signals eminating from three satellites.

GPS to track potential ground and air targets before they are flagged as hostile and destroyed. These weapons systems pass the GPS coordinates of targets to in-flight precision-guided munitions (aka "smart bombs"). Even 155-mm howitzer shells can have embedded GPS receivers and are able to withstand accelerations of 12,000 g. Pretty amazing. The GPS satellites themselves carry nuclear detonation detectors, which form the backbone of the U.S. monitoring system for the Nuclear Nonproliferation Treaty.

Civilian examples of GPS? Plenty! Its ability to determine absolute location allows GPS receivers to perform as surveying tools or as aids to commercial navigation. Portable GPS receivers are now very common (Fig. 10.5), just look in your car. Heck, they even help land aircraft in the dark! The capacity to determine relative movement enables a GPS receiver to calculate both local velocity and orientation, which is useful in tracking ships, cars, teenagers, you name it. Being able to synchronize clocks to exacting standards is critical in large communications and observation systems. A prevalent example is a CDMA cellular phone system. Each base station has a GPS timing receiver to synchronize its spectra spreading codes with other CDMA base stations, to facilitate better intercell handoff and support hybrid GPS–CDMA positioning for emergency calls and other applications. Whether you realize it or not, GPS is now rapidly moving into your cell phones, and in 2008 all major vendors announced GPS-enabled GSM/WCDMA handsets. GPS survey equipment has revolutionized plate tectonics by directly measuring the motion of faults in earthquakes. And let's not forget the burgeoning role of GPS in cars – no need to get lost anymore! Want to know how GPS actually works? Read on.

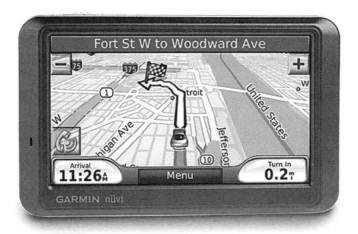

Technical Aside #1: Space Weather

Weather in space? You bet. So is it rainy or sunny out today in orbit? Space weather, in the context of GPS, actually refers to the highly dynamic and ever-changing environmental conditions found in Earth orbit. Because the GPS constellation resides in orbit, space weather is a big deal, as it is in fact for any electronic system orbiting the Earth [the International Space Station (ISS), the Space Shuttle, TV and telephone broadcast satellites, you name it]. Fundamentally, space weather is a direct consequence of the interaction of the Sun, which continuously streams gobs of highly energetic (10s to 1000s of mega electron volts) particles (the "solar wind") from its fusion furnace, with the Earth's own magnetic field, which subsequently traps certain types of energetic particles (e.g., electrons and protons) around the Earth, forming the so-called Van Allen radiation "belts". A familiar consequence of the Van Allen belts are the Northern and Southern Lights (the famous *auroras*), as shown in Fig. 10.6. Beautiful, yes, but a major disaster for the electronics on a satellite! It can get worse. A so-called "coronal mass ejection" (CME – sounds ominous!) from a solar flare can easily take down all the satellites it finds in orbit. Suddenly your cell phone goes blank and your GPS stares at you while it searches furtively to reacquire its lock. Trust me, astronauts don't want to be "outside" walking around during a CME either! You may know that solar "events" are correlated with sunspot activity and come in 11-year cycles. Hold on tight. We are headed back into a solar maximum in the next few years, and the consequences for our satellites have lots of very bright people quite worried. A direct consequence of space weather is that conventional microelectronic components developed for us on the ground cannot be simply flown into space.

(continued)

Technical Aside #1: Space Weather (*continued*)

Reason? The ubiquitous radiation (which by the way cannot be easily shielded – tough to fly lead vests for everything!) experienced during each orbit will shift the characteristics of the requisite transistors, eventually killing the electronic components and then the systems. No electronics, no GPS, no communication satellites, (aka comm-sats), no nothing. Instead one has to "radiation harden" the electronics, a very complex and highly costly endeavor, but required for doing business in space. Even then, a sufficiently strong solar event can still take down a satellite if the weather turns especially sour. The severity of the effects of the space weather depends on the exact orbit your "bird" (satellite-speak) is circling in, and, as mentioned, GPS occupies MEO. So where is that exactly? Well, MEO is the region of space around the Earth above low Earth orbit (LEO), which is located at up to 2,000 km (1,243 mi) above the Earth's surface, and below geostationary orbit (GEO), which is located 35,786 km (22,236 mi) up. For reference, the ISS is parked in LEO about 320 km up. GEO is nice because the satellite then sits above the Equator, and thus from the ground it remains fixed in the sky as the Earth rotates (hence the name). Nice feature for communications coverage. (Trivia Alert! GEO was actually first popularized by science fiction author Arthur C. Clarke in 1945 as a useful orbit for communications satellites, long before we ever put anything up in space). The GPS constellation actually sits in MEO about 20,200 km (12,552 mi) up. Why there? Well, MEO with 24 birds gives the optimal planetary coverage. Yep, it's still pretty high up. Lots of nasty weather. The orbital period (how fast the satellites circle the globe) of GPS satellites in MEO is about 12 h.

Figure 10.6   An orbital view of the Van Allen radiation belts from the Space Shuttle *Discovery* (STS-39) in 1991 (courtesy of NASA).

A Brief History of GPS

The earliest origins of GPS can be traced to similar ground-based radio navigation systems that were developed by the allied forces during World War II [e.g., LORAN (long-range navigator) and the Decca Navigator systems]. Ironically, further inspiration for GPS came just after the Soviet Union's launch of Sputnik in 1957 [1]. As a team of exasperated (read: just-scooped) U.S. scientists led by Dr. Richard B. Kershner were lazily monitoring Sputnik's orbital radio transmissions, they made an interesting discovery regarding that the annoyingly periodic "beep-beep-beep" signal beamed back by Sputnik. Lightbulb moment: Kershner and company realized that they could actually pinpoint the satellite location in its orbit by measuring the transmitted signal at several known places on the Earth and then applying some math. Now imagine doing this in reverse. That is, if you know the precise location of several transmitting satellites, then you can determine the precise location of the person receiving the signals on the Earth. Result? A location finder! Hummm . . . The first satellite navigation system was deployed in 1960 by the U.S. Navy and was called "Transit." Using a constellation of five satellites, Transit could provide a navigational fix to a ship approximately once every hour. In 1967, the U.S. Navy developed the Timation satellite, which carried the first highly accurate clock into space – a key developmental step to GPS. In the 1970s, the ground-based Omega Navigation System became the first global radio navigation system. Onward ho to GPS. In 1972, the U.S. Air Force Central Inertial Guidance Test Facility (at Holloman Air Force Base) conducted developmental flight tests of two prototype GPS receivers over the White Sands Missile Range, using ground-based pseudo-satellites. The first experimental Block-I GPS satellite was finally launched in February 1978 (then made by Rockwell, now a part of Boeing, and currently manufactured by Lockheed Martin and Boeing). In 1983, after a Soviet interceptor aircraft shot down the inadvertently off-course civilian airliner KAL 007 (I recall this vividly – very sad day) in restricted Soviet airspace, killing all 269 people on board, U.S. President Ronald Reagan announced that the GPS system would be made available for civilian use (especially safety) once it was completed. By 1985, 10 additional Block-I GPS satellites had been launched to validate the GPS concept. They worked! In December of 1993, the GPS system was finally switched on. Hooray! By January 17, 1994, a complete constellation of 24 GPS satellites was in orbit, and full GPS operational capability was declared by NAVSTAR in April 1995. Not that long ago. In 1996, recognizing the emerging importance of GPS to civilian users, U.S. President Bill Clinton issued a policy declaring GPS to be an official "dual-use" system, and in 1998, U.S. Vice-President Al Gore announced plans to upgrade GPS with two new civilian signals for enhanced user accuracy and reliability, particularly with respect to aviation safety. On May 2, 2000, "selective availability" (SA) was officially discontinued as a result of the 1996 executive order, allowing civilian users to globally receive the pristine (nondegraded) GPS signal available to the military (as you can easily imagine, there are some subtleties, still, as to what can and cannot be used in DoD vs. commercial GPS receivers – refer to the subsequent GPS transmission frequencies sidebar discussion). In 2004, the U.S. government signed an agreement with the European Community establishing cooperation related to GPS on Europe's planned GALILEO system. The most recent GPS satellite launch was on December 20, 2007, and the oldest GPS satellite still in operation was launched on July 4, 1991, and became operational on August 30, 1991. Just keeps on ticking![2] The rest, as they say, is history.

(continued)

---

[2] You know . . . Timex – "Takes a licking, and just keeps on ticking!"

A Brief History of GPS (*continued*)

Some accolades for the GPS visionaries? You bet. It is pleasing to note that many folks received their due recognition for helping develop the marvel that is now GPS. On February 10, 1993, the National Aeronautic Association selected the entire GPS team as winner of the 1992 Robert J. Collier Trophy, the most prestigious aviation award in the United States. This team consists of researchers from the Naval Research Laboratory, the U.S. Air Force, the Aerospace Corporation, Rockwell International Corporation, and the IBM Federal Systems Division. The citation accompanying the presentation of the trophy honors the GPS team "for the most significant development for safe and efficient navigation and surveillance of air and spacecraft since the introduction of radio navigation 50 years ago." Now that's a big deal! The 2003 National Academy of Engineering Charles Stark Draper prize was awarded to Ivan Getting, president emeritus of The Aerospace Corporation and engineer at the Massachusetts Institute of Technology, and Bradford Parkinson, professor of aeronautics and astronautics at Stanford University. Roger L. Easton received the National Medal of Technology on February 13, 2006, for his developmental role.

10.2 Nuts and bolts

But what exactly IS GPS and how does it actually work? Well, the GPS is composed of three separate but highly coordinated components: a "space segment" (the satellites themselves); a "control segment" (ground control of the satellites); and a "user segment" (you and me), as depicted in Fig. 10.7.

GPS Control, User, and Space Segments

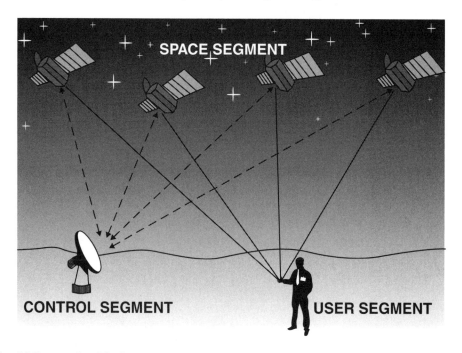

Figure 10.7   Artist's conception of the three components of GPS: the space segment, the control segment, and the user's segment.

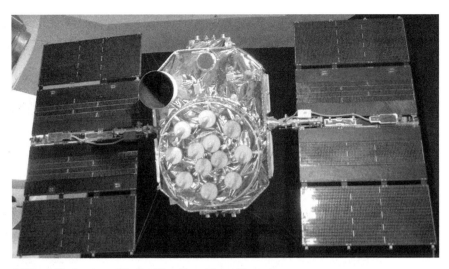

Figure 10.8 A GPS satellite (courtesy of the San Diego Air and Space Museum).

The *space segment* comprises the orbiting GPS satellite constellation, aka the GPS space vehicles (SVs), one of which is shown in Fig. 10.8. The GPS design originally called for 24 SVs, 8 each in three circular orbital planes, but has since been modified to six planes with four satellites each. The orbital planes are centered on the Earth, not rotating with respect to the distant stars. The six planes have approximately 55° inclination (tilt relative to Earth's Equator) and are separated by 60° right ascension of the ascending node (angle along the Equator from a reference point to the orbit's intersection). The SV orbits are arranged so that at least six satellites are always within line of sight from virtually anywhere on the Earth's surface. Orbiting at an altitude of approximately 20,200 km (12,600 mi) with an orbital radius of 26,600 km (16,500 mi), each SV makes two complete orbits each day (Fig. 10.9). As of September 2007, there are 31 actively broadcasting GPS satellites in the constellation (Fig. 10.10). The additional seven satellites, although not essential, are used to improve the precision of GPS receiver calculations by providing redundant measurements.

The *control segment* consists of the ground-based monitoring and control of the GPS SVs. The flight paths of the satellites are tracked by U.S. Air Force monitoring stations in Hawaii, Kwajalein, Ascension Island, Diego Garcia, and Colorado Springs, along with additional monitoring stations operated by the National Geospatial-Intelligence Agency (NGA). The satellite tracking information is sent to the U.S. Air Force Space Command's master control station at Schriever Air Force Base in Colorado Springs, which contacts each GPS satellite regularly with a navigational update by using the ground antennas at the monitoring stations. These control updates synchronize the atomic clocks on board each satellite to within a few nanoseconds (billionths of a second) of each other and adjust each satellite's internal orbital calculator based on current space weather information, and various other inputs. Often a slight readjustment in a particular

Figure 10.9   Artist's conception of a GPS satellite in Earth orbit (courtesy of NASA). (See color plate 32.)

Figure 10.10   Artist's conception of the GPS satellite constellation.

Figure 10.11   Some examples of handheld GPS receivers (courtesy of NASA).

SV's orbit is mandated. Ironically, satellite maneuvers within the constellation are not especially precise by GPS standards. So to change the orbit of a satellite, the satellite must first be taken off-line and marked as "sick," so the user receivers on the ground will delete it from their current location calculations. The SV orbital maneuver is then performed, and the resulting orbital changes are tracked from the ground, and when complete the satellite is once again marked healthy. Good to go.

The *user segment* represents us, the end users of GPS. The GPS receivers we use come in a wide variety of formats, from devices integrated into cars, planes, cell phones, watches, and artillery shells, to handheld dedicated devices such as those shown in Fig. 10.11. They may also include a display for providing location and speed information to the user. A receiver is often described by its number of "channels," and this signifies how many satellites it can monitor (track) simultaneously. Originally limited to four or five channels, GPS receivers have steadily improved over the years so that, as of 2006, receivers typically have between 12 and 20 channels available. You'll never be lost!

Each GPS satellite continuously broadcasts a "navigation message" at 50 bps, giving the time of day, the GPS week number, and satellite health information (all transmitted in the first part of the message); an "ephemeris" (transmitted in the second part of the message); and an "almanac" (later part of the message). The messages are sent in "frames," each taking 30 s to transmit 1,500 bits. Read: pretty darn slowly.

The first 6 s of every frame contain data describing the satellite clock and its relationship to GPS system time. The next 12 s contain the ephemeris data, giving the satellite's own precise orbit. The ephemeris is updated every 2 h and is generally valid for 4 h, with provisions for updates every 6 h or longer in nonnominal conditions (yep, think space weather). The time needed to acquire the ephemeris is becoming a significant element of the delay in the position fix, because, as GPS receivers become more capable, the time to lock onto the satellite signals shrinks, but the ephemeris data still require 30 s (worst case) before they are received because of the low data transmission rate (something to work on). The almanac consists of coarse orbit and status information for each satellite in the constellation, an ionospheric model, and information to mathematically relate GPS-derived time to the universal time standard "Coordinated Universal Time" (UTC – see Geek Trivia sidebar discussion). A new portion of the almanac is received for the last 12 s in each 30-s frame. Each frame contains 1/25 of the almanac, so 12.5 min are required for receiving the entire almanac from a single satellite.

The almanac serves several purposes. The first is to assist in the acquisition of satellites at power-up by allowing the receiver to generate a list of visible satellites based on stored position and time, whereas an ephemeris from each satellite is needed to compute position fixes by using that satellite. The second purpose is for relating time derived from the GPS system (called "GPS time") to UTC. Finally, the almanac allows a single-frequency receiver to correct for ionospheric error by using a global ionospheric model. Importantly, each satellite transmits only its own ephemeris data, but also transmits an almanac for ALL of the other GPS satellites.

In GPS-speak, one refers to a GPS "cold start" vs. a "warm start." What gives? A GPS cold start is when you haven't turned on your GPS unit in quite some time (say a few days). All of the ephemeris–almanac data the unit previously received are clearly now out of date, and the GPS unit is essentially blind. Consequently, it will take your GPS unit some time, maybe even a minute or two (because of the slow transmission rates from the SVs), to "find itself." In a warm start, however, you might have turned off the unit for only a few minutes (while you went into the store, say), and thus the ephemeris–alamanac data, although invalid, is still close-enough-for-government-work (sorry), and your GPS unit should thus be able to find lock in only a few seconds by using "last known values." That is, the GPS cleverly remembers where it last was and will search for itself there first before "crying uncle" and bringing in a brand-new set of SV-transmitted data to locate itself. Cool.

Geek Trivia: Coordinated Universal Time

Coordinated Universal Time (UTC) is a high-precision atomic time standard (atomic time?! – see sidebar discussion on atomic clocks) and is utilized by GPS. UTC has uniform seconds defined by International Atomic Time, with leap seconds introduced at irregular intervals to compensate for the Earth's slowing rotation. The addition of leap seconds allows UTC to closely track Universal Time (UT), a time standard based, not on the uniform passage of seconds, but on the Earth's angular rotation (a logical fit for GPS given its nature). Various time zones around the world are expressed as positive or negative offsets from UTC. Local time is UTC plus the time zone offset for that location, plus an offset (typically +1) for daylight savings time, if in effect. On January 1, 1972, UTC replaced Greenwich Mean Time (GMT) as the *de facto* reference time scale (bet you didn't know that!). UTC is usually referred to by military and civilian aviators as "Zulu (Z) time".

Geek Trivia: Atomic Clocks

An "atomic clock" is key for a GPS to work as advertised and is simply a particular type of unbelievably accurate time piece (your wind-up grandfather clock, beautiful as it is, is just not going to cut it. Tough to put on a satellite too!). Atomic clocks use an atomic resonance frequency standard, based on an isotope such as cesium-133 [1]. The first atomic clock was demonstrated in 1949 at the U.S. National Bureau of Standards (NBS) (now NIST, the National Institute of Standards and Technology), and the first highly accurate (modern) cesium-133 atomic clock was built by Louis Essen in 1955 at the National Physical Laboratory in the United Kingdom. From the very beginning, atomic clocks have been based on the "hyperfine transitions" in hydrogen-1, cesium-133, and rubidium-87. Things have progressed. What was once desk sized is now chip sized (no surprise to us studying the science of miniaturization). In August 2004, NIST scientists demonstrated the first chip-scale atomic clock (Fig. 10.12), only millimeters across and using only 75 mW of power. Sweet! Since 1967, the SI has defined a "second" of time as the duration of precisely 9,192,631,770 cycles of radiation corresponding to the transition between two energy levels of the ground state of the cesium-133 atom. Whew! This definition makes the cesium-133 oscillator (the guts of our most common atomic clocks) the world's main standard for time and frequency measurements. Interestingly, other important units such as the volt and the meter themselves rely on this definition of the second. Read: Atomic clocks are pretty important . . . and guess what – each GPS satellite has one on board, chirping away 24/7!

Figure 10.12   A chip-scale atomic clock (courtesy of NIST).

Each GPS satellite transmits its navigation message with at least two distinct codes: the "coarse–acquisition" (C/A) code, which is freely available to the public, and the "precise" (P) code, which is encrypted and used in many military applications. The C/A code is a 1,023-long pseudo-random-number (PRN) code (just what it sounds like) transmitted at 1.023 million numbers/s, so that it repeats itself every millisecond. Each satellite has its own C/A code so that it can be uniquely identified and received separately from the other satellites transmitting on the same frequency. The P-code is a 10.23 million numbers/s PRN code that repeats every week. When the "anti-spoofing" mode is on in your receiver, the P-code is encrypted by the "Y-code" to produce the "P(Y) code," which can be decrypted only by units with a valid decryption key (read: DoD – off limits to you and me). Both the C/A and P(Y) codes impart the precise time of day to the end user – us. Importantly, all of the GPS SVs transmit their codes at the same frequency, using "spread-spectrum" techniques (just as in CDMA cell phones – refer to Chap. 3). This is important because it enables each GPS receiver to track MANY satellites (10–20) at the same time, while using the same set of electronics to receive those signals. Big cost and complexity savings. What transmission frequencies are actually used by GPS and why? Refer to the sidebar discussion.

Technical Aside #2: What Transmission Frequencies Does a GPS Actually Use?

Curious what frequencies GPS actually uses? Well, they are in the range of frequencies known as the "L-band," from 950 MHz to 1,450 MHz, which itself is a part of the UHF (ultra-high-frequency) broadcast band, which ranges from 3 MHz to 3 GHz, the same as for TV and most cell phones. Why L-band? Well, it ultimately traces back to optimal EM transmission windows. That is, these particular EM frequencies will generally experience the smallest attenuation when transmitted through the atmosphere (think longest path length, ergo best transmission fidelity at lowest transmitted power). EM transmission windows will be discussed more in Chap. 11. Here are the actual GPS signals [1]:

- L1 (1,575.42 MHz): L1 ("L" is for L-band) is a mix of multiple things, including (a) the Navigation Message (NM), (b) the C/A code, (c) the encrypted (read: DoD) precision P(Y) code, and (d) the new L1C to be used on coming Block III GPS satellites. L1 is the workhorse RF signal for consumer GPS receivers.
- L2 (1,227.60 MHz): L2 contains the P(Y) code, plus the new L2C code used on the Block IIR-M and newer satellites. Ironically, L1 is just outside the formal L-band frequency range. L2 is generally used by the DoD for improved accuracy.
- L3 (1,381.05 MHz): L3 is used by the Nuclear Detonation (NUDET) Detection System (NDS) payload to signal detection of nuclear detonations. This monitoring is to help enforce the various nuclear test ban treaties. Hot topic (pun intended!).
- L4 (1,379.91 MHz): L4 is currently dedicated for use for future ionospheric correction schemes.
- L5 (1,176.45 MHz): L5 has been proposed for use as a civilian safety-of-life (SoL) signal in emergency situations. L5 falls into an internationally protected frequency range for aeronautical navigation, thus having the utility that there is little or no interference to worry about. The first Block II-F satellite to provide this L5 signal was launched in 2008.

So how does the GPS receiver actually determine its location? A typical GPS receiver calculates its position by using the broadcast signals from four or more GPS satellites. Four satellites are needed because the process needs a very accurate local time, more accurate than any normal clock can provide, so the receiver can internally solve for time as well as for position. In other words, the receiver uses four measurements to solve for four variables: x, y, z, and t (think back – four equations, four unknowns). These values are then turned into more user-friendly forms, such as latitude–longitude or location on a map, and then displayed to the user. Each GPS satellite has an atomic clock and continually transmits messages containing the current time at the start of the message, parameters to calculate the location of the satellite (the ephemeris), and the general system health (the almanac). The signals travel at a known speed – the speed of light through outer space, and slightly slower through the atmosphere, subject to local space weather. The receiver uses the arrival time to compute the distance to each satellite, from which it determines the position of the receiver using a little geometry and trigonometry. Go math! You should have paid more attention during high school trig! Although four satellites are required for normal operation, fewer may be needed in some special cases. For example, if one variable is already known (for example, a seagoing ship knows its altitude is 0), a receiver can determine its position by using only three satellites (Fig. 10.4). In addition, receivers can cleverly use additional clues (e.g., Doppler shift of satellite signals, last known position, dead reckoning, inertial navigation, etc.) to give "degraded" answers (but still useful if you're lost) when fewer than four satellites are visible.

So just how accurate is a good GPS receiver? The position calculated requires the current time, the position of the satellite, and the measured delay of the received signal. The position accuracy is primarily dependent on the satellite position and signal delay. To measure the delay, the receiver compares the bit sequence received from the satellite with an internally generated version. By comparing the rising and trailing edges of the bit transitions, modern electronics can measure signal offset to within about 1% of a bit time, or approximately 10 ns for the C/A code. Because GPS signals propagate basically at the speed of light, this represents an error of about 3 m (the math is easy, try it). This is the minimum error possible when only the GPS C/A signal is used. Position accuracy can be improved by use of the higher-chip-rate P(Y) signal. Assuming the same 1% bit time accuracy, the high-frequency P(Y) signal in a high-end (read: bigger bucks, DoD) GPS system can in principle result in an accuracy of about 30 cm. Pretty amazing. At present, a low-cost civilian GPS location fix is typically accurate to about 15 m (50 ft). With various GPS "augmentation" systems (e.g., WASS, a wide-area augmentation system [6]; or differential GPS) your unit can provide additional real-time correction data, yielding up to five times better resolution. A WASS-capable GPS receiver, for instance, which is actually pretty routine these days, is good to about 3-m accuracy about 95% of the time. Wow! And ... GPS resolution is getting better all the time. FYI – the

U.S. government is already working on GPS Block-III, a 3G (read: more accurate still) GPS satellite constellation. Stay tuned.

10.3 Where are the integrated circuits and what do they do?

Okay, let's have some fun and pull off the covers of a handheld GPS receiver and see what lurks inside doing all this magic! So what electronic gizmos must be present to do all this fun stuff? Well, a vanilla GPS receiver is basically a sophisticated radio receiver (much like a cell phone) and is composed of an antenna, a RF "front-end," power management circuitry, a digital signal processor, and a highly stable "clock." A complete GPS receiver would thus be composed of the following parts:

- #1 *Antenna:* The antenna on a GPS receiver is a carefully designed piece of (electrically conductive) metal that is precisely tuned to efficiently receive the modulated RF signals coming to the radio receiver from the transmitting satellite (e.g., L1 = 1,575.42 MHz). For those in-the-know, the GPS signal is actually circularly polarized, and one can thus gain an advantage by using a non-monopole, slightly fancier, antenna. Most GPS units do this.
- #2 *Low-Noise Amplifier:* Given that the transmitted power from the GPS satellites is VERY weak (they are a long way away!), we need to first receive the signal via the antenna, and then immediately amplify (boost) the signal amplitude in order to properly decode it. A low-noise amplifier (LNA) is the workhorse here (just as in a cell phone). Why low noise? Well, we don't want to corrupt the GPS signal during the amplification process. Eliminating spurious background noise (think: hiss, static) is thus required. In modern GPS receivers, the LNA is often incorporated directly into the RF downconverter to save space and cost.
- #3 *RF Bandpass Filter:* Filtering the RF signal coming into the receiver is also needed and is typically performed immediately after amplification. We use a RF "bandpass" filter, which allows our L1 signals to come right through, but strongly attenuates (damps out) any spurious (noisy) components contained in the signals that are closely spaced (in frequency) to our desired GPS signal, but are unneeded and unwanted. Think of RF filtering as a way to "clean up" the received RF signal, improving downstream signal fidelity (read: improved accuracy).
- #4 *Power Management Circuitry:* Like a cell phone, GPS receivers are often battery operated, and thus need to carefully manage their power drain to extend battery life. For instance, the GPS receiver has built-in electronic intelligence to infer when a given component or subsystem in the GPS unit can be powered down and put into "sleep" mode, thus saving battery life. Unlike for a cell phone, GPS only listens, and does not transmit, making it generally MUCH more power efficient.

- #5 *RF Downconverter:* Just as with a cell phone, the usefulness of a GPS receiver begins and ends with the quality of its radio. A GPS RF "downconverter" contains "frequency-translation" electronic circuitry to convert the (high) RF of the EM satellite signal (e.g., L1 = 1,575.42 MHz) to an "intermediate frequency" (IF), say 220 MHz, that is more easily manipulated by the subsequent electronics of the digital signal processor of the GPS unit. With GPS, unlike for cell phones, this RF-to-IF conversion is only a one-way process needed to "receive" the RF signal and then manipulate it.

- #6 *Display:* The display is the visual interface between you and the GPS receiver (cheaper models may simply print out the coordinates rather than show a detailed map). A GPS display in a car, say, consists of a color liquid-crystal display (LCD) for showing those pretty maps. A piece of electronics called a "display driver" is used to interface between the receiver and the LCD, just as with a cell phone.

- #7 *Reference Clock:* A "crystal oscillator" is often used to build a stable electronic timepiece; aka a reference clock. Yep, real crystal; but not a semiconductor – typically a quartz crystal. A crystal oscillator is an electronic circuit that exploits the mechanical resonance of a rapidly vibrating crystal of piezoelectric material to create an electrical signal with a very precise and very stable frequency. When a quartz crystal is properly cut and mounted, it can be made to distort in an electric field by applying a voltage to an electrode near to or even on the crystal. This property of crystals is known as piezoelectricity. When the field is removed, the crystal will generate an electric field as it returns to its previous shape, generating a voltage. In essence, the quartz crystal behaves like a mini electronic circuit composed of an inductor, capacitor, and resistor, with a precisely defined "resonant" or vibrational frequency (the well-known RLC resonator). Quartz crystals for oscillators are manufactured for clock frequencies ranging from a few 10s of kilohertz to 100s of megahertz, and more than 2 billion such quartz crystals are manufactured annually (Fig. 10.13)! Stable clocks are needed to keep track of time (duh), as needed in quartz-crystal-powered wristwatches (no more wind-up springs); to provide a stable clock signal for digital ICs (think GPS); and to stabilize the frequencies in radio (e.g., cell phone) transmitters. A little crystal oscillator history? Sure! Piezoelectricity was discovered by Jacques and Pierre Curie in 1880 [1]. Paul Langevin first investigated quartz resonators for use in sonar systems during World War I (for hunting U-boats). The first crystal-controlled oscillator, using a crystal of Rochelle salt, was built in 1917 and patented in 1918 by Alexander M. Nicholson at Bell Labs. Walter Cady built the first quartz-crystal oscillator in 1921, a precursor for many good things to come. Good ideas stick around. Crystal oscillators are absolutely ubiquitous in all modern (timed) electronic systems and hence a key building block of our Communications Revolution. Yep, pretty important.

- #8 *Digital Signal Processor:* The digital signal processor (DSP) is essentially a specialized microprocessor that serves as the brains of the GPS receiver,

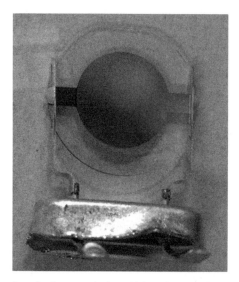

Figure 10.13   An under-the-covers look at a high-performance quartz crystal oscillator.

functioning as a "command and control center," if you will, and importantly doing LOTS of math very rapidly in the background. As in a cell phone, the DSP likely has its own SRAM and DRAM on-chip for its immediate storage needs.

Good. So THAT is what really lurks inside a GPS receiver! Yep, pretty sophisticated. Some visuals? Sure! Let's wrap things up by looking under the covers of a Garmin zūmo 550 GPS receiver [7]. Kids – don't try this at home! Are you a fan of black leather? (Careful, don't answer that!) Is your favorite movie *Easy Rider*? Do you own a Harley? No? Well, do you fantasize about owning a Harley? Are you addicted to the uncanny sweet rumble coming from an exhaust pipe? If so, the Garmin zūmo 550 GPS receiver is meant for you! The zūmo was actually designed by bikers (at Garmin!), for bikers, and is "motorcycle friendly," making it easy to operate while "in the saddle." It has lots of preloaded maps and navigation features that will give you all the freedom a biker needs. Your zūmo comes preloaded with City Navigator NT street maps, route planning software, and a points-of-interest (POIs) database, including motels, restaurants, gas, ATMs, and . . . leather shops(?). For the biker purist? Perhaps not. For the weekend biker road warrior? Definitely! Your zūmo is waterproof; has a glove-friendly, high-brightness, touch-screen LCD display; is voice prompted (speak loudly!); comes with a custom bike mount; runs 4 h off a lithium battery; and is of course vibration (read: pothole) rated, with built-in bug shields (joking). Heck, it even has an on-board MP3 player for your tunes.

Figure 10.14 shows a pristine zūmo 550 GPS unit, an exploded view of what's really inside (Fig. 10.15), and then front and back views with the covers pulled down and the individual ICs indicated (Figs. 10.16 and 10.17). You can see that a GPS receiver has the look and feel of a cell phone's innards, with lots of

Figure 10.14   The Garmin zūmo 550 GPS unit. (Image produced with the permission of Garmin. Copyright 2008, Garmin Ltd. or its subsidiaries, all rights reserved.)

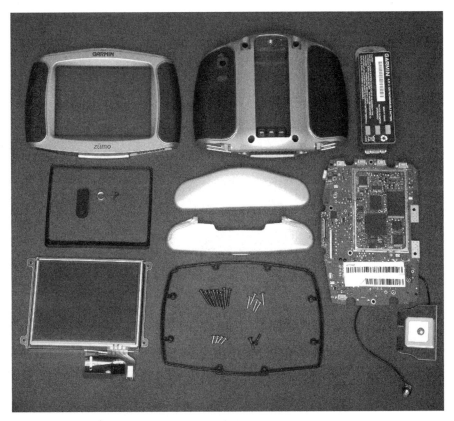

Figure 10.15   Exploded view of a naked Garmin zūmo 550. (Image produced with the permission of Garmin. Copyright 2008, Garmin Ltd. or its subsidiaries, all rights reserved.) (See color plate 33.)

Programmed
Map / Voice
Prompt
Database

Programmable
Logic Device

GPS Patch
Antenna

Serial Computer
Interface

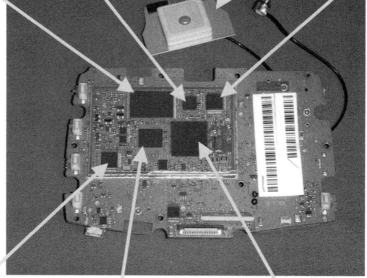

XM RF
Transceiver

Low Power DDR
RAM

Microprocessor

Figure 10.16   Front-side view of the IC board of a Garmin zūmo 550. (Image produced with the permission of Garmin. Copyright 2008, Garmin Ltd. or its subsidiaries, all rights reserved.)

GPS
Processor

Digital Signal
Processor

8 GByte Memory

Stereo Digital
to Analog
Converter

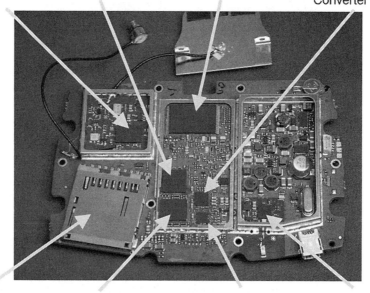

SD Card Slot

Combination
FLASH/SRAM

High Power Stereo
Audio Amp

Bluetooth
Transceiver

Figure 10.17   Back-side view of the IC board of a Garmin zūmo 550. (Image produced with the permission of Garmin. Copyright 2008, Garmin Ltd. or its subsidiaries, all rights reserved.) (See color plate 34.)

sophisticated close packing and form-factor minimization. One of the present quests in the GPS business is to implement the entire GPS on a single piece of silicon, and that day is rapidly approaching. The goal – make GPS receivers so small and so cheap that they can be placed on virtually everything. You can run but you can't hide! There are some obvious implications of this trend for modern society (both good and bad), which will be discussed in our last chapter.

GPS – modern marvel of engineering. For those pundits who might be tempted to pooh-pooh the costs of space exploration, GPS is a nice example of the many technological fruits of that investment. Go, go, GPS, go!

References and notes

1. See, for instance, Wikipedia: *http://en.wikipedia.org*.
2. See, for instance, *http://www.gps.gov*.
3. "GPS Guide for Beginners," Garmin, Inc., 2000.
4. B.W. Parkinson, *Global Positioning System: Theory and Applications*, American Institute of Aeronautics and Astronautics, Washington, D.C., 1996.
5. W. Sweet, "No payoff for GALILEO navigation system," *IEEE Spectrum*, pp. 48–49, January 2008.
6. See, for intsance, *http://www8.garmin.com/aboutGPS/waas.html*.
7. I am deeply indebted to Garmin for volunteering to dissect their zūmo 550 GPS unit and provide me (and you!) with some cool pics. You can visit Garmin at *http://www.garmin.com/*.

11 Let There Be Light: The Bright World of Photonics

Alfred Sisley's "The Effect of Snow at Argenteuil" (1874). (See color plate 35.)

Winter Is Here.
The Impressionist Paints Snow.
He Sees That, in the Sunlight
The Shadows of the Snow Are Blue.
Without Hesitation,
He Paints Blue Shadows.
So the Public Laughs,
Roars With Laughter.

Théodore Duret (art critic)

1874. France. The Impressionists are afoot, radicals all: Monet, Sisley, Renoir, Pissarro, Morisot, Bazille. Scandalous! The artistic revolution that is today instantly recognizable as French Impressionism centered on a valiant attempt to record a "truer" visual reality in terms of the myriad transient effects of juxtaposed light and color. Impressionism was most definitely a calculated rejection of the prevailing stiff formalism of the "good art" propagated by the French Academy. Their astounding results endure, bearing witness to their "impressions" of nature. Blue snow? You bet! Peruse for a moment the chapter opening figure showing Sisley's (1839–1899) "The Effect of Snow at Argenteuil" (1874) – see color plate 35. Or better yet, take a winter hike and see for yourself. Given the proper sunlight, snow is indeed most definitely blue.

So what exactly does French Impressionism have to do with semiconductors? Oh... more than you might think! It is perhaps tempting to infer that our 21st-century Communications Revolution is being driven solely by the world of micro/nanoelectronics. Not so. Let there be dancing light and color. Enter "photonics," a reasonably recent but now vital subdiscipline of electrical engineering that uses radiant energy (aka photons; in common parlance, light) to do its business. What the electron is to electronics, the photon is to photonics. Why bother? Good question. For one thing, photons can travel much faster than electrons (some numbers in a moment), meaning ultimately that digital data transmitted photonically can travel longer distances and in a fraction of the time required for performing the same data transmission electronically. And... often it can do it cheaper. That's not all. Visible-light and IR beams, for instance, unlike electric currents, can pass through each other without interacting, preventing undesirable signal interference. Nice feature! Interestingly, as we will shortly see, the field of photonics today is largely coincident with semiconductors, and the merger of semiconductor-based photonics with semiconductor-based electronics is often referred to as "optoelectronics." Each of us uses optoelectronic components hundreds of times daily. Skeptical? Think TV remote control; think CD player; think digital camera; think DVD player; think laser pointer; think traffic lights; think... gulp... the Internet! Yep, it's time to know a little bit more about how all this photonics stuff works. Read on.

11.1 Let there be light!

Care for a mathematical translation of a well-known Biblical verse? You bet! Remarkably, the behavior of light can be completely understood by the four mathematical equations shown in Fig. 11.1. These are "Maxwell's equations," first published in 1861 and named for Scottish mathematician and physicist James Clerk Maxwell (1831–1879).[1] These four equations compactly describe the exceptionally complex interrelationship among electric field, magnetic field, electric charge, and electric current, all as functions of both position and time. They are universally considered one of humankind's greatest intellectual triumphs, and certainly one of the most useful. One-hundred and fifty years later,

---

[1] In point of fact, the form of Maxwell's equations shown here dates not to 1861 and Maxwell, but rather to Oliver Heaviside and Willard Gibbs. In 1884 they used newly developed concepts from vector calculus (unknown in 1861 to Maxwell) to write the equations in the compact form we know today.

$$\nabla \cdot \vec{E} = \frac{\rho}{\epsilon_0}$$

$$\nabla \cdot \vec{B} = 0$$

$$\nabla \times \vec{E} + \frac{\partial \vec{B}}{\partial t} = 0$$

$$\nabla \times \vec{B} - \frac{1}{c^2} \frac{\partial \vec{E}}{\partial t} = \mu_0 \, \vec{J}$$

Figure 11.1   A mathematical translation of Genesis 1:3.

no self-respecting undergraduate student in electrical engineering or physics has failed to study these four equations in detail. The fact that the human mind can conceive of a mathematical framework, a mathematical language of nature if you will, to so elegantly and compactly represent such a varied and complex physical phenomena, is truly staggering and worthy of some serious reflection. You may recall that one of Maxwell's equations, Poisson's equation (the first equation shown in Fig. 11.1), also comprises one of the semiconductor's fundamental "equations of state" (Chap. 5). It would not be an exaggeration to say that all of electronics and photonics (read: virtually all of modern technology at some level) at minimum touches, and more often directly rests upon, Maxwell's bedrock equations. They are that important.

So what exactly do Maxwell's equations say about the visible light that you and I see with? Well, if you speak vector calculus and know a little bit about solving partial differential equations (no worries if you don't!) you can nicely demonstrate that Maxwell's equations have a solution that can be expressed in terms of a traveling sinusoidal plane wave, with the electric- and magnetic-field directions (vectors) orthogonal both to one another and their direction of propagation, and with the two fields in phase, traveling at a speed c given by (Fig. 11.2):

$$c = \frac{1}{\sqrt{\mu_0 \, \epsilon_0}}, \tag{11.1}$$

where $\epsilon_0 = 8.85419 \times 10^{-12}$ F/m is called the "permittivity of free space" (vacuum) and $\mu_0 = 4\pi \times 10^{-7}$ H/m is called the "permeability of free space," both well-defined and measurable fundamental constants of nature. Maxwell discovered that this quantity c is simply the speed of light traveling in a vacuum (aka free space), thus proving formally that visible light is just another form of electromagnetic (EM) radiation (all EM radiation moves at velocity c in vacuum). The accepted value (in SI units) for $c = 2.99792458 \times 10^8$ m/s. Or, $1, 079, 252, 848.8$ km/h. Or, $186, 282.397$ mi/s. Is this a big number or a small number? Well, moving at the speed of light, you could circumnavigate the equator of the Earth approximately 7.5 times before 1 s elapsed. Yep, cookin'!

There are some subtleties associated with c to be aware of. The speed of light when it passes through a transparent or translucent material medium, like glass or water, or, as we will soon see, a silica optical fiber, is actually a little slower than its speed in vacuum. The ratio of c to the observed velocity in the medium in question is called the "refractive index," an important term in optics that you may have heard. The index of refraction of glass, for instance is about 1.52. This is why light bends, for example, when it travels from air to water, distorting the real position of Mr. Bass to the anxious angler hovering above. It is also

An EM Wave: aka "Light"

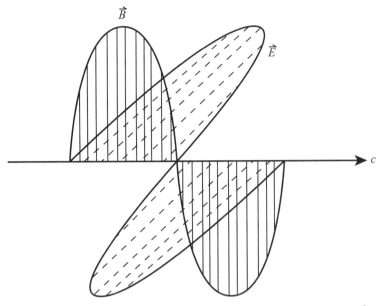

Figure 11.2 A cartoon representation of a beam of light, comprising undulating sinusoidal electric ($\vec{E}$) and magnetic ($\vec{B}$) fields traveling at velocity c.

the reason why a plastic lens can be cleverly used to bend light and correct for near-sightedness by using soft contact lens. Interestingly, Maxwell's equations, as originally written, were by definition consistent with Einstein's Theory of Special Relativity (1905).[2] Einstein's Theory of General Relativity (gravity) explains how a gravitational potential can affect the apparent speed of distant light in a vacuum (the origins of so-called "gravitational lensing," a hot topic these days in astronomy). Locally, however, light in a vacuum always passes an observer at a rate equal to c, and thus c represents the upper bound, the fundamental speed limit, of the universe. The hunt for experimentally determining c is a fascinating story in its own right, but starting with Michelson is 1926, the world homed in on a solid number pretty rapidly (Fig. 11.3) [1].

Geek Trivia: The Speed of Light

So, why the letter c for the speed of light? Although c is now the universal symbol for the speed of light, the most common symbol in the 19th century was actually an uppercase V (velocity), which Maxwell began using in 1865 [2]. Interestingly, the V notation was adopted by Einstein for his first few papers on relativity from

(continued)

[2] Famously, Einstein's quest that ultimately culminated in his highly counterintuitive (e.g., objects get more massive the faster they travel) Theory of Special Relativity began with his imagining what the universe would look like if he could ride along on the back of a beam of light moving at speed c.

Date	Investigators	Method	c (m/s)
1926	Michelson	Rotating mirror	299,796 ± 4
1935	Michelson, Pease, and Pearson	Rotating mirror in vacuum	299,774 ± 11
1940	Hüttel	Kerr cell	299,768 ± 10
1941	Anderson	Kerr cell	299,776 ± 6
1950	Bol	Cavity resonator	299,789.3 ± 0.4
1950	Essem	Cavity resonator	299,792.5 ± 3.0
1951	Bergstrand	Kerr cell	299,793.1 ± 0.2
1951	Alakson	Radar	299,794.2 ± 1.9
1951	Froome	Microwave interferometer	299,792.6 ± 0.7

Figure 11.3   Historical experiments establishing the speed of light in vacuum.

Geek Trivia: The Speed of Light (*continued*)

1905. The origins of the use of the letter c can be traced to a paper of 1856 by Weber and Kohlrausch. In that paper they defined and measured a quantity denoted by c that they used in an electrodynamics force law equation. It soon became known as Weber's constant and was later shown to have a theoretical value equal to the speed of light times the square root of two. In 1894 physicist Paul Drude modified the usage of Weber's constant so that the letter c became the symbol for the speed of EM waves. Slowly but surely, the c notation was adopted by Max Planck, Hendrik Lorentz, and other influential physicists in their papers. By 1907, Einstein himself had switched from V to c in his own papers, and c finally became the standard symbol for the speed of light in vacuum for electrodynamics, optics, thermodynamics, and relativity. So why choose c? Weber apparently meant c to stand for "constant" in his force law. There is evidence, however, that physicists such as Lorentz and Einstein were accustomed to a then-common convention that c could be used as a mathematical variable for velocity. This usage can be traced back to classical Latin texts in which c stood for "celeritas," meaning "speed." There you have it!

11.2 Spectral windows

Great. So visible light is actually a wave train of undulating orthogonal electric and magnetic fields moving along in phase at the speed of light. Cool! But visible light comprises only a tiny portion of the entire EM spectrum, across a remarkably narrow band of frequencies to which our retinal light detector cells can respond. Recall from Chap. 1 that frequency (f) and wavelength (λ) are related by

$$\lambda = \frac{c}{f}. \qquad (11.2)$$

The complete EM spectrum of EM waves one encounters in the universe ranges from audio frequencies (a few hertz = tens of thousands of kilometer wavelengths) to x-ray frequencies (1000s of petahertz = tenths of nanometer of wavelength: 1 PHz = 1,000,000,000,000,000 Hz). This frequency range of the EM radiation world is shown in Fig. 11.4.

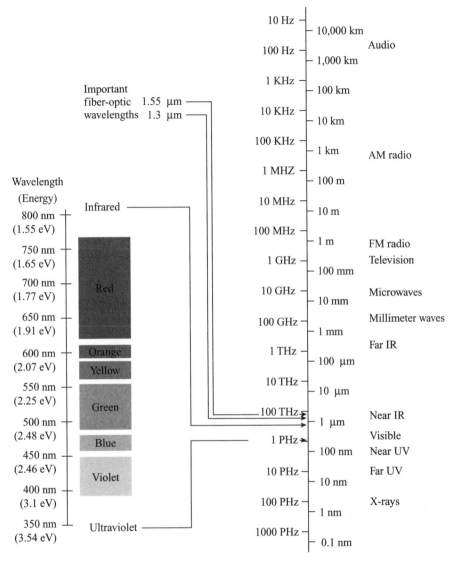

Figure 11.4 The visible spectrum and where it sits within the context of the universal distance, frequency, and energy scales.

In dimensions of wavelength (crest-to-crest distance between the undulating EM waves), visible light roughly spans 0.4–0.7 μm or 400–700 nm in micro/nano dimensions. The "colors" of visible light span the rainbow, obviously, from violet (0.4 μm) to red (0.7 μm), according to the elementary school mnemonic "ROYGBIV" (red–orange–yellow–green–blue–indigo–violet). We know, however, that nature provides a simple equivalence between energy (E) and wavelength of all EM radiation, given by

$$E = \frac{h\,c}{\lambda}, \tag{11.3}$$

where h is Planck's constant (6.63×10^{-34} J s). Conveniently, if we work in units of electron volts for energy (think energy bandgap in semiconductors!) and

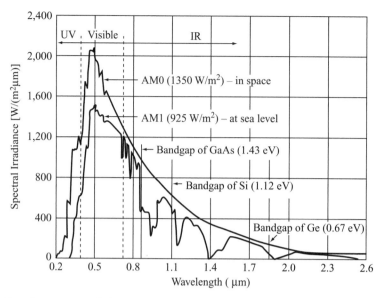

Figure 11.5 The solar spectrum at both sea level (AM1) and in Earth orbit (AM0). Superposed are the visible spectrum boundaries and the bandgaps of several important semiconductors.

microns for wavelength, we can reexpress this in an easy-to-remember form:

$$E \text{ (eV)} = \frac{1.24}{\lambda \text{ (µm)}}. \tag{11.4}$$

For instance, blue–green light has a wavelength of about 500 nm, corresponding to about 2.48 eV of energy. It is important to note at this stage that the energies associated with visible light are in the range 1.5–3.0 eV. Food for thought: The energy bandgaps of the useful semiconductors span a range from about 0.5 eV to about 4 eV. As we will soon see, this matters a lot in the grand scheme of things.

Here's a question for you. We know that EM radiation in a vacuum propagates at the speed of light – always. But you and I clearly do not live in a vacuum. Given that we are usually interested in sending EM radiation from point A to point B (think information flow), and that path most often happens on Earth with its complex atmosphere, does this affect the way we send or receive EM radiation? Answer? It does. Profoundly. Imagine the EM radiation generated by the helium fusion furnace inside the Sun. The energy that actually makes it to you and me on the surface of the Earth depends on the wavelength (energy) of the EM radiation. Oh yeah? Consider Fig. 11.5, which plots the "solar irradiance" (essentially the amount of energy deposited per unit area) at two locations: AM0 (in Earth orbit, above the atmosphere) and AM1 (at sea level, below the atmosphere). Three things can be gleaned from a cursory look at this plot: (1) the bulk of the solar EM radiation reaches us in the visible spectrum, conveniently lighting our world; (2) not much short-wavelength (high-energy) UV EM radiation reaches us, protecting our DNA from too much damage and mutation; but (3) lots of longer-wavelength (lower-energy) IR EM radiation does reach us, providing warmth for the planet. Excellent! You can also clearly see that the solar spectrum is profoundly altered by the atmosphere (the difference between the AM0 and AM1 curves). In particular, the extreme brightness in the visible range is muted by the atmosphere; otherwise we'd have to wear sunglasses 24/7; and the withering UV radiation is effectively damped out by the ozone layer (this is why it is essential to life!). The other "dips" we see in the sea-level solar spectrum are associated with absorption by the various chemical components in the atmosphere (e.g., by water vapor or oxygen). In between these strong absorption regions lie the so-called transmission "windows," spectral windows if you will. The existence of spectral windows has major implications for the design of satellite systems. For instance, if I put a satellite into orbit and want it to look downward at the surface of the Earth for remote imaging, I definitely don't want to center my detector at the 1.4-μm wavelength. Why? Refer to Fig. 11.5. No signal will get through the atmosphere. If, instead, I use visible light, say 0.5 μm or even 1.2 μm, however, my satellite can "see" the surface of the Earth just fine. Said another way, the atmospheric "loss" (attenuation) of EM radiation is a strong function of wavelength. Global communications satellites that beam Internet or telephone or TV signals back and forth through the atmosphere a gazillion times a second must obviously choose their transmission frequencies very, very carefully, for just this reason.

Key take-away? When transmitting EM radiation from point A to point B when we are NOT in a vacuum, an intelligent choice of wavelength, energy, and frequency of that EM radiation is the key to success. Another nonvacuum EM transmission medium that is highly relevant to photonics is the optical fiber. The ubiquitous silica fibers used in optical communications systems (think long-haul Internet backbones) have complicated, but ultimately very well-defined, loss characteristics, as illustrated in Fig. 11.6. In particular, an EM pulse sent down an optical fiber experiences minimum loss at two particular wavelengths; 1.3 μm and 1.55 μm, corresponding to energies of 0.95 eV and 0.80 eV, respectively (see Fig. 11.7). There is serious financial incentive to choose these lowest loss transmission wavelengths, because they will determine how many kilometers my optical signal can travel before it gets too small to detect, directly determining the number of (expensive) amplifier blocks (aka "repeaters") that must be inserted between points A and B to regenerate the attenuated optical signal. Big bucks at stake! So what does this have to do with semiconductors? Lots, actually. Keep reading.

11.3 Getting light in and out of semiconductors

For building a long-haul fiber-optic communications system, wouldn't it be cool if we could use light of just the right wavelength to be centered at the precise

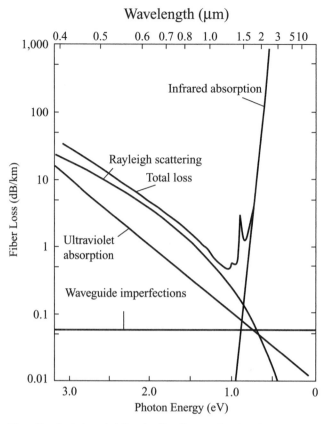

Attenuation (loss) characteristics of a silica fiber as a function of energy and wavelength.

minimum loss point that nature provides for the optical fiber? You bet it would! And it turns out we can. Question: What color would that light need to be? Answer: Invisible, because 1.55 μm is in the near-IR region of the spectrum![3]

As you may recall from Chap. 5, "bandgap engineering" in devices (electronic or photonic) is all about using different semiconductors in clever ways to change the bandgap energy to our advantage [3]. Here is a concrete example from the photonics world of why and how this is done. Figure 11.8 shows the various families of chemically compatible semiconductors that can be used for just this purpose – think of it as the bandgap-engineering materials landscape. As an example, I can lattice-match (same lattice constant) $In_{0.53}Ga_{0.47}As$ to an InP substrate (wafer), producing an energy bandgap with a corresponding wavelength quite close to 1.55 μm. For a more complete look, see Fig. 11.9, which superposes the various available material systems onto a wavelength scale. The vertical bars for each material system in question represent the bounds of viable material composition that can be achieved in practice when attempts are made to vary the bandgap. For instance, in the InGaAs material system, we can span 0.6 μm to

---

[3] Unless you happen to be the *Predator*, who sees quite nicely in the IR region, thank you – this is why Dutch (Arnold Schwartzenegger) had to cover his body with river mud (war paint?!) to effectively hide himself from his crustacean-like alien nemesis (ugly cuss!) in this 1987 science fiction classic.

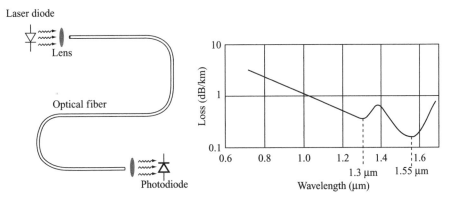

Figure 11.7 Optical-fiber communications system, showing the fiber attenuation (loss) characteristics. Note the importance of 1.30 μm and 1.55 μm as transmission wavelengths in a modern silica optical fiber (inspired by [5]).

3 μm (part visible light, part IR EM radiation) just by changing the In and Ga mole fractions. This clearly puts in play the optical-fiber sweet spots of 1.3 and 1.55 μm. Observe that if we are instead more interested in the visible spectrum (e.g., to build an visible laser – patience, we'll get to how in a second), we might consider GaAsP for red light (think laser pointer!) or perhaps InGaN for blue light. As shown in Fig. 11.10, we can simply change the mole fraction of As and P in the $GaAs_{1-y}P_y$ material system to tune to the precise hue of red, or even green, we might actually desire (primary colors are particularly useful for building up other color systems). These are clearly powerful tuning knobs in the photonics world.

But how exactly do we generate light by using a semiconductor? Good question. Let's look first at what happens when we shine light on a semiconductor. It's all about band structure (if you need a refresher, look back at Chap. 5 for a moment).

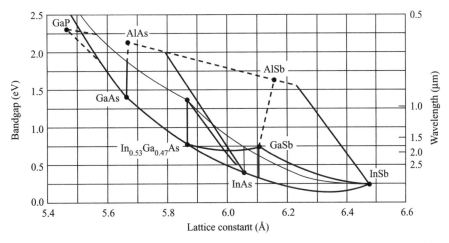

Figure 11.8 The bandgap-engineering landscape. Energy bandgap is plotted as a function of the lattice constant of the semiconductor crystal.

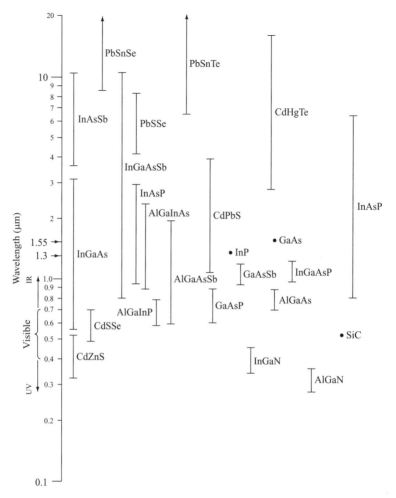

Superposition of technologically important optical wavelengths with a variety of available semiconductors. When ranges of wavelengths are indicated, bandgap engineering by varying the composition (i.e., component mole fractions) is implied. Note the various material options for 1.30-μm and 1.55-μm optical fibers.

11.3.1 Optical absorption

As depicted in Fig. 11.11, first consider the energy band diagram of the semi-conductor in question. We know there is some density of electrons and holes present in equilibrium in the conduction and valence bands. Question. Intuitively, how much energy does it take to boost an electron from the valence band to the conduction band? Answer? E_g. Imagine now shining a photon (light) of energy $E = h\nu > E_g$ on the semiconductor. If an electron in the valence band absorbs the energy associated with that photon, then it has sufficient energy to be boosted up into the conduction band, where it is now available to move in response to an electric field (think current flow, the bread and butter of the electrical engineering world). This process is called "optical absorption" for obvious reasons – we absorb the incident photon, creating a free carrier. What happens if the incident

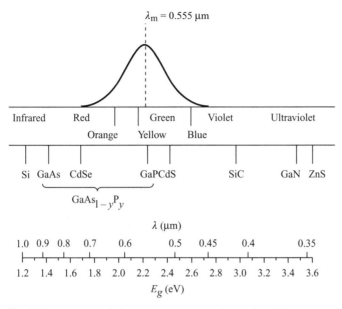

The visible spectrum and the standard response of the retina of the human eye, overlayed with several semiconductors of varying energy bandgaps.

photon energy is smaller than E_g? Clearly, there is not enough energy to boost the carrier from the valence band to the conduction band. Read: Nothing happens. In this sense the E_g of a semiconductor can be thought of as a "color filter" of sorts – light with energy below E_g does nothing, whereas light with energy above E_g does lots. Because energy and wavelength are reciprocally related, this is equivalent to saying that certain wavelengths (colors) are absorbed, and others are not. A color filter. AND, we can easily control what energy we want by changing the bandgap of the material by means of bandgap engineering.

Let's make this optical absorption process a bit more concrete. Consider a slab of semiconductor, and now imagine shining light (with $E > E_g$) on its surface. Two things can happen: (1) the photon is absorbed, generating free carriers

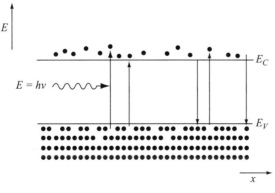

The energy band diagram of a semiconductor under photon illumination, illustrating the optical absorption mechanisms.

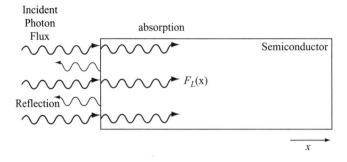

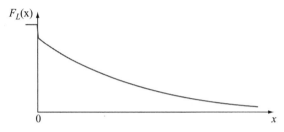

Figure 11.12   Incident, absorbed, and reflected photon fluxes as functions of position for an illuminated semiconductor.

(electron–hole pairs); or (2) the photon might be reflected from the surface (that is, the surface acts like a mirror to the incident radiation) (see Fig. 11.12). The flux of photons as we move inside the semiconductor starts at a maximum value at the surface and then decreases in an exponential fashion as we move inside the volume of the semiconductor, as more and more photons are absorbed (if you are mathematically inclined you might attempt to solve the simple differential equation to show this). We define the "absorption coefficient" (α) of this process to be the distance at which the photon flux decreases by $1/e$. The larger the absorption coefficient, the more photons we absorb, producing lots of "photogenerated" carriers for our use. Absorption coefficient data for several important semiconductors are shown in Fig. 11.13. Note that the absorption process stops once we go below the E_g, as expected. Hence the wavelength of the light we are interested in utilizing will, in essence, determine the material required.

11.3.2 Optical emission

Although we will shortly see some very important examples of photonic devices that exploit optical absorption, we are often more interested in the exact opposite process: optical emission. In this case, if an electron is present in the conduction band and then recombines with a hole in the valence band, it loses its energy ($E = E_g$) in the recombination process (conservation of energy in action). Under particular circumstances, though, that energy exchange can PRODUCE a photon of energy $E = E_g$. That is, light is emitted in the recombination event. This is

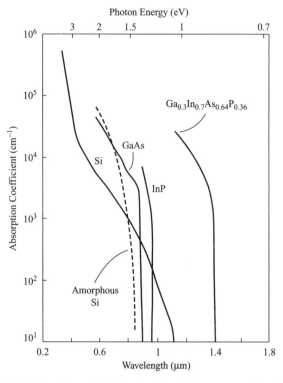

Figure 11.13 Optical absorption coefficient as a function of photon energy and wavelength for a a variety of semiconductors. Note the differences in shape between the indirect (e.g., Si) and direct (e.g., GaAs) semiconductors.

called "spontaneous emission." In optical absorption we trade a photon for an electron, whereas in optical emission, we trade an electron for a photon. As with absorption, the semiconductor acts as a color filter of sorts, because the photon will have a wavelength roughly corresponding to the magnitude of the energy bandgap of the material used.

11.3.3 Direct vs. indirect bandgap materials

There is one last subtlety related to photon absorption and emission that we need to consider. Fact: It is impossible to get significant optical emission (read: large number of emitted photons) from silicon, whereas we can easily get efficient optical emission from GaAs or InP. Why? Gulp... It's all in the band structure! Recall from Chap. 5 the fundamental difference between indirect bandgap and direct bandgap semiconductors. As illustrated in Fig. 11.14, a direct bandgap material (e.g., GaAs) has its principal bandgap (maximum in E_V and minimum in E_C) at the origin (zero point) in "k space" (crystal momentum), whereas an indirect bandgap material (e.g., Si) has its principal bandgap located away from the origin. Why does this matter? Sigh... truthfully, you would have to delve into the dark depths of quantum mechanics to answer that completely, and that topic is well beyond even a simple "Caffeine Alert" sidebar discussion. Suffice

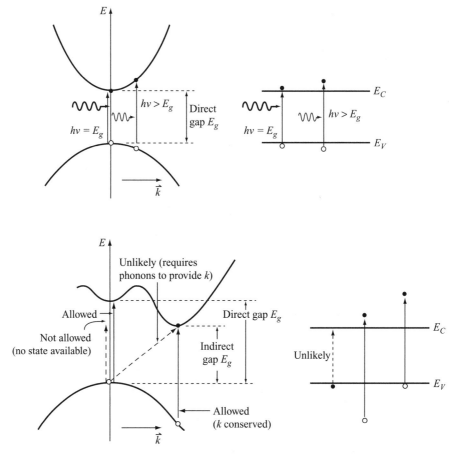

The differences in optical absorption for direct vs. indirect energy bandgap semiconductors (inspired by [5]).

it to say, though, that nature (through quantum mechanics) imposes so-called "selection rules" for energy exchange between carriers located in different k states. In an indirect semiconductor, for an electron to be boosted from the top of the valence band ($k = 0$) to the bottom of the conduction band, this cannot happen without a secondary mechanism to shift the k state of the electron. A crystal-lattice vibration (aka Mr. Phonon) can in principle do this for us, but even so, it is inevitably a very inefficient process because it involves two distinct steps. In a direct bandgap semiconductor, however, no secondary k-shifting process is required, and the energy transfer can readily occur, making optical absorption very efficient. As you might imagine, exactly the same game is involved in the optical emission process – inefficient in indirect bandgap materials, but very efficient in direct bandgap materials (Fig. 11.15). This is the fundamental reason why lasers or light-emitting diodes (LEDs) are built from direct bandgap III-V materials, and not from silicon. It does NOT mean we cannot get light out of silicon, but simply that the two-step emission process is just too inefficient to easily utilize in practical photonic devices. Clearly the same can be said for optical absorption, for identical reasons. Hence, III-V materials are more efficient than silicon.

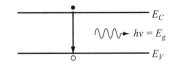

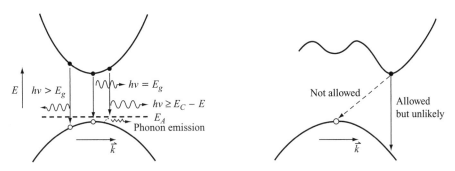

Figure 11.15 Comparison of photon emission processes in both direct and indirect energy bandgap semiconductors (inspired by [6]).

11.4 Photodetectors and solar cells

Now that we are armed with a bit of knowledge of basic optical processes in semiconductors, let's use this to build some remarkably useful photonic devices. Let's start with devices that directly exploit optical absorption: photodetectors and solar cells. First, photodetectors.

11.4.1 Photodetectors

A photodetector is a simple little object; nothing more than a *pn* junction designed specifically to detect photons (hence the name). Why might we want to do this? Well, at the low end of the sophistication ladder, a photodetector makes a nice light sensor (ever wonder how your porch light knows how to turn itself on at night and off at daylight?!). Or we might want to detect the signal from your TV remote control (this operates in the IR and hence is conveniently invisible to your eye). Or at the highest level of sophistication, we might want to detect the encoded data in a modulated laser pulse coming out of a fiber-optic line in a long-haul optical transmission system. A photodetector is a *pn* junction with a twist. As shown in Fig. 11.16, we create a circular *pn* junction and take great care to remove the metal layers from above the collecting junction (see sidebar discussion on why this is required). The target collection point for the incident photons is the space-charge region of the *pn* junction, where band-bending (hence electric field) occurs. We can maximize the physical volume of that collection region by building a so-called *p-i-n* junction; that is, by inserting an undoped (intrinsic $=i$) region inside the *pn* junction (Fig. 11.17). If we now reverse-bias the *p-i-n* junction, an electron–hole pair generated during the absorption of the incident photon will simply slide down (electrons) or up (holes) the band edges, generating

Photodetector

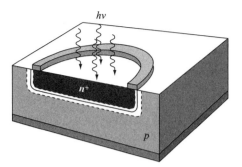

Figure 11.16   The basic structure of a *pn* junction photodiode.

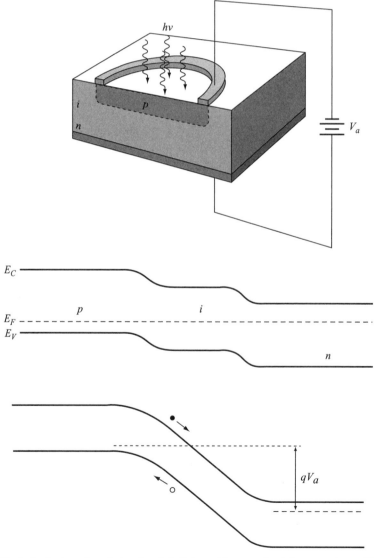

Figure 11.17   The basic structure of a *p-i-n* junction photodiode, and its energy band diagram under equilibrium condition and under reverse-bias (use) conditions (inspired by [5]).

useful current (refer to the band diagram in Fig. 11.17). Turn on the light, you get a photogenerated current; turn off the light, and the current goes away. Presto, a light-triggered switch! We're in business. Because this p-i-n photodetector can be made tiny, with minimal parasitic R and C, it can respond to light very, very rapidly (think billionths of a second). Clearly, given the preceding discussion, the wavelength of the light we are trying to detect will determine the type of semiconductor (bandgap) used to make the photodetector guts. And of course, a direct bandgap material will still give the best efficiency. Silicon, for instance, will work just fine for detecting sunlight, but not as well for detecting the signal from your TV remote control.

> ### Geek Trivia: Why Do We Remove Metals From The Tops of Photodetectors and Solar Cells?
>
> Question: Why on earth do we have to remove the contact metals from the top of our photodetectors (and solar cells)? Answer: If we don't, those pesky metals will absorb or reflect all of the incident photons, yielding little or no light collection within the p-i-n junction. Bummer. But why? Consider. Metals tend to be bright and shiny, suggesting that the incident light is efficiently reflected from their surface. And this is true. But even for silver (think silvered mirror!), the most reflective metal out there, only about 88% of the visible light that hits the surface is reflected. Metals also happen to be very efficient absorbers of visible and IR EM radiation. As we have seen, to absorb a photon, any material has to have available electrons that can absorb the incident photon energy, boosting them to a higher energy state. In metals, this is trivial. Metals do not have an energy bandgap (the conduction and valence bands overlap), and thus metals, unlike semiconductors, are composed of a veritable sea of free electrons that can act as efficient photon absorbers (this is also why metals have high electrical and thermal conductivities). This is the same reason, by the way, that materials like glass are transparent. As insulators, free electrons are very hard to come by, and thus optical absorption is difficult. No photon absorption to visible light equals transparency (glass does absorb UV light). Astute readers such as yourself might then logically wonder: How then do we make colored glass? Answer: Glass has to be "colored" by adding a small quantity of one of the transition metals (read: lots of electrons). Cobalt produces a blue glass, chromium produces a green glass, and traces of gold give a deep-red color. Moral to the story: Dielectrics on top of the photodetector–solar cell are fine; metals are a no-no.

11.4.2 Solar cells

Closely related to photodetectors are solar cells. Given the serious constraints we humans currently face in the arena of clean energy generation, solar power is something everyone should know about, because it will play an increasingly important role in 21st-century global energy production. As shown in Fig. 11.18 (color plate 36), there are large regions of the Earth's surface that enjoy tons of

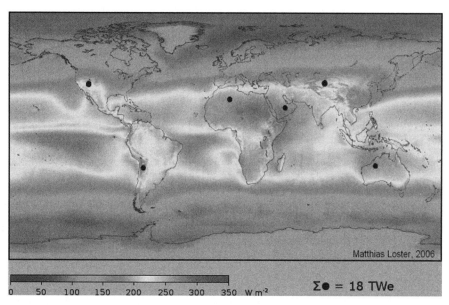

Matthias Loster, 2006

0   50   100   150   200   250   300   350   W m⁻²   Σ● = 18 TWe

Figure 11.18   The solar irradiance map of the Earth's surface, indicating solar "sweet spots" (courtesy of NASA). (See color plate 36.)

bright sunshine. Free photons for dependable energy production by use of solar cells – consider: 18 TW of cumulative solar energy hitting the surface on the various continents!

Solar cells have a characteristic look recognizable to all. A reminder: Figure 11.19 shows the ubiquitous solar panel, and Fig. 11.20 an individual solar

Figure 11.19   A solar panel from the largest photovoltaic solar power plant in the United States (2007), located at Nellis Air Force Base. The solar arrays will produce 15 MW of power at full capacity. (Courtesy of the U.S. Air Force. Photo taken by Senior Airman Larry E. Reid, Jr.) (See color plate 37.)

Figure 11.20 Solar cell array from a solar panel (courtesy of the U.S Department of Energy).

cell array used to populate such panels. Depending on where you live, solar power is increasingly being embedded into homes and buildings (Fig. 11.21) and serves as the principal power source for all satellite systems (Fig. 11.22), and is thus clearly integral to the global communications infrastructure.

The operation of solar cells is quite simple actually. Refer to Fig. 11.23, which shows the cross section of a solar cell. Yep, it's a *pn* junction and looks pretty similar to a photodetector (you obviously have to minimize the metal coverage over the collecting junction – see sidebar discussion). Just as with the photodetector, the more incident light intensity the solar cell sees (sorry), the larger the photogenerated current available for our use. Figure 11.24 shows

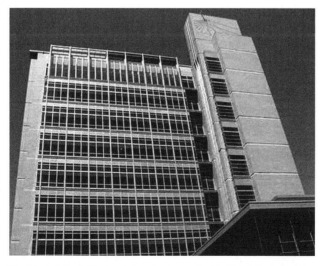

Figure 11.21 The Alfred A. Arraj U.S. Courthouse, in Denver, Co. The first federal courthouse to get PV glazing. (Courtesy of NREL).

Figure 11.22 **Artist's rendition of a space-based solar panel powering a satellite system (courtesy of NASA).**

the current–voltage characteristics of a solar cell. The fourth quadrant is solar cell land. The total current in the solar cell can be written as

$$I_{\text{solar cell}} = I_{\text{dark}} + I_{\text{light}}, \tag{11.5}$$

where I_{dark} is the normal voltage-dependent pn junction diode current (Chap. 8),

$$I_{\text{dark}} = I_0 \, (e^{qV/kT} - 1), \tag{11.6}$$

and I_{light} is negative and independent of the junction voltage. If we short circuit the solar cell (connect a wire between the anode and cathode of the pn junction, such that the voltage across the junction is zero), a current flows, even though no voltage is applied! This is I_{SC}, for short-circuit current. If, instead, we now open circuit the solar cell (that is, $I_{\text{solar cell}} = 0$; disconnect the wires running to the anode and cathode of the pn junction) and solve for the resultant voltage (V_{OC}

A Generic Solar Cell

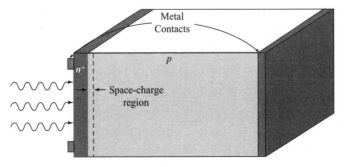

Figure 11.23 **The basic structure of a solar cell.**

Solar Cell Land

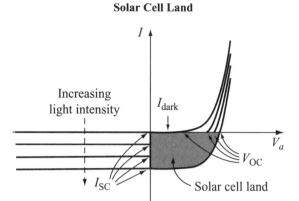

Figure 11.24

The current–voltage characteristics of a solar cell under illumination.

for open-circuit voltage), we find

$$V_{OC} = \frac{kT}{q} \ln\left(1 + \frac{I_L}{I_0}\right). \tag{11.7}$$

Moral: Simply park a solar cell out in the Sun, and a small, but well-defined, voltage magically appears across it. All thanks to optical absorption in semiconductors. Connect a "load" to the solar cell (e.g., a light or a battery or even the power grid), and current will now flow, generating "free" electrical power (Fig. 11.25). In essence, the solar cell is acting as a neat little optical-to-electrical transducer, and this process is called the "photovoltaic effect" (literally photovoltage);

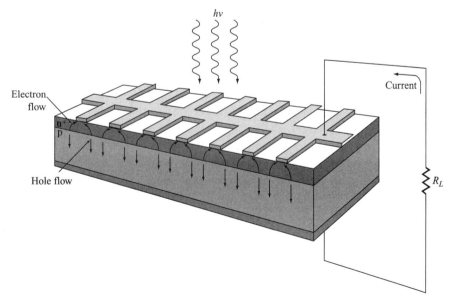

Figure 11.25

An illustration of the use of interdigital "fingers" in solar cell design. Also shown are the bias and electron and hole flow contours under illumination (inspired by [5]).

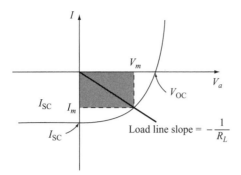

Figure 11.26 The current–voltage optimization plane used in solar cell design. Also shown are the resistor load line and its intersection on the *I–V* curve. Generally, the larger the V_m–I_m box, the better the solar cell.

"PV" for those in-the-know.[4] The characteristic look of solar cells has to do with the use of interdigitated metal "fingers" to minimize metal coverage while providing good electrical contact to the cell (Fig. 11.25).

Important point: Because the solar spectrum has significant IR content (Fig. 11.5), building solar cells in low-cost silicon is viable, even with their limited efficiency compared with that of III-V materials. Working with silicon has obvious cost advantages.

So what exactly makes for a good solar cell? It is all about power-generation efficiency. How much electrical bang for the sunlight buck. As shown in Fig. 11.26, the maximum voltage (V_m) and maximum current (I_m) are defined as the side lengths of a rectangle in the fourth quadrant of the current–voltage characteristics, which are bounded by V_{OC} and I_{SC}. The maximum output power of the solar cell is simply $P_m = V_m I_m$, and the cell power-conversion efficiency is then given by

$$\eta = \frac{P_m}{P_{L,i}} 100(\%), \tag{11.8}$$

where $P_{L,i}$ is the incident optical power from the Sun. A key solar cell figure of merit is the so-called "fill-factor" (FF), which is defined by

$$\mathrm{FF} = \frac{I_m V_m}{I_{SC} V_{OC}} \tag{11.9}$$

and is essentially a measure of how well the rectangle fills the fourth quadrant of the current–voltage characteristics. From this result, we can write, finally,

$$\eta = \mathrm{FF} \frac{I_{SC} V_{OC}}{P_{L,i}}. \tag{11.10}$$

What is a good efficiency number? Well, the higher the better, obviously. The theoretical limit for a silicon solar cell is about 26%, and the best achieved to date is in the range of 22%, using some fancy micromachining techniques (think

[4] Life as we know it rests on natural power transducers (consider the magical optical-to-chemical conversion process found in plants; namely, photosynthesis). No photosynthesis, no plants; no plants, no us.

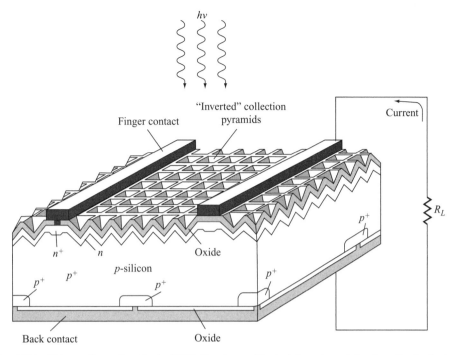

hv

Finger contact

"Inverted" collection
pyramids

Current

R_L

n^+ n

Oxide

p^+

p-silicon

p^+ p^+

p^+

p^+

p^+

Back contact

Oxide

Figure 11.27 A high-efficiency silicon solar cell using MEMS-like micromaching techniques to maximize absorption (inspired by [4]).

MEMS meets solar cells!) to maximize incident photon collection (Fig. 11.27). That is, 1.0 W of free optical power gives me 0.22 W of electrical power. For direct bandgap III-V solar cells, the record number (2007) is currently 42.8%, and the holy grail in this context is a 50% efficient cell. We're getting close. Clearly the low-cost volume manufacturing of silicon-based (either single-crystal or polysilicon) solar cells is highly attractive for the terrestrial power-generation market, where massive "solar farms" are the end goal, whereas the much-more-expensive III-V solar cells might be used in critical applications such as satellite systems, where power needs are relatively small but launch weight and system efficiency are everything. In many ways, PV is the ultimate green power source, at least if we disregard the carbon footprint associated with fabricating them (more on this in a later chapter). With the rising cost of exhaustible (and environmentally unfriendly) fossil fuel sources, it is likely only a matter of time until PV-generated energy becomes a key cog in the global electric power grid. All possible through the magic of semiconductors and micro/nanoelectronics fabrication. Pretty cool, huh?!

11.5 CCD imagers, CMOS imagers, and the digital camera

Remember back in the dark ages when you actually used photographic "film" to take your pictures? Think way back! That 37-lb single-lens reflex (SLR) monster (yes, I still have my beloved Canon A1 fully auto-SLR!) and that menagerie of bulky film canisters. Unless you were a pro, post-photo-shoot, you had to drive to the film-processing center across town and come back an hour later to rifle through your instant-Pulitzer classics, only to discover 2 decent shots among your roll of 36 of ASA 400. And, no, they didn't offer you a discount on those bad pics! Did you print 2-for-1? Great, now you have 68 bad pics to shove in a drawer!

The photographic industry has been revolutionized by microelectronics technology. Consider the differences only a few years make. A sleek little lightweight, palm-of-the-hand, ten-Mega-pixel digital camera to rival SLR image quality. Instant viewing so you can trivially store your prizewinners or erase your dogs before printing. Put 500 pics on a single thumbnail-sized CompactFlash or Mini-SD card (see sidebar Geek Trivia discussion on file formats). Carry the digital image to the store for a fancy print (complete with digital retouching!), or just do it at home on your own printer; or better yet, send them via the Internet to your extended family or even post them on Facebook so your worldwide circle of "friends" can instantly partake.[5] Consider: It is virtually unacceptable today for most people under the age of 25 to not have a digital camera integrated directly into their cell phones. Grrrr. Boy, times have changed! How on earth did this happen? Read on.

Geek Trivia: Digital Image File Formats

So what's the deal with the file extensions on those digital images your camera spits out – "*birthday-surprise.jpg*"? Well, *jpg* stands for JPEG (in geek-speak, JAY-peg) and is a commonly used method of data compression for photographic images [2]. The name JPEG stands for "Joint Photographic Experts Group," the name of the committee that created the compression standard. The group was organized in 1986 and finally issued the JPEG standard in 1992 (approved in 1994 as ISO 10918-1). JPEG is distinct from MPEG ("Moving Picture Experts Group"), which produces data compression schemes for video files. The JPEG standard specifies both the "CODEC," which defines the algorithm by which an image is first compressed (shrunk) into digital bits ("1s" and "0s") and then decompressed (expanded) back into an image, and the file format used to contain those compressed bits. Why do it? Well, just to make image files much smaller, so that they can be more effectively stored and especially transmitted from point A to point B. Alas, there is a price to be paid for doing this. The compression method used in JPEG is "lossy," meaning that some visual quality is lost by the compression algorithm. Image files that use JPEG compression or decompression are commonly called "JPEG files." JPEG is the file format most commonly used for storing and transmitting digital photographs on the Internet, but its compression algorithm is actually not well suited for line drawings and other textual or iconic graphics, and thus other compression formats (e.g., PNG, GIF, TIFF) also exist. For color pics from your digital camera, however, think JPEG.

---

[5] E-mail etiquette tip #1: Please do not e-mail-attach 47 individual 2M jpg pics and send them to Mom and Dad; they will not respond well to "say-cheese" next time you see them!

Figure 11.28
CCD imager IC with the package cover removed (courtesy of NASA).

11.5.1 CCD imagers

Behold the ubiquitous charge-coupled device (CCD) imager IC (Fig. 11.28). Not much to look at, perhaps, but a magical means, nonetheless, to capture a high-resolution digital image. Why does digital matter so much? As always, a DIGITAL image (bits, "1s" and "0s") can be stored, manipulated, and transmitted very, very easily using conventional IC technology. Interestingly, the imaging process in a CCD begins in the analog domain (as all sensing functions inevitably must, because we live in an analog world). That is, we use a *pn* junction of collected photogenerated charges from incident photons (at above bandgap energies!) reflected from the object that we are attempting to image (read: taking a picture of). Yep, you have to keep the metals off the top of the photon collection node, just as with a photodetector or a solar cell. The amount of charge generated is obviously proportional to the intensity of the incident light (number of incident photons).

To build an imager, the question then becomes this: How do we "read out" the charge that is generated by the image in the 2D array (pixels) of the image sensor? Well, a CCD basically acts like an analog "shift register," enabling photogenerated analog signals (i.e., electrons) to be transported (shifted) point by point out of the 2D pixel array through successive storage stages (on capacitors) under the control of a clock signal that turns "on" and "off" a "transfer gate" (Figs. 11.29 and 11.30). That is, when light exposure from the image is complete, the CCD transfers each pixel's photogenerated charge packet sequentially to a common output node, which then converts that charge to a voltage (yes, using a transistor amplifier!), and then sends it off-chip for data processing by the guts of the digital camera. Out pops Mr. JPEG file of the pic you took! In this sense, a

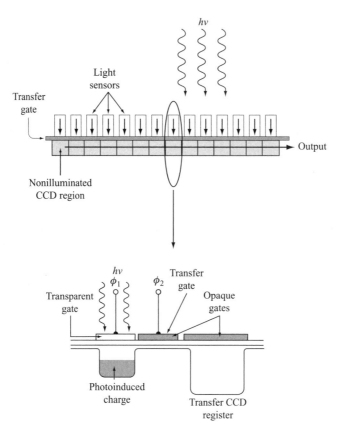

Figure 11.29 Illustration of the operational principles of a CCD for optical imaging (inspired by [5]).

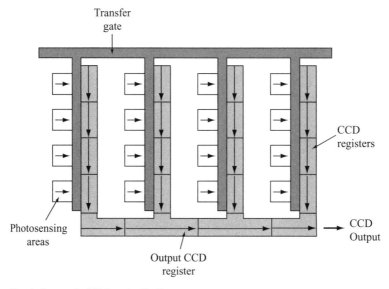

Figure 11.30 The pixel array of a CCD-based optical imager.

Digital
Camera

CCD
Image Sensor

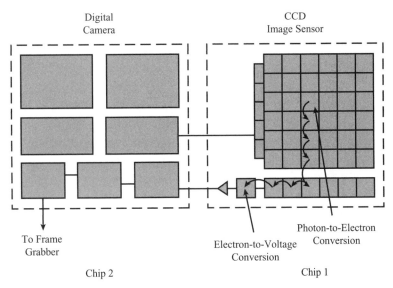

To Frame
Grabber

Electron-to-Voltage
Conversion

Photon-to-Electron
Conversion

Chip 2

Chip 1

Figure 11.31   Block-level diagram of a CCD imager, showing the CCD imager chip and the associated processing circuitry on a companion circuit board.

CCD-based camera actually consists of two distinct pieces: the CCD imager itself, and the data processing engine that manipulates it (Fig. 11.31). For those who may be curious for a closer look (don't get nervous!), a circuit-level schematic of a SINGLE CCD pixel is shown in Fig. 11.32. A decent digital camera might have a 8-megapixel (8M) = 8,000,000 or even 10M-pixel CCD imager IC inside them. Pretty cool.

CCDs can actually be used for many, many different things, including as a form of memory, and as a signal delaying technique in analog signal processing

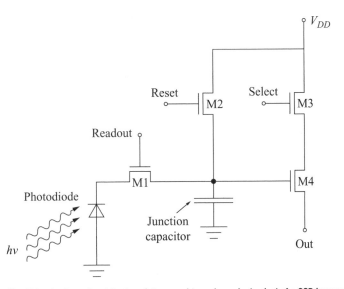

Figure 11.32   Circuit-level schematic of the transistors used to make a single pixel of a CCD imager.

(see sidebar historical aside). Today, however, CCDs are most widely used for extracting collected photogenerated charges from 2D pixel arrays (cameras). This imaging application of CCDs is so predominant that, in common parlance, "CCD" is often incorrectly used as a synonym for a type of image sensor, even though, strictly speaking, the term "CCD" refers only to the manner by which the 2D image is read out from the imager sensor chip. Today, CCDs are ubiquitous in digital photography, video cameras, astronomy, optical scanners, satellite-based remote imaging systems, night-vision systems, and a wide class of medical imaging applications, to name but a few [7]. A well-made CCD can respond to 70% of the incident light (a quantum efficiency of about 70%), making them far more efficient than photographic film, which captures only about 2% of the incident light. The advantages for optical astronomy should be obvious.

So how exactly does one then get a color image with a CCD? Good question! Digital color cameras generally use a "Bayer mask" over the CCD imager [2]. Each square of four pixels has one filtered red, one filtered blue, and two filtered green (the human eye is more sensitive to green – the center of the visible spectrum – than to either red or blue). The result is that image intensity is collected at every pixel, but the color resolution is lower than the intensity resolution. Better color separation can be achieved with so-called "three-CCD" (3-CCD) imagers and a prism that splits the image into red, green, and blue components. Each of the three CCDs is then optimized to respond to a particular color. Some high-end digital video camcoders use this technique. Another advantage of 3-CCD imagers over a Bayer mask CCD approach is that it has higher quantum efficiency (and therefore better light sensitivity for a given camera aperture). This is because in a 3-CCD imager most of the light entering the aperture is captured by a sensor, whereas a Bayer mask absorbs a high proportion (perhaps 2/3) of the light falling on each CCD pixel. Needless to say, ultra-high-resolution CCD imagers are expensive, and hence a 3-CCD high-resolution still-camera is beyond the price range of most of us amateurs. Some high-end still-cameras use a rotating color filter to achieve both color fidelity and high resolution, but these multishot cameras are still rare and can photograph only objects that are not moving. Clearly CCD cameras are still evolving. Stay tuned!

Historical Aside: The CCD

A little CCD history? Sure! The CCD was invented in 1969 by Willard Boyle and George E. Smith at AT&T Bell Labs [7]. Boyle and Smith were working on the "picture phone" and particularly on the development of what is now referred to as semiconductor "bubble memory" (ultimately a commercial dead end). Merging these two initiatives, Boyle and Smith conceived of the design of what they termed "Charge 'Bubble' Devices." The essence of the design was in figuring out how to transfer collected charge from point A to point B along the surface of a semiconductor. The CCD thus actually started its life as a sophisticated memory device. Unfortunately, one could "inject" charge into the CCD at only one point and then read it out. It was soon clear, however, that the CCD could also receive charge at many multiple points simultaneously by means of photoabsorption, providing a clever (and new) means to generate images. By 1970 Bell Labs researchers were able to capture black and white images by using 1D linear arrays, and the modern CCD was born. Several companies, including Fairchild Semiconductor, RCA, and Texas Instruments, picked up on the invention and began their own development programs. Fairchild was the first to market and by 1974 had a linear 1D 500-pixel CCD and a 2D 100 × 100 pixel CCD. Under the leadership of Kazuo Iwama, Sony also started a large development effort on CCDs, primarily for application in camcorders, a huge success, and the rest is history! In January 2006, Boyle and Smith received the Charles Stark Draper Prize (read: a big deal) by the National Academy of Engineering for their work on CCDs.

11.5.2 CCD vs. CMOS imagers

A recent battleground in the digital imaging world centers on the use of traditional CCD imagers vs. so-called "CMOS imagers," the new-kid-on-the-block. Both CCDs and CMOS imagers are very closely related, because they both use MOSFETs and *pn* junctions to do their thing. Both accumulate photogenerated charge in each pixel proportional to the local illumination intensity. When exposure is complete, the CCD (Fig. 11.31) transfers each pixel's charge packet sequentially to a common output node, converts the charge to a voltage, and sends it off-chip for processing. In a CMOS imager (Fig. 11.33), however, the charge-to-voltage conversion takes place in each pixel. This difference in image readout techniques has significant system design implications. In a CMOS

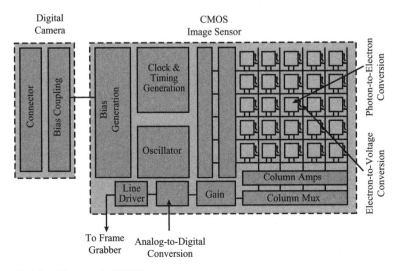

Figure 11.33   Block-level diagram of a CMOS imager.

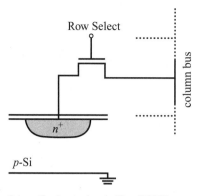

Figure 11.34 Schematic of a passive pixel in a CMOS imager array.

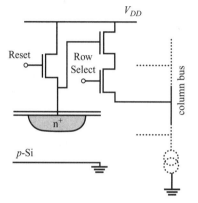

Figure 11.35 Schematic of an active pixel in a CMOS imager array.

imager, the pixel itself is of a simpler architecture and can be either a bare-bones "passive pixel" (Fig. 11.34), or a more sophisticated "active pixel" (Fig. 11.35), the so-called "smart pixel" one oftens hears about these days in camera circles. In a CMOS imager system, the imager itself does far more of the on-pixel data processing BEFORE it is sent to the brains of the camera. CMOS imagers tend to be more tightly coupled to the evolutionary trends in CMOS technology (Fig. 11.36) and leverage those scaling advances by the computing industry to produce larger, faster, cheaper pixel arrays. Current pixel sizes are sub-5 μm^2 at the state-of-the-art.

There are of course trade-offs, as with all things in life. In general, the relative new-comer CMOS imagers offer superior integration levels (higher pixel density), lower power dissipation, and reduced system form factor, at the expense of image quality (particularly under low-light conditions), and system design flexibility [7]. CMOS imagers are thus becoming the technology of choice for high-volume, space-constrained applications for which image quality requirements are low, making them a natural

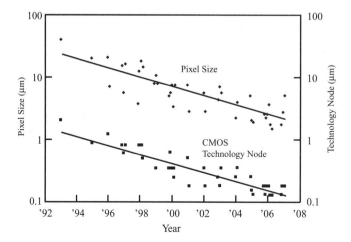

Figure 11.36 Trends in pixel size of CMOS imagers compared with CMOS technology node evolution.

fit for security cameras, PC videoconferencing, wireless handheld device video-conferencing, bar-code scanners, fax machines, consumer scanners, toys, biometrics, and some automotive in-vehicle uses. CCD imagers, however, offer superior image quality and flexibility at the expense of system size. They remain the most suitable technology for high-end imaging applications, such as digital photography, broadcast television, high-performance industrial imaging, and most scientific and medical applications. Furthermore, the flexibility of CCDs means users can achieve greater system differentiation with CCDs than with CMOS imagers. For the foreseeable future, there will be a significant role to be played by BOTH CCD imagers and CMOS imagers in an ever-changing market landscape.

11.6 LEDs, laser diodes, and fiber optics

We have thus far talked about photonic devices that exploit the absorption of photons in semiconductors to create extremely useful functionality: to detect incoming light (photodetectors); to transduce light into electricity (solar cells); and to take pictures (CCD and CMOS imagers). We might term such applications "passive photonics." But wouldn't it be cool if we could cleverly coax light OUT of semiconductors to build some light-generating marvels? We can, and your day-to-day existence is bathed in a myriad of examples, often so ubiquitous as to (sadly) go virtually unnoticed. Care for some examples? How about your TV remote control, that bane of modern family life and the butt of many couch-potato jokes (pun intended – okay, I'm not so proud of that use of our photonic gadgets!)? Or the monster jumbo-tron screen at the nearest football stadium or basketball arena, or gazillions of other types of less massive displays? Or the traffic signals on the roads? Or even the taillights in your car? Even flashlights! And how could you think of doing a powerpoint presentation without your laser pointer? The list is virtually inexhaustible. Yep, all of these clever gadgets use . . . gulp . . . semiconductor technology, and let's term these types of light-*emitting* photonic devices "active photonics." Definitely something you should know a little about.

11.6.1 LEDs

Let's start simple: The unassuming LED; aka the "light-emitting diode." Figure 11.37 shows a printed circuit board, and some common off-the-shelf discrete (individual) LEDs you might have come across. Observe that they all have a similar look – two metal leads and some sort of shaped, transparent plastic package encasement, often colored. No coincidence, as we shall see. Simply put, a LED is a semiconductor *pn* junction diode (hence, two terminals) that emits "incoherent" (the various emitted photons are NOT in phase with each other), fairly broad-spectrum (i.e., over a finite range of wavelengths) EM radiation (near-UV, visible, or near-IR) when electrically forward biased. Said another way, construct a *pn*

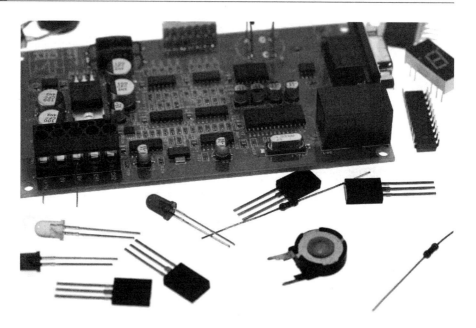

Figure 11.37   A variety of discrete LEDs. (Courtesy of agefotostock.com, © W. Zerla).

junction of the appropriate material, electrically bias it to turn it on, and presto, light comes out! As previously discussed, photons generated from spontaneous emission by electron–hole recombination in semiconductors at near-bandgap energies are called "electroluminescence" – literally, light from electricity! – and are the physical basis for the operation of LEDs [5]. A LED typically has only a tiny light source (think small junction area – 100s of microns on a side), with extra optics (a lens) added within the package itself to shape the outgoing radiation pattern (hence the "domed" shaped packages seen in Fig. 11.37). The color of the emitted light from a LED, as you can easily imagine by what you now know regarding photon emission from semiconductors, depends on the type and composition of the semiconductor used. A direct bandgap material will give the highest efficiency (most light out),[6] and we can tune the "color" of the light from near-UV (invisible to the human eye), to individual colors within the visible spectrum, to near-IR (also invisible), simply by carefully tuning the composition of the semiconductor (aka bandgap engineering).

LEDs have come a long way in recent years![7] Figure 11.38 plots a key LED performance metric – lumens (brightness) of light output per watt of electrical power input – essentially LED efficiency, as a function of evolving time [9].

---

[6] An exception to this rule of needing a direct bandgap semiconductor for efficient LEDs is GaP, an indirect bandgap material. In this case, we create "isoelectronic traps" within the indirect material (e.g., nitrogen in GaP), to "trick" the photon-producing recombination event into thinking it occurs in a direct bandgap system. Nitrogen-doped GaP can thus yield a decent green LED (565 nm).

[7] When I was an undergraduate student at Georgia Tech (nope, won't tell you when!), the best-available scientific calculator was the TI-55, complete with a cool BRIGHT red LED numeric display. VERY slick. Downside? Well, the LEDs used so much power that to get through a 3-h final exam, you had to bring a backup 9-V transistor battery, else you might have to finish that 30-step calculation by using long division! Fortunately, LED times have since changed . . . considerably.

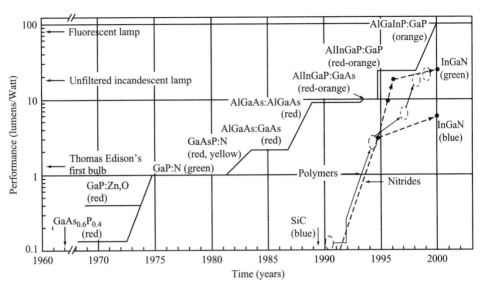

Figure 11.38 Historical trends in LED performance (optical output power per unit electrical input power – lumens per Watt) (inspired by [9]).

For reference, both standard incandescent and fluorescent bulbs are also shown. It may surprise you to learn that today LEDs can be far brighter than standard lightbulbs and even surpass those efficient screw-in fluorescent bulbs that are the latest rage for energy conservation. This is why they make good traffic signals and taillights and brake lights. It took a while, though, to hone the technology and to make some advances in materials fabrication. Although visible LEDs began as red, virtually all of the visible colors now exist, as well as near-IR and near-UV LEDs, for occasions when we don't want to "see" the light output (e.g., when using a remote control). Blue LEDs, which require wide-bandgap semiconductors [see Eq. (11.4)], began with SiC, and have since migrated to the GaN (gallium nitride) material system for blues and greens (Fig. 11.39).

Figure 11.40 shows a typical light emission spectrum of a LED and the now-obvious connection among the light emission energy (wavelength; color), the semiconductor bandgap, and the electron–hole populations generating the light by means of recombination events. This is an important connection to make. A typical LED has an emission spectrum that might be 50–100 nm wide about the designed wavelength; fine for generating colors for displays, but with its own set of drawbacks for many applications, as we will see. Not surprisingly, the photon emission in an LED is omnidirectional (spills out everywhere) and can be blocked by metal layers, just as with photodetectors and solar cells, as illustrated in Fig. 11.41. To concentrate the emitted light, metal is removed from the majority of the emission surface, and the LED is mounted in a metallic, optically reflective "cup" and then packaged with an epoxy "dome" (see Fig. 11.42) that will act as a concentrating lens for the light (think magnifying glass). Presto – bright LEDs to cover the rainbow (Fig. 11.43).

GaN blue LED Christmas lights! (Courtesy of agefotostock.com.)

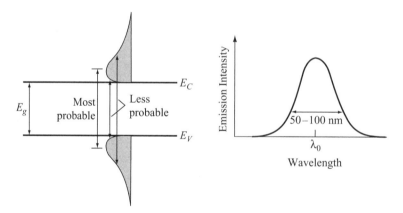

Figure 11.40 Photon emission characteristics resulting from recombination events between the electron and hole populations in the energy bands of a semiconductor.

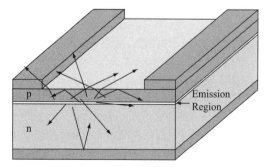

Figure 11.41 Illustration of the outgoing photon directions in a LED. Note that surface emission is relatively inefficient.

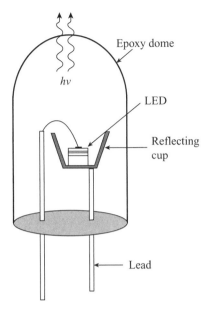

Epoxy dome

hv

LED

Reflecting cup

Lead

Figure 11.42 An illustration of how a discrete LED is packaged.

One can get fancier. Simple question. For a given LED, how do we improve its efficiency? Well, for a given bias applied, we would like to get more light out. Bandgap engineering can help us here. As shown in Fig. 11.44, for instance, we might embed a region of semiconductor with a narrow (direct) bandgap within two regions of semiconductor of a larger bandgap. The result is that the narrow-bandgap "quantum well" (think of a well for trapping electrons and holes) can produce more light because the carrier density trapped in the well and made available for recombination is greatly enhanced. More recombination, more photons for a given bias; more photons, higher efficiency; higher efficiency, a brighter LED. This approach will also tend to tighten up the spectrum of the emission (a narrower range of wavelengths, often a good thing).

Clearly these improvements come at added cost, however, because the fabrication is now more complex.

A hot topic these days in the LED world is the realm of so-called "solid-state lighting" (SSL). The goal of SSL is essentially to replace home or office conventional incandescent and fluorescent lighting with LEDs (hence solid-state). Why? LEDs can produce brighter, more reliable, and far more energy-efficient

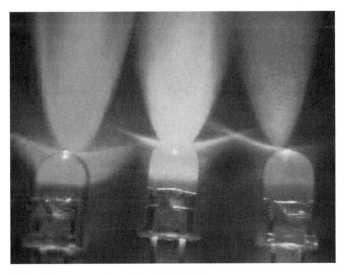

Figure 11.43 LEDs emitting primary colors. (See color plate 38).

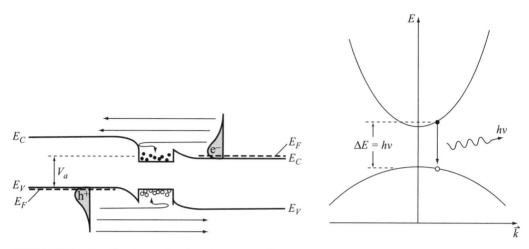

Figure 11.44   Use of bandgap engineering to improve the efficiency of LEDs (inspired by [5]).

lighting.[8] Clearly a good thing for Mother Earth. There are also clever games to be played. For instance, with solid-state lighting we could in principle electrically tune the emission spectrum of the LED source with the hour of the day to affect our biorhythms. Perhaps brighter white in the morning to help us wake up, followed by a subdued hue in the evening to help us chill out?! LED-based mood control might sound a bit scary, but the potential productivity boost has already sparked such investigations and may well be within near-term reach with LEDs. One of course has to make the LEDs cheap enough to enable pervasive SSL, and that is why this goal remains a goal at present; albeit one being actively pursued by many around the world.

For SSL we clearly need to produce something approaching white light. So how would we make a white LED? Well, we could do this in a brute-force manner: Add the light from a blue LED to a red LED and a green LED. But this is actually not a very efficient use of LEDs. Instead, most "white" LEDs in production today are based on an InGaN–GaN bandgap-engineered structure, which emits blue light of a wavelength between about 450 nm and 470 nm [2]. These GaN-based LEDs are then covered by a yellowish phosphor coating usually made of cerium-doped yttrium aluminum garnet (Ce^{3+}:YAG) crystals that have been powdered and bound in a type of viscous adhesive to cover the LED. The LED chip emits blue light, part of which is then efficiently converted to a broad spectrum centered at about 580 nm (yellow) by the phosphor layer. Because yellow light stimulates the red and green receptors of the eye, the resulting mix of blue and yellow light gives the appearance of white, and the resulting shade is often called "lunar white" (think of staring at the full Moon on a crisp, winter evening – that color). Pretty clever, huh? Developing the holy grail of white LEDs

[8] The estimated time to failure (ETTF) for LEDs can be as high as 1,000,000 h. Fluorescent bulbs, however, are typically rated to about 30,000 h, and incandescent bulbs to only about 1,000–2,000 h.

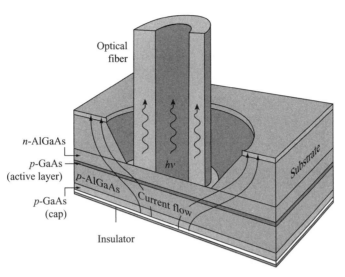

n-AlGaAs

p-GaAs
(active layer)

p-AlGaAs

p-GaAs
(cap)

Insulator

Optical
fiber

hv

Substrate

Current flow

Figure 11.45   The coupling of an LED into an optical fiber for use in an optical communications system (inspired by [5]).

remains a heavily researched topic (major bucks for the winner via SSL). One recent method for producing white LEDs uses no phosphors at all (a big plus, because it is expensive), but is instead based on epitaxially grown zinc selenide (ZnSe) deposited on a ZnSe substrate. This LED simultaneously emits blue light from its active region and yellow light from the substrate. Result? Looks white to your eye!

So what can we use LEDs for besides building jumbo-trons, SSL, and remote controls? Well, suppose we wanted to couple the light emitted from a LED into an optical fiber to build a short-haul communications system. Not a problem. As illustrated in Fig. 11.45, we can etch out a cavity in the guts of the LED and mate the cleaved optical fiber directly to the LED to maximize the transmission of the generated photons into the fiber line. Now, by electrically modulating the voltage on the LED, we can directly modulate the light traveling down the fiber line. On–off, on–off. A simple Morse code for signaling "1s" and "0s" down an optical line! Quite useful.

LEDs – fun stuff. Despite the fact that LEDs have so many nice features, yielding many cool apps, they do have their drawbacks: (1) the light is fairly hard to get out of the LED; (2) the emitted light is incoherent (i.e., the photons are of random phase); (3) the light is fairly broadband (a range of wavelengths is present); and (4) the light intensity is fairly modest. What is a photonics geek to do? Humm . . . build a laser? You bet!

11.6.2 Semiconductor laser diodes

Consider the ubiquitous laser pointer (Fig. 11.46). Lasers (see sidebar discussion on the naming of the laser) are used in virtually all fields of science and

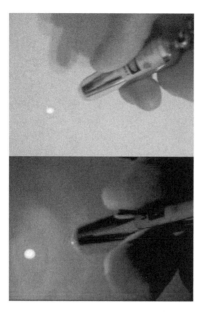

Figure 11.46

A red laser pointer uses a diode laser. Operation in both the daylight (top) and at night (bottom) are shown.

engineering, examples of which include (1) spectroscopy (analyzing the composition of an object based on the light it emits); (2) ranging and targeting (determining distance and setting direction); (3) materials processing (yep, lasers play a prominent role in semiconductor fab); (4) nuclear fusion; (5) medical imaging and instrumentation; (6) industrial cutting and welding; (7) radar; and (8) long-haul optical communications [11]. And . . . sigh . . . laser pointers. The list of practical laser uses is virtually endless. A nice thing to have invented!

But why exactly are lasers so darn useful? Well, the light emitted from lasers is fundamentally different from that of other light sources (including LEDs) in several important ways. A laser emits light in a focused, narrow, low-divergence (doesn't spread out moving from point A to point B) beam, with a well-defined and tunable wavelength (color if visible, else in the IR or even UV). The emitted photons are coherent (in-phase), and laser light can thus reach extremely high intensity, and even be switched on and off very, very quickly (think femtoseconds: 10^{-15} s). Yep, a darn useful invention.

Geek Trivia: Who Coined the Acronym LASER?

The acronym "LASER" was first coined by Gordon Gould, then a Ph.D. student at Columbia University, in his 1959 conference paper titled "The LASER, Light Amplification by Stimulated Emission of Radiation" [11]. An image of his notebook entry naming the laser is shown in Fig. 11.47. Actually, Gould intended "–aser" to be a suffix, and used with the appropriate prefix to indicate the portion of the EM spectra of emitted light from the given device (e.g., x-ray laser = "xaser," ultraviolet laser = "uvaser," etc.). Although none of the other terms became popular, "laser" stuck. Gould's notebook entries even included possible applications for the then-infant laser, such as spectrometry, interferometry, radar, and nuclear fusion. He continued working on his idea and filed a patent application in April of 1959. The U.S. Patent Office denied his application and awarded a patent instead to Bell Labs in 1960. This sparked a legal battle that ran for 28 years, with scientific prestige and significant bucks at stake. Gould won his first minor laser patent in 1977, but it was not until 1987, 28 years after the fact, that he could claim his first significant patent victory, when a federal judge ordered the government to issue patents to him for the optically pumped laser and the gas discharge laser [2]. Sometimes persistence pays off!

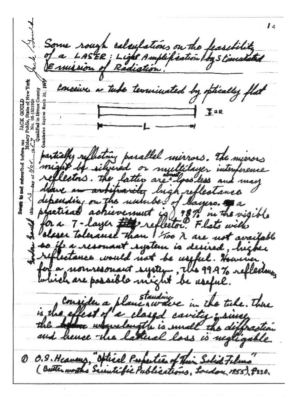

Figure 11.47

Photograph of a lab journal entry by Gordon Gould, graduate student at Columbia University, coining the acronym LASER.

11.6.2.1 OPTICAL GAIN

So how do lasers do their thing? Well, to produce a working laser, two things are required: (1) optical gain, and (2) a feedback mechanism. Big words, but not so hard to digest. We'll focus our discussion on semiconductors because we are after diode lasers. Let's look at optical gain first. In general, optical gain exists when the number of photons leaving the semiconductor is larger than the number of photons entering (yep, opposite of loss). There are actually three optical processes in semiconductors: absorption (photodetector, solar cell), spontaneous emission (LEDs), and one new one I have yet to mention – stimulated emission (Fig. 11.48). In general, all three optical processes can and do happen in semiconductors. Note that absorption of a photon boosts an electron from the valence band to the

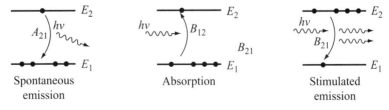

Figure 11.48

The march toward lasers. An illustration of the differences among spontaneous emission, absorption, and stimulated emission of electrons between two energy levels. For semiconductors, E_1 and E_2 would be E_V and E_C, respectively.

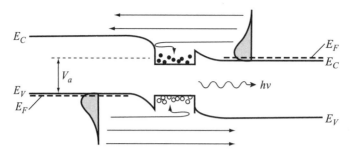

Figure 11.49   A double-heterojunction diode laser, showing population inversion at high bias currents (inspired by [5]).

conduction band (higher energy). Spontaneous emission emits a photon as that high-energy electron recombines with a hole (loses its energy).

If, by some trick, however, we can put lots of electrons into a high energy state, then something counterintuitive can occur. Here goes ... If a photon of the correct energy (in our case, near E_g) enters a system with lots of electrons in a high energy state, then one can theoretically show (okay, Einstein did this first!) that stimulated emission can actually be favorably compared with absorption. In this case the incident photon is not absorbed (as we might naively expect), but instead causes an electron to recombine, producing another photon, in phase with the incident photon! That is, one photon in produces two photons out – optical gain! Putting lots of electrons into the conduction band sets the stage here, and is called "population inversion," because electrons would really prefer to be in the their lowest-energy configuration (the valence band). Fortunately for us, a carefully crafted *pn* junction under forward bias naturally produces a population inversion at high-injection (current) levels. If, for instance, we sandwich a narrow (10s of nanometers) direct bandgap region between two heavily doped wider direct bandgap regions (aka a double heterojunction), then under high current flow there will be lots of electrons and lots of holes confined to the narrow small-bandgap region. Yep, population inversion (Fig. 11.49). Incident photons (with an energy near the bandgap energy) created by spontaneous emission can now themselves induce stimulated emission. Note that this laser diode structure is basically the same as that used for an efficient LED, albeit the laser diode is generally much more heavily doped (to create a larger population inversion). At low bias (before inversion is significant), a laser diode thus acts like a normal LED, governed by spontaneous emission within the semiconductor. The larger the bias on the diode, the larger the population inversion, and the stronger the stimulated emission becomes.[9] At some "threshold current" stimulated emission becomes dominant,

[9] You might wonder how population inversion occurs in a non-semiconductor laser system; e.g., a ruby laser or a gas laser. In that case, one has to optically "pump" (stimulate) electrons to go to high energy, often using a "flash lamp" (just what it sounds like), coaxing electrons to high energy and then allowing them to pile up at some high lifetime energy level in the system, which produces a recombination bottleneck of sorts (for ruby, metal impurity levels are introduced to accomplish this). With enough pumping, population inversion is achieved, and stimulated emission can then commence.

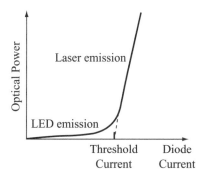

and the stage is set for lasing. The current–voltage characteristics of a diode laser look as shown in Fig. 11.50. One obviously has to continuously supply current to preserve the population inversion and stimulated emission. Clearly, reducing the laser threshold current is highly prized, because it will allow the laser to burn less power (think longer battery life).

Figure 11.50   Optical output power as a function of bias current for a laser diode, with the threshold current for lasing indicated.

Historical Aside #1: Who Really Invented the Laser?

It all starts with theory – go figure! John von Neumann actually carried out the first documented theoretical treatment of a laser in 1953, as documented in [12]. The first working laser had nothing to do with semiconductors, however, and was demonstrated by Theodore H. Maiman on May 16, 1960 [13] at Hughes Research Laboratories in Malibu, California, beating out several competing research teams, including those of Charles Townes at Columbia University, Arthur Schawlow at Bell Labs, and Gordon Gould at a company called TRG (Technical Research Group) [14]. Maiman used a solid-state flashlamp-pumped synthetic ruby crystal to produce red laser light at 694-nm wavelength. Maiman's ruby laser, however, was capable of only pulsed-mode operation, not continuous wave (CW); that is, it could not run continuously. Later in 1960, Ali Javan, working with William Bennet and Donald Herriot, made the first gas laser by using a combination of helium and neon (think red). The ubiquitous low-cost, shoe-box-sized lasers most students encounter in their labs at school are He–Ne gas lasers.

The idea of the semiconductor diode laser was proposed by Basov and Javan, and on Sunday, September 16, 1962, Gunther Fenner, a member of the team headed by Robert N. Hall at General Electric, operated the first semiconductor diode laser [15]. Within about 30 days, researchers at three other laboratories independently demonstrated their own versions of the semiconductor diode laser (talk about a race to the wire!): Nick Holonyak, Jr. and co-workers at General Electric (Fig. 11.51) [17]; Marshall I. Nathan and co-workers (Fig. 11.52) at IBM [16]; and Robert Rediker and co-workers at MIT Lincoln Laboratory. Hall's diode laser was made of GaAs and emitted at 850 nm (in the near-IR). As with the first gas lasers, these early semiconductor diode lasers could be used only in pulsed mode, and in fact only when cooled to liquid nitrogen temperatures (77.3 K). In 1970, Zhores Alferov in the Soviet Union, and Izuo Hayashi and Morton Panish of Bell Labs, independently developed diode lasers capable of CW operation at room temperature, using a bandgap-engineered heterojunction design to improve efficiency, paving the way for their eventual commercialization. The rest is history!

Figure 11.51   The first visible laser diode photographed in its own light. It was produced by Nick Holonyak, Jr. and co-workers, then at the GE Laboratory in Syracuse (courtesy Nick Holonyak, Jr. and Russ Dupuis).

11.6.2.2 FEEDBACK

Alas, engaging stimulated emission is only step one. In step two, we need to impose a feedback mechanism to help things out. Recall that spontaneously emitted photons travel in random directions. Stimulated emission, hence optical gain, can occur, however, in only the localized region of population inversion. Optical feedback is used to produce additional photon amplification within the laser, ensuring that photons experience maximal time within the gain region of

Figure 11.52   IBM scientists (from left-to-right: Gordon Lasher, William Dumke, Gerald Burns, Marshall Nathan, and Frederick Dill, Jr.) observe their new GaAs laser diode. The picture was taken on November 1, 1962; (Courtesy of International Business Machines Corporation. Unauthorized use is not permitted.)

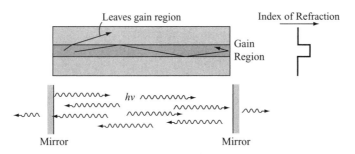

Figure 11.53 By confining the photon emission to a region of one semiconductor between two semiconductors with a higher index of refraction, the emitted photons can be directed parallel to the semiconductor surface with high efficiency. The resultant optical-gain characteristics of semiconductor depend on photon wavelength. The cleaved edges of the semiconductor crystal can be used as embedded mirrors to reflect generated photons.

the system. How? Well, we physically embed the laser medium within what is called a *Fabry–Perot* cavity – a fancy word for two partially reflective mirrors, arranged in a way to reflect the photons back and forth repeatedly across the gain medium (Fig. 11.53). Here's the kicker. In a two-mirrored Fabry–Perot cavity, the photons bouncing back and forth establish what are called "resonant modes," meaning that only certain wavelengths within the cavity are actually amplified. Why? Well, think of the photons as EM waves, now bouncing back and forth between the mirrors. Some wavelengths constructively interfere with each other and grow in magnitude (the same as with sound or water waves), whereas some destructively interfere, decreasing in magnitude. Much like a standing wave. At particular wavelengths ("modes") determined by the cavity size, the photon intensity can grow because of this "cavity resonance" (Fig. 11.54). So how exactly do we embed our teeny, weeny little laser diode into a two-mirrored Fabry–Perot cavity? VERY carefully! Fortunately, nature is on our side here. Because the diode is made of a semiconductor of a given index of refraction, if we cleave (break) the two ends of the diode they will act like very nice little mirrors (the index of

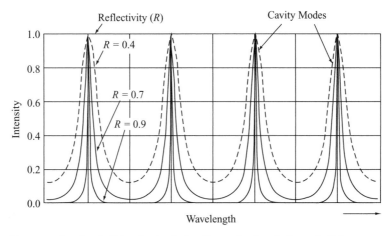

Figure 11.54 The resonant modes of a Fabry–Perot optical cavity as a function of mirror reflectivity.

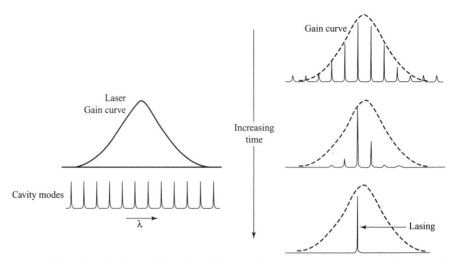

Figure 11.55 Overlay of the optical-gain characteristics of the diode laser with the resonant modes of a Fabry–Perot optical cavity as functions of wavelength. As time elapses, the first resonant mode to achieve optical gain will dominate the emission spectrum of the laser diode.

refraction of air is not the same as for a semiconductor)! Moral: Snap the edges of the diode at a precise dimension needed for a given target cavity resonance wavelength and, presto, you have a resonant cavity ideal for building a laser.

Next stop, the diode laser. If we now imagine overlaying the optical-gain characteristics of the diode as functions of wavelength on the resonant modes of the Fabry–Perot cavity, as shown in Fig. 11.55, lasing occurs as follows: Bias the diode to high current levels such that population inversion and stimulated emission start to occur within the semiconductor. Optical energy (photons) in the various cavity modes (wavelengths) will now start to grow. Continue to crank the diode bias current and let some time elapse, and an increase in the optical intensity in the cavity modes occurs. When the energy in one (or a few) of the resonant modes reaches the condition of optical gain (i.e., crosses the semiconductor gain curve), lasing onsets, and all of the additional optical energy is then dumped into that one (or few) resonant modes (wavelengths) where gain is now present (photons are being actively generated). The intensity of the light emission grows dramatically. The originally broad optical output spectrum (LED-like) now narrows to a single (or few) wavelengths or modes, the principle characteristic of laser light. A question to see if you are paying attention? Sure! If the laser is embedded inside a double-mirrored cavity, how on earth does the laser light get out? Well, one end is mirrored (depositing a metal on the surface will do this nicely), but the other is only *partially* mirrored, with some reflectivity (R) preserved to pull out a fraction of the light bouncing back and forth during lasing. Note that this arrangement will produce a highly directional laser beam, because the optical gain occurs in only one particular direction within the cavity parallel to the gain region of the diode (Fig. 11.56). Welcome to the world, Mr. Laser!

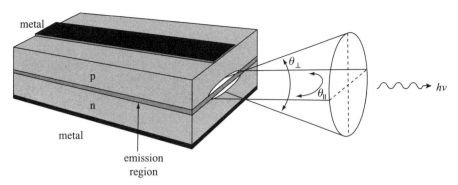

Figure 11.56 Illustration of the structure of an edge-emitting laser diode (inspired by [5]).

11.6.3 Laser perks, drawbacks, and future directions

The now-famous diode laser has some obvious perks compared with other types of lasers (see sidebar discussion for popular myths and misconceptions in the laser world): (1) Because the diode itself is made with standard microelectronics (microphotonics?) fabrication techniques, with the resonant cavity built into the diode body, diode lasers can be tiny, an obvious advantage for things like portable CD players or laser pointers (Fig. 11.57). (2) This size advantage translates into major cost advantages, because a single semiconductor wafer can yield thousands of diode lasers, just as it does with their microelectronic IC cousins (Fig. 11.58). (3) The diode laser requires only current to make it work – no messy flash lamps – and hence it integrates very well with conventional microelectronic circuits and systems requiring a small form factor. Any downside to diode lasers? You bet; life is full of trade-offs! (1) The output power of diode lasers is fairly limited (these are not the lasers that punch holes through inch-thick metal slabs!). (2) We burn quite a bit of current to produce the lasing; meaning diode lasers are not

Figure 11.57 A packaged GaAs laser diode (courtesy of NASA). (See color plate 39.)

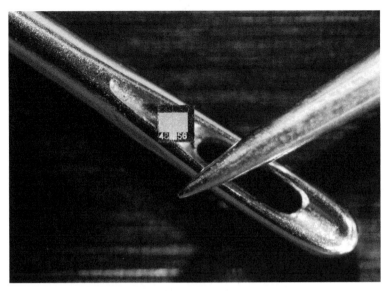

Figure 11.58 An unpackaged GaAs laser diode, showing the size of the semiconductor die (courtesy of NASA).

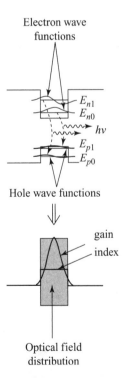

Electron wave functions

E_{n1}
E_{n0}

hv

E_{p1}
E_{p0}

Hole wave functions

gain

index

Optical field distribution

Figure 11.59   Using bandgap engineering to improve the efficiency and tunability of laser diodes. The influence of quantum-well width and depth on the gain characteristics is illustrated (inspired by [5]).

terribly efficient. (3) Once we fix the material we fix the wavelength, so tuning the wavelength of a fabricated diode laser is not so simple. Finally, (4) edge emission is the rule here, and the diodes have to be cleaved to produce the requisite resonant cavity, so building any kind of laser that emits from the *surface* of the semiconductor wafer is tough (there are good reasons for wanting to do this – can you offer some insight?). To attempt an end-around to solve these types of diode laser limitations, current research in this field is largely centered on (1) reducing threshold current by clever bandgap-engineering techniques to improve carrier confinement, population inversion, or both (e.g., using quantum wells or index grading – Figs. 11.59 and 11.60); (2) improving the level of electrical tunability; and (3) producing diode laser structures that permit surface rather than edge emission (the vertical-cavity surface-emitting laser – VCSEL, Fig. 11.61). Perhaps the final frontier in the diode laser world is to produce a silicon-based laser, which would have profound implications for the merger of silicon-based electronics and the

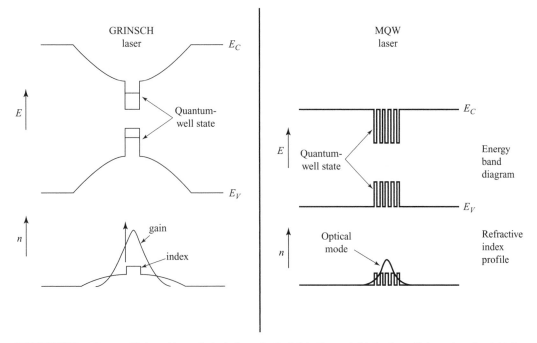

Figure 11.60   More sophisticated laser diode designs. On the left is the graded Index laser diode, and on the right, the multiple-quantum-well laser diode (inspired by [5]).

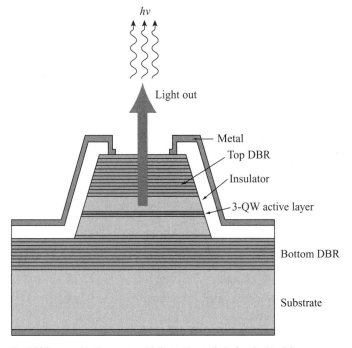

Figure 11.61   The VCSEL, arguably the most sophisticated laser diode (inspired by [5]).

photonics world. A silicon-based diode laser at first glance is seemingly impossible, given our previous discussion on the requisite direct bandgap needed for efficient photon production. Some serious cleverness is required for tricking Mother Nature, and we are not there yet, but externally pumped lasing in silicon has already been demonstrated, and the race for a true silicon laser diode is officially on [18]. Stay tuned!

Myth Busters #1: Popular Misconceptions About Lasers

The representation of lasers in popular culture, and especially in science fiction and action films, is very often misleading, if not downright wrong [2]. Some examples? Sure! Unlike in air, a laser beam would not be visible to the naked eye in the near-vacuum of space, because there is not sufficient matter in space to produce the light scattering needed to actually see the beam. Forget those old-but-still-cool *Star Wars Episode IV* scenes of laser (aka a "blaster") strikes from speedy Tie Fighters impacting on the Millenium Falcon's hindquarters as she limps away from Vader and the Death Star. Science fiction films also often depict laser weapons propagating at only a few meters per second, slowly enough to see their progress, in a manner reminiscent of conventional tracer ammunition (admittedly, it definitely makes for better eye-candy). Clearly, however, a laser beam, being an EM wave, travels at the speed of light and would thus appear to the naked eye to instantly connect muzzle to impact point. Okay, *Star Wars* geeks, go ahead and argue that blasters aren't really lasers, and thus move VERY slowly! Fine!

Action movies (e.g., *Mission Impossible*) often depict laser-based security systems whose beams can be illuminated with dust or powder and then defeated by a clever combination of mirrors to redirect the beams. Such security systems almost NEVER use visible lasers, and putting enough dust into the air to make the beam visible would likely trigger the alarm anyway (the Mythbusters on the Discovery Channel proved this!). Several glaring laser misconceptions can be found in the 1964 James Bond film *Goldfinger*. Think back. Bond, played by – the one and only – Sean Connery, faces a smokin'-hot laser beam approaching his groin, inch-by-inch melting the solid gold table to which Bond is strapped. *Goldfinger's* director, Guy Hamilton, found that a real laser beam would not show up on camera so it was inserted after filming as an optical effect (crudely by today's standards, but cut 'em some slack, it was 1964!). The table was also precut and then coated with gold paint, and the faux-laser melting effect was achieved by a man sitting below the table armed with an oxyacetylene torch (Bond should have been more worried by the torch!). Goldfinger's laser even makes a now-campy ominous whirring electronic sound, whereas a real laser would have produced a neat, fairly heat-free, and definitely silent, cut.

11.6.4 Fiber optics

An important piece in our photonics pie is the creation of a medium for moving crisp and low-divergence laser-diode-generated light, now cleverly encoded with

Figure 11.62   The ubiquitous fiber optic! (Courtesy of NASA.)

digital information, from point A to point B, say, Atlanta to San Francisco. Clearly we can shine lasers through the air, but NOT over the needed distances because of the ever-present attenuation characteristics of the atmosphere (not to mention the curvature of the Earth!). Instead, we use thin glass (silica) fibers – the ubiquitous optical fiber – to do our magic for us (Fig. 11.62). Let's just scratch the surface (pun intended).

Here's a question for your consideration: If we do want to move lots of bits from Atlanta to San Francisco, why not just use a metal wire? Why fiber? Several compelling reasons, actually: (1) Laser light moving in an optical fiber experiences far smaller attenuation (loss) in transmission over long distances than an electrical signal in a metal wire (provided, obviously, that our laser wavelength is properly tuned to the lowest loss point of the fiber attenuation characteristics). Lower loss means greater transmission distances without intermediate amplification, and this directly translates to much lower cost. (2) The laser light is not subject to electromagnetic interference (EMI) from the plethora of electrical noise generators out there (lightning, motors, radios, TVs, cell phones, you name it). This means that we can transmit the digital information without worrying a whole lot about creative shielding to suppress EMI. Read: much lower cost. (3) The bandwidth of an optical fiber (in this context, how much information can be stuffed down the pipe) is MUCH larger than an electrical wire. Read, it is FAR cheaper to send large amounts of information on optical fibers.

Optical fibers are not as simple as they look [19]. There are in fact many, many different types of fibers in use, depending on the exact application needs, and these are basically classified as follows: "single-mode" fibers (these have the lowest loss and thus constitute virtually all "long-haul" fiber systems currently deployed); "multimode fibers" (lossier and hence used only for "short-haul" apps such as inside buildings or the home, but larger and simpler to package and use than single-mode fibers); and the new-kid-on-the-block, "plastic" fibers (their cores are fancy polymers, not glass, and again are lossier than single-mode fibers but with their own set of advantages for short-haul systems). Interestingly, in low-loss single-mode fibers, digital data can be encoded on multiple wavelengths of the laser light, and we can then send multiple wavelengths down the fiber *at the same time*, greatly enhancing the information-carrying capability (aka the data bandwidth). Clearly, we need some creativity for multiplexing the beams when we "launch" them onto the fiber (combining multiple beams without losing information) and demultiplexing the beams (separating them back out) at the end of the fiber post-transmission, but that is actually pretty routine today, and long-haul fiber backbones for Internet traffic, say, use up to 40 different wavelengths in a single fiber, yielding a $40\times$ bandwidth adder. Pretty cool!

Fine. So what exactly is an optical fiber? (See sidebar discussion on the history – it is predictably interesting). Well, we have a solid silica (glass) "core" of about 8-μm diameter (single mode) or 50 μm (multimode), and doped with germanium (typically). This silica core is surrounded by an undoped solid silica "cladding" of a differing index of refraction to ensure that the laser light stays focused down the core even when the fiber bends. To accomplish this, the core has a higher index of refraction than the cladding. Finally, a protective coating surrounds the fiber and is typically made of UV-cured acrylate. This coating seals the fiber from the environment. The typical diameter of a coated fiber is about 245 μm, pretty darn thin, and because of that, yep, it's quite flexible. Presto – the optical fiber (Fig. 11.63). For a telecommunications fiber line, we "bundle" fibers together and encase the whole bundle in plastic (Fig. 11.64). Clever games can be played in the cladding layer to improve the transmission properties of the fiber – think refractive index engineering – as illustrated in Fig. 11.65. Fiber types in use today are many and varied, including graded-index optical fibers; step-index optical fibers; non-dispersion-shifted fibers (NDSFs); nonzero dispersion-shifted fibers (NZDSFs); dispersion-shifted fibers (DSFs), birefringent polarization-maintaining fibers (PMFs); and, more recently, photonic crystal fibers (PCFs).

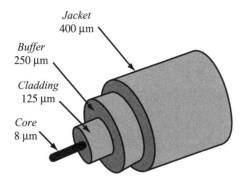

Figure 11.63 Cross-sectional view of a fiber-optic line.

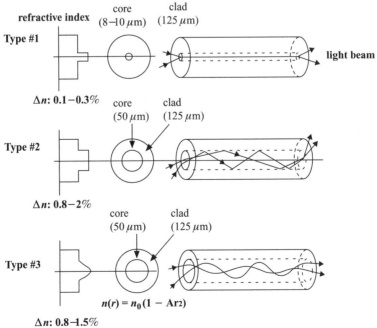

Figure 11.64 Fiber-optic bundle.

Figure 11.65 Different approaches to optical-fiber design (inspired by [18]).

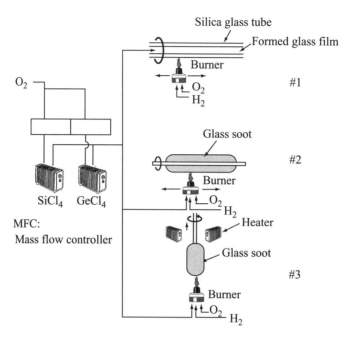

Figure 11.66 The fabrication of silica optical fibers (inspired by [18]).

Fabrication of optical fibers is a highly refined art form [19]. Standard fibers for telecommunications are made by first constructing a large-diameter "preform" with a carefully controlled refractive index profile (undoped cladding around doped core), and then "pulling" the preform into long, thin optical fibers. The preform is commonly made by using three chemical-vapor-deposition (CVD) steps (the same CVD as in semiconductor fab!), as illustrated in Fig. 11.66. The preform is then placed in a device known as a "drawing tower" (just what it sounds like), where the preform tip is heated to red hot, and the optic fiber is then carefully pulled out as an ultra-thin "string." By measuring the resultant fiber width as it is pulled, the tension on the fiber can be controlled to maintain a precise fiber thickness. Now wind the cooled fiber onto a large spool, and you are in business. On to coating and bundling.

At the state-of-the-art, laser light propagating in single-mode optical fibers can be modulated at data rates as high as 40 Gb/s, and each fiber can carry many independent data channels, each using a different wavelength of light (this is called "wavelength-division multiplexing"). If, for instance, we put 128 laser wavelengths traveling down a single fiber at 40-Gb/s (40 billion bits per second!) data rates, that translates to an aggregate bandwidth of 1,500,000,000,000 bits/s (1.5 Tb/s) in a single fiber! Imagine for a second plugging that baby into the back of your laptop. Suddenly you go from your-now-wimpy 10–100 Mb/s (generic cable modem bandwidth) to a blazing 1.5 Tb/s, with a single fiber connection. HEAVEN! This is a predictably dynamic field, and the race is now on to reach

100-Gb/s data rates with an increasing number of multiplexed wavelengths. Stay tuned; "you ain't seen nothing yet"![10]

Historical Aside #2: Origins of the Optical Fiber

The light-guiding principle exploited in today's optical fibers was actually first demonstrated by Daniel Colladon and Jaques Babinet in the 1840s, with Irish inventor John Tyndall offering public displays using crude fiber-lighted water fountains 10 years later [2] (maybe the origin of those tacky fiber Christmas trees?!). Although folks dabbled in using fibers to guide light for the medical and dental fields, it didn't get really interesting until people started thinking about lasers – ultimately, a marriage made in heaven! In 1952 physicist Narinder Singh Kapany began conducting experiments that eventually led to the invention of the optical fiber [20]. Development then focused on creating fiber "bundles" for image transmission for use in medicine (you know; for looking in those hard-to-reach corners and crevices within the body!). The first fiber-optic semiflexible gastroscope (not fun, but MUCH better than exploratory surgery!) was patented by Basil Hirschowitz, C. Wilbur Peters, and Lawrence E. Curtiss, researchers at the University of Michigan, in 1956. In the process of helping to develop the gastroscope, Curtiss produced the first glass-clad fibers. In 1965, Charles K. Kao and George A. Hockham of the British company, Standard Telephones and Cables, were the first to suggest that light attenuation in fibers was caused by impurities (which could in principle be removed), rather than fundamental physical effects such as light scattering. They speculated that optical fibers could ultimately be a practical medium for communications systems, provided the attenuation could be reduced to below about 20 dB/km. A global race ensued! This attenuation level was first achieved in 1970 by researchers Robert D. Maurer and co-workers at Corning Glass Works. They demonstrated an optical fiber with 17-dB attenuation per kilometer by doping silica glass with titanium. A few years later they produced a fiber with only 4-dB/km loss by using germanium oxide as the fiber core dopant. Such low attenuations ushered in the optical-fiber telecommunications industry, ultimately enabling the present telephone system and of course . . . drumroll please . . . the Internet. Read: a BIG deal. Today, attenuation in optical cables is far less than what can be achieved in copper-wire electrical cables, leading to long-haul fiber connections with repeater distances in the range of 500–800 km. The erbium-doped fiber amplifier, with light amplification now cleverly incorporated directly within the fiber, was co-developed by teams led by David Payne of the University of Southampton and Emmanuel Desurvire at Bell Laboratories, in 1986. This was a key development in telecommunications, because it dramatically reduced the cost of long-haul fiber systems by reducing (or even eliminating) the need for optical–electrical–optical "repeaters" (detect light, decode the data, and convert them to an electrical signal, amplify them – yep, with transistors! – and reencode them, and relaunch them down the fiber). The latest craze? Well, in 1991, the emerging field of "photonic crystals" led to the development of photonic crystal optical fibers, which guide light by means of diffraction from a periodic structure, rather than total internal reflection (i.e., fundamentally differently). The first photonic crystal fibers became commercially available in 1996 [21]. Photonic crystal fibers can be designed to carry higher power than conventional fibers, and their wavelength-dependent properties can be manipulated to improve their performance in many applications.

---

[10] Not bad English, just a famous Bachmann-Turner Overdrive (BTO – 1973 Canadian rock band) lyric.

Figure 11.67   The ubiquitous CD/DVD.

11.7 CDs, DVDs, and Blu-ray

For another wonderful example of photonics in action, you need look no further than your portable CD player, or its close cousin, the DVD player, lurking inside your laptop or perhaps connected to your home theater with surround-sound. Ah yes, the ubiquitous CD (Fig. 11.67). Workhorse of the digital data storage industry. Friend to the audiophile. Care to guess how many CDs are out there in the world? From 1982 to 2007, only 25 years – 200 billion! Wow ... that's a lot of CDs. Moral: Must have been a good idea! The compact disc (CD for short) is simply an optical disc used to store digital data (see the sidebar on CD history). In short, a CD represents cheap, portable memory. They make nice Frisbees too! Yep, the CD player that reads (writes, or both) all those CDs/DVDs hinges completely on a semiconductor laser diode. No semiconductors, no CDs or DVDs. No tunes. No movies. Bummer.

CDs were originally developed for storing digital audio files (to replace those nasty, hissy, popping LP vinyl records) but have since morphed into many different venues, including data storage (CD-ROM), write-once ("burnable") audio and data storage (CD-R), rewritable media (CD-RW), SACD, VCD, SVCD, PhotoCD, PictureCD, CD-i, and Enhanced CD – you name it, the list goes on and on. Whew! And of course, CDs have now evolved into DVDs (read: audio to video – movies).

A standard audio CD consists of from 1 to 99 stereo tracks stored by 16-bit encoding at a sampling rate of 44.1 kHz per channel [2]. Early CDs could hold 74 min of stereo sound (they needed to hold the content of a vinyl LP record – if you are under the age of 25, you may not know what I am talking about!), but today 80-min audio CDs are more common.

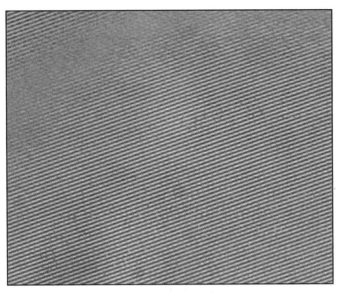

Figure 11.68 A blowup of the surface of a CD, showing the optical data track ("groves") on which data are encoded.

Some CD specifics? Sure! The main features of a generic CD include [2]

- Track pitch (track-to-track spacing): 1.6 μm (see Fig. 11.68)
- Disc diameter: 120 mm
- Disc thickness: 1.2 mm
- Inner radius program area: 25 mm
- Outer radius program area: 58 mm
- Center spindle hole diameter: 15 mm
- Disc scanning velocity: 1.2 m/s (constant linear velocity); approximately 500 rpm at the inside of the disc and 200 rpm at the outside edge. Pretty darn fast![11]

The "program area" on a CD (where we put all those bits of data) is 86.05 cm^2 (try the math), and thus the end-to-end length of the recordable 1.6-μm-wide track spiral is 86.05 cm^2/1.6 μm = 5.38 km! Miles! With a spiral scanning speed of 1.2 m/s, the playing time of a CD is thus 74 min, or around 650 megabytes (MB) of data (for a CD-ROM).

Fine, but what exactly is a CD and how does it work? Well, a CD is actually made from a 1.2-mm-thick disc of almost pure polycarbonate plastic and weighs a meager 16 g [2]. A thin layer of aluminium is first applied to the surface to make it reflective and then is protected by a thin film of lacquer. The lacquer is normally spin-coated (think photoresist) directly on top of the reflective layer. On top of that surface, the label is eventually applied by screen printing or offset printing.

The data on a CD are stored as a series of tiny indentations (aka "pits"), encoded in a tightly packed spiral track molded into the top of the polycarbonate layer. This

---

[11] You might logically infer that the CD player must therefore accurately slow down the disc's rotational speed as it traverses the disc diameter during playback.

Figure 11.69   A DVD-R/W drive with the cover removed.

is the same basic idea behind the vinyl LP record, which uses undulated groves cut into a piece of plastic, but with many advantages. The areas between the CD pits are known as "lands." Each pit is approximately 100 nm deep by 500 nm wide, and varies from 850 nm to 3.5 μm in length. The spacing between the tracks, the "pitch," is 1.6 μm (much smaller than on an LP record). The digital data encoded on the CD are then read by focusing a 780-nm wavelength (0.78 μm = near-IR, slightly reddish to the eye; made from AlGaAs) semiconductor diode laser through the bottom of the polycarbonate layer. The change in height between pits and lands results in a difference in intensity in the reflected light. By measuring the intensity change with a semiconductor photodiode (go figure!), the data can be then read back from the disc. A tiny, precisely tuned wavelength semiconductor diode laser is key to building a small and cheap CD player (Figs. 11.69 and 11.70). Cleverly, the pits and lands are much closer to the label side of a CD, so that any defects or dirt or nasty fingerprints on the clear (exposed) side of the CD will conveniently remain out of focus during playback. Say goodbye to scratch-induced LP "clicks" and "pops." Cool!

Recordable (burnable) CDs (CD-Rs) are injection-molded with a "blank" data spiral. A photosensitive dye is then applied to the CDs, after which the discs are metalized and lacquer-coated as before. Two semiconductor lasers are now needed to do business. The "write laser" of the CD player changes the color of the dye as it "burns" the track encoding the data, whereas a normal "read laser"

Figure 11.70 A laser diode read/write head from a CD-R/W drive.

is used to read data, just as it would in a normal CD. Clever. CD-R recordings are designed to be permanent, but over time the dye's physical characteristics will degrade (don't leave them in the Sun in your hot car!), eventually causing read errors and data loss. Bummer. A CD-R's designed lifespan is from 20 to 100 years, depending on the quality of the discs (no pirated copies allowed), the quality of the write laser, and CD-R storage conditions. However, in practice degradation of CD-Rs in as little as 18 months under normal storage conditions is observed and is often affectionately known as "CD rot."

Rewritable CDs (CD-RWs) are re-recordable (one step fancier) CDs that use a metallic alloy instead of a dye. The write laser in this case is used to heat and alter the properties (e.g., amorphous vs. crystalline) of the alloy and hence change its reflectivity.

DVDs ("Digital Versatile Disc"; originally "Digital Video Disc" – Fig. 11.71)? Well, same basic idea, but the laser wavelength is shorter, enabling a higher track density and hence much more data to be stored (think movie vs. audio). The DVD, introduced in December 1995, has a storage capacity of 4.7 GB, about six times as much data as a CD holds, giving it enough capacity to hold a standard-length feature film. How? Well, a DVD uses a 650-nm wavelength (red; made from AlGaInP, but shorter than the CD's 780 nm) diode laser, enabling a smaller laser spot on the media surface (1.32 μm for DVD versus about 2.11 μm for a CD). Tighter tracks. And writing speeds for DVDs are faster ($1\times = 1350$ kB/s vs. 153.6 kB/s for a CD).

The latest buzz in the CD/DVD industry? Blu-ray! The Blu-ray Disc (aka Blu-ray or BD) uses a 405-nm semiconductor laser (blue! – made from GaN), yielding a storage capacity of 50 GB, six times the capacity of a DVD! The first BDs were released on February 12, 2008.

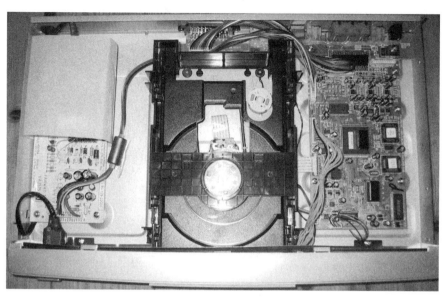

Figure 11.71 A DVD player with the cover removed.

A Brief History of CDs

The CD is the hugely successful by-product of a hugely unsuccessful technology – the video "LaserDisc" (if you are over the age of 40 you may vaguely recall these) [2]. In 1979, Sony and Philips teamed up and formed a joint task force of their brightest minds to design a new digital audio disc platform (nominally to replace vinyl LP records). This CD task force, led by Kees Immink (Philips) and Toshitada Doi (Sony), built on early research into semiconductor lasers and primitive laser-readable optical discs, which had begun at Philips in 1977. After a year or so of work, the task force produced the now-famous CD "Red Book," the CD standard. Philips contributed the general manufacturing process of CDs, based (not surprisingly) on its video LaserDisc technology, and its patented Eight-to-Fourteen Modulation (EFM) scheme for encoding or decoding digital data. EFM offers both a long playing time (needed for a record replacement) and an excellent tolerance against defects such as scratches and fingerprints on the discs. Sony contributed their error-correction method to further improve the audio fidelity. Jointly the Philips/Sony task force defined the CD sampling frequency, playing time, disc diameter, you name it. According to Philips, the CD was thus "invented collectively by a large group of people working as a team." Nice thought.

The first CD commercially released rolled off (pun intended) the assembly line on August 17, 1982, at a Philips factory in Langenhagen, Germany. The first CD music title? Hold your breath... ABBA's *The Visitors* (1981). Gulp. These new CDs and of course Sony's requisite CD player (the Sony CDP-101), where the real money is made, reached consumers on October 1, 1982, in Japan, and early in 1983 in the United States and elsewhere. This 1982 watershed event is often viewed as the birth of the digital audio revolution (hence, the digital video revolution). From the very beginning the audio CD was very enthusiastically received, especially in the early-adopting classical music and audiophile communities (read: excellent fidelity, smaller and easier to handle and store than LPs, and no more ugly hisses and pops between Mozart concertos).

As the price of the CD players rapidly dropped (read: manufacturing economy-of-scale), the CD began to gain popularity in the (much larger) rock music market (college students have less cash!). "I'll take CD Trivia for $100, Alex." The first musical artist to sell a million copies (platinum) on CD. "Who was Dire Straits, with their 1985 album *Brothers in Arms?*" You are correct.

Although the CD was conceived as a modest evolutionary replacement of the LP record, it soon became the workhorse of the digital data storage industry. Moral – good ideas always morph into better ideas when clever people are involved. In June 1985, the CD-ROM (read-only memory) was released, and, in 1990, the CD-R (CD-Recordable) was introduced, also developed by Sony and Philips; go figure – it pays handsomely to be first.

Whew! Deep breath. Exhale. Good. We are now at the end of our photonics road. The end of the beginning, really, of this fascinating field. In many ways photonics is every bit as exciting as micro/nanoelectronics, and certainly just as important in the scheme of life on Earth as it will play out in the 21st century. One important takeaway is that these two seemingly diverse fields are actually highly synergistic, leveraging each other's fabrication tools and manufacturing techniques, and of course both squarely resting on nature's great gift of semiconductors and bandgap engineering. Let there be light . . . indeed!

References and notes

1. F. A. Jenkins and H. E. White, *Fundamentals of Optics*, 4th ed., McGraw-Hill, New York, 1976.

2. See, for instance, Wikipedia: *http://en.wikipedia.org*.

3. M. J. Madou, *Fundamentals of Microfabrication: The Science of Miniaturization*, 2nd ed., CRC Press, Boca Raton, FL, 2002.

4. R. F. Pierret, *Semiconductor Device Fundamentals*, Addison-Wesley, Reading, MA, 1996.

5. B. L. Anderson and R. L. Anderson, *Fundamentals of Semiconductor Devices*, McGraw-Hill, New York, 2005.

6. D. A. Neamen, *Semiconductor Physics and Devices*, 3rd ed., McGraw-Hill, New York, 2003.

7. D. Litwiller, "CCD vs. CMOS: Facts and fiction," in *Photonics Spectra*, January 2001.

8. A. Theuwissen, "CMOS image sensors: State-of-the-art and future perspectives," in *Proceedings of the 37th European Solid-State Device Research Conference*, pp. 21–27, 2007.

9. R. D. Dupuis and M. R. Krames, "History, development, and applications of high-brightness visible light-emitting diodes," *IEEE Journal of Lightwave Technology*, Vol. 26, pp. 1154–1171, 2008.

10. See, for instance, *http://www.ieee.org/organizations/pubs/newsletters/leos/feb03/diode.html*.

11. R. G. Gould, "The LASER, light amplification by stimulated emission of radiation," presented at The Ann Arbor Conference on Optical Pumping, June 1959.

12. J. von Neumann, *IEEE Journal of Quantum Electronics*, Vol. 23, p. 658, 1987.

13. T. H. Maiman, "Stimulated optical radiation in ruby," *Nature (London)*, Vol. 187, pp. 493–494, 1960.

14. J. Hecht, *Beam: The Race to Make the Laser,* Oxford University Press, New York, 2005.

15. R. N. Hall, G. E. Fenner, J. D. Kingsley, T. J. Soltys, and R. O. Carlson, *Physical Review Letters,* Vol. 9, p. 366, 1962. The editor of *Physical Review* received the manuscript on September 24, 1962.

16. M. I. Nathan, W. P. Dumke, G. Burns, F. H. Dill, Jr., and G. Lasher, *Applied Physics Letters,* Vol. 1, p. 62, 1962. This paper was received on October 6, 1962.

17. N. Holonyak, Jr. and S. F. Bevacqua, *Applied Physics Letters*, Vol. 1, p. 82, 1962. This paper was received on October 17, 1962.

18. S. K. Moore, "Laser on silicon," *IEEE Spectrum*, p. 18, November 2006.

19. M. Yamane and Y. Asahara, *Glasses for Photonics,* Cambridge University Press, New York, 2000.

20. W. A. Gambling, "The rise and rise of optical fibers," *IEEE Journal on Selected Topics in Quantum Electronics*, Vol. 6, pp. 1084–1093, 2000.

21. J. Hecht, *Understanding Fiber Optics*, 4th ed., Prentice-Hall, Englewood Cliffs, NJ, 2002.

12 The Nanoworld: Fact and Fiction

Gerd Binnig and Heinrich Rohrer looking at their baby: the world's first scanning tunneling microscope. (Courtesy of International Business Machines Corporation. Unauthorized use is not permitted.)

知之为知之，不知为不知，是知也。
————孔子

> *To Know That We Know*
> *What We Know,*
> *And That We Do Not Know*
> *What We Do Not Know:*
> *That Is True Knowledge.*
>
> K'ung Ch'iu (Confucius)

Some 2,500-year-old words of wisdom we would do well as a global community to embrace [1]. Case in point. We know that people who breath in smoke (micron-sized particles) end up with deposits of those smoke particles in their lungs (sadly, often leading to lung cancer). Roll-your-eyes obvious today. BUT, change those microparticles to nanometer-sized particles, and the inhaled nanoparticles are

Figure 12.1   This image of a clean gold (Au) surface was made using a scanning tunneling microscope, and the individual gold atoms are clearly visible. Nanoscale "surface reconstruction" causes the surface atoms to deviate from their natural bulk crystal structure and arrange themselves in columns several atoms wide with regularly spaced pits between them. Nanoscale self-assembly in action. (Courtesy of Erwin Rossen, Technical University Eindhoven.)

instead rather mysteriously deposited into the brain, bypassing the lungs. At least in lab rats [2]. We think we know the answer; but alas we do not. Moral to the story? Scale can change everything, often in very counterintuitive ways. Welcome to the nanoworld. Savior of civilization or lurking monster ready to gobble us up lest we misstep? Perhaps a little of both. Read on.

12.1 Nanotech, nanobots, and gray goo

Nano? From the Latin word *nanus*, literally "dwarf." We're not talking Gimli here.[1] Nano – 1,000 millionth; one billionth; 0.000000001; 1×10^{-9}. Pretty darn tiny. Nanometer (nm), length scale of modern electronics. Nanosecond (ns), time scale of modern electronics. But also, Nanovolt (nV). Nanomole (nmol). Nanoliter (nl). Nanohenry (nH). Nanogram (ng). Yep, nano as a free-for-all prefix is most definitely in vogue. Nanomaterials; nanoprobes; nanomedicine; nanorobots. Gulp. Welcome to nanotechnology!

A Google search on the term "nanotechnology" yields (on July 7, 2008) . . . gulp . . . 16,000,000 hits! Clearly a clue that there is something big time going on here [2,3]. Ah yes, nanotechnology. Kids and grandmas, and you and I, have all heard the term, but alas nanotechnology has proven very tough to nail down solidly, and signifiant confusion exists on what is and what is not possible with nanotechnology. Nanoscale self-assembly of atoms? Yep; you just have to look very closely, as shown in Fig. 12.1. Nanotransistors? Most definitely. They

---

[1] You know, Gimli; *The Lord of the Rings*. Dwarf of Middle Earth, creation of J.R.R. Tolkien. Dwarfs – fabled builders of the Mines of Moria, where Gandalf fought the Balrog.

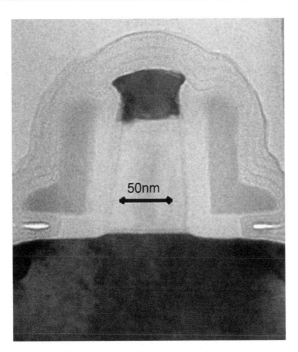

Figure 12.2 A cross-sectional view of a FET from a 90-nm CMOS technology. The active transistor region is 50 nm, 50 billionths of a meter (courtesy of Intel Corporation).

are already inside your laptop (Fig. 12.2). Nanolightbulbs? Uh-huh; see Fig. 12.3. Stronger-than-steel, lighter-than-cotton nanomaterials? Already here (Fig. 12.4). Nanomedicine for fighting cancer? Very soon. Nanorobots? Well . . . don't hold your breath (see sidebar discussion).

Nanobots and Gray Goo

Nanorobots (aka nanobots, nanoids, nanites) are machines or robots with a dimensionality at or at least close to the nanoscale. Nanobots would in principle enable complex interactions with and manipulations of other sythesized nanoscale objects. The possibilities are nearly endless and are fun to think through (careful, you may be labeled a nanogeek). A nanobot that makes on-site welding repairs (nanobot house calls?) to defects in trillion-element nanoelectronic chips? A nanobot that delivers drugs to specific cells to fight cancer? A nanobot that goes in and selectively trims a stand of DNA for some specific gene manipulation? Borg nanoprobes from *Star Trek Next Generation*? Nanogenes in the *Doctor Who* episode "The Empty Child"? Nanites in Asimov's *I, Robot*? The (supposedly) nanoenabled T-1000 in *Terminator 2: Judgment Day*? You get the idea. The good, the bad, and the ugly. How about nanobots that self-replicate to form nanobot "swarms"? Humm . . . that doesn't sound so good. Ah yes, the infamous nano-gray-goo problem. "Gray goo" is nanogeek-speak for an end-of-the-world scenario involving nanotechnology run amuck (literally), in which out-of-control, self-replicating nanobots get loose, and gobble up all living matter on Earth (yep, you and me and Fido), all the while replicating more of themselves, *ad infinitum* [4]. Yikes! Wistful sigh . . . Alas, nanobots are at present only scientifically hypothetical (a polite word for fanciful; okay, pure science fiction!), and are likely to remain that way, at least in the near future, despite what you may hear in the media.

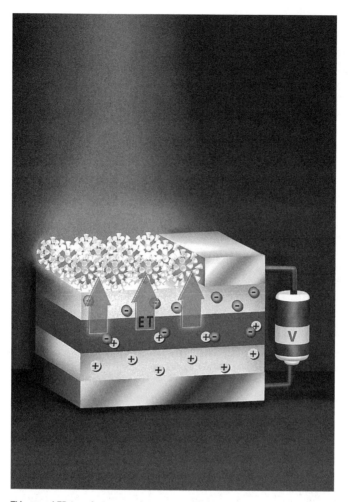

Figure 12.3 This nano-LED transfers energy from nanoscale layers of quantum wells to nanocrystals ("quantum dots") above them, prompting the nanocrystals to emit visible light (courtesy of Marc Achermann, Los Alamos National Laboratory).

Not surprisingly, gazillions of venture capital (VC) bucks are lined up to grow this nanobeast into start-ups and make a ton of money, and everyone is rushing to jump on the nanobandwagon – CEOs, lawyers, lawmakers, professors, you name it. Heck, if you are a decent-sized university today and you don't have a building named "John Doe Nanotechnology Research Center," you are seriously behind the times and soon to be relegated to the out-of-date dinosaur bin of academia.

As might be imagined, when big VC bucks, politics, and government all get intertwined, there are some pretty bold claims being trumpeted regarding the prospects of nanotechnology. From a 2001 study, the National Science Foundation predicted that nanotechnology "stuff" would reach a value of "$1 trillion by 2015" (trillion – 1,000 billion – a million million). Ergo, we can't afford to miss this boat! Result: Major government infusion of research dollars. A good thing. More recently, however, it has been boldly proclaimed that by 2014 nanotechnology will be a $3T industry [5]. That is enough dough to drive a major country's

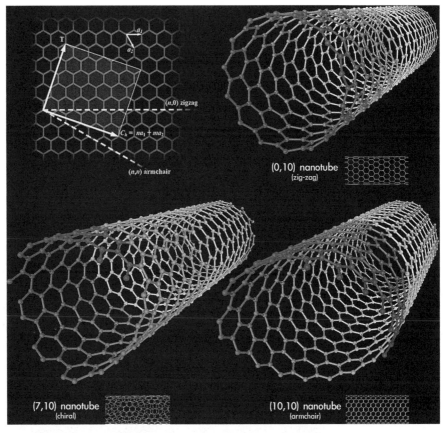

Figure 12.4   The three families of single-walled carbon nanotubes. (Courtesy of Michael Ströck.)

economy several times over, and, as you can easily imagine, VC fund managers are currently licking their lips with glee. But these types of enormous market projections basically count as nanotechnology the entire value of any product that has nano-anything in it. For instance, if a $100 item has 2 nanocents worth of a nanoparticle coating, then that is treated as $100 in nanotechnology, a slight stretch (read: at least potentially misleading if blindly interpreted) [5].

Despite all this wild-eyed hype surrounding nanotechnology, present commercial applications (beyond semiconductor electronics – more on this in a moment) almost exclusively center on the very limited use of so-called "first-generation" (passive) nanomaterials to improve the shelf-life, functionality, appearance, or performance of very familiar everyday products: lotions, foods, clothes, makeup, batteries, paper, paint, you name it (Fig. 12.5). That is, add a little nanosomething to an existing product to improve it. Example nanomaterials in present use include titanium dioxide and zinc oxide nanoparticles used in sunscreens and cosmetics; silver nanoparticles used in food packaging, clothing, and household appliances as a bacterial disinfectant; zinc oxide nanoparticles in various surface coatings, paint pigments, and varnishes; and cerium oxide nanoparticles, which act as fuel catalysts for better combustion. The "Project on Emerging Nanotechnologies"

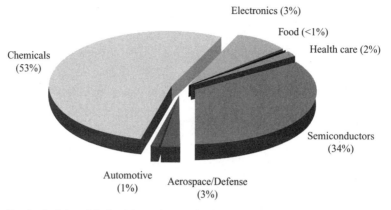

Figure 12.5   The pie-chart view of the breakdown of nanotechnology market share in 2007.

was established in April 2005 as a partnership between the Woodrow Wilson International Center for Scholars and the Pew Charitable Trusts and actually tracks the global inventory of consumer products that contain nanomaterials [6]. This is a good thing. Hold on tight – the product categories currently include appliances, automotive, coatings, electronics and computers, food and beverage, goods for children, health and fitness, and home and garden. Yep, nearly everything affecting your daily life.

Nanoliterature?

You bet! Start with some science fiction masters. Both Robert A. Heinlein, in his 1940 novella *Waldo*, and Arthur C. Clarke, in his 1956 short story "The Next Tenant," described highly miniaturized machines that could operate on a micron scale – i.e., microbots – obvious anticipations of nanobots.

Enter legendary bongo-playing physicist Richard Feynman. His famous lecture, "There's Plenty of Room at the Bottom," presented at an American Physical Society meeting on December 29, 1959, set the stage for scientific nanotech discussion [7]. Feynman put the nanoworld under serious scrutiny, entertaining the possibility of direct manipulation of individual atoms as a means for improving chemical synthesis, as a means to develop far smaller transistor circuits, and a way to improve microscope resolution; all remarkably prescient, given that these three predictions have since come true. Feynman also presented the "weird possibility" of building a tiny, swallowable surgical robot (yep, enter nanobots), to be developed as follows: Make a set of 1/4-scale manipulator hands slaved to the operator's hands and then use those hands to construct 1/4-scale machine tools analogous to those found in a machine shop. This set of miniaturized tools would then be used by the small hands to build and operate 10 sets of 1/16-scale hands and tools, and so forth, ultimately culminating in massively parallel (think billions) tiny nanofactories capable of performing complex operations on the nanoscale. Slick. Clearly, as the sizes got ever smaller, the tools would have to be modified because the relative strength of various forces in play would change with the system dimensionality (a major theme in present-day nanotech).

For example, gravity becomes far less important to a nanotool, whereas surface tension would become much more important, etc. Feynman concluded by issuing two challenges to the scientific community: (1) build a tiny motor, and (2) write the information from a book page on a surface 1/25,000 smaller in linear scale. In true Feynman fashion, he offered prizes of $1,000 for each challenge. His motor challenge was quickly met by a meticulous craftsman who used fairly conventional tools; but in 1985, Tom Newman, a Stanford graduate student, successfully reduced the first paragraph of *A Tale of Two Cities* by 1/25,000, using photolithographic techniques, and collected the second "Feynman Prize." The world was suddenly abuzz with nanospeak.

Of course, all sorts of writers have since embraced the nanoidea: e.g., Robert Silverberg's 1969 short story "How It Was When the Past Went Away" describes nanotech loudspeakers, with a thousand speakers per square inch. Enter nanotech for John Q. Public. K. Eric Drexler, in his controversial 1986 book, *Engines of Creation: The Coming Era of Nanotechnology*, popularized the idea of nanotechnology for the masses. Among other things, Drexler took Feynman's idea of a billion tiny nanofactories and added the notion that they could figure out how to self-replicate. Starting to sound scary? Crichton's 2002 novel *Prey* exploited this general notion of a colony of molecular-sized self-replicating nanobots that developed some smarts (aka artificial intelligence) and decided they wanted to run the show! Not good for inferior humans – gray-goo here we come! The nanotech reading list is long and growing – presto, nanoliterature.

Some nanoexamples you might currently have in your house? Sure! Here's one courtesy of Nano Care Technology, Ltd. (in Hong Kong) [6]. Antibacterial tableware, whose active ingredient is a silver nanoparticle coating on the utensils that kills (they say in 10 min!) a host of bad-boys, including spirillums, virosis hepatitis, salmonela, and golden staphylococcus (golden?!). Eating and cooking with nanocoated tableware prevents the spread of disease. Very nice!

Here's another everyday nanoobject, courtesy of Eastman Kodak (in the United States). KODAK Ultima Picture Paper, for printing hard copies of those beautiful pics you took, uses a nine-layer composition engineered with ceramic nanoparticles so that the paper better resists the degrading effects of heat, humidity, light, and ozone (ah, city life). The net? Longer lasting pictures.

Okay, one more. Live from the kitchen via Shemen Industries, Ltd. (in Israel). Canola Active Oil. Nanooil?! Yep, they use something called NSSLs (nanosized self-assembled structured liquids), aka nanodrops, that enable good-for-you stuff like vitamins and minerals, which are often insoluble in oil, to be incorporated directly into the liquid to nourish us. The vitamin-toting nanodrops are added to the cooking oil, do their thing, and then pass right on through the digestive system (no visuals, sorry).

Alright, alright, just one more. But this is the last one! Courtesy of Toshiba Corporation (in Japan). Their new lithium ion battery uses a (unnamed) nanoparticle coating to prevent organic liquid electrolytes from corrupting the negative electrode of the battery during recharging (read: much longer battery life). The nanoparticles quickly absorb and store lots of lithium ions without causing any deterioration in the electrode itself during recharge. Sweet!

Take-away? There is clearly a whole lot riding on developing a firm under-standing of what nanotechnology actually is and what it is not; what we can do now with it, and what we MAY be able to do with it in the future. Would you like the facts and my own take on the future of nanotech? You got it! Read on.

12.2 Say what you mean and mean what you say: Nanotech definitions

Okay, but what exactly IS nanotechnology? Good question! Loosely defined, nanotechnology is the science, engineering, and application of materials of nanometer-scale dimensionality. When I say nanometer scale, I mean objects in the range of say 0.1–100 nm in size. Nanogames to be played at a "few" molecular diameters [for reference, the diameter of a water molecule (H_2O = H—O—H) is about 0.29 nm; the double helix of a DNA molecule is about 2.5 nm in diameter]. Want to be in-the-know? Don't say nanotechnology, say "nanotech."

Let's also differentiate between several common nanotech terms you'll hear touted: "top-down" vs. "bottom-up" nanotech, leading to "evolutionary" vs. "rev-olutionary" nanotech. The terms "top-down" and "bottom-up" originally referred to strategies of knowledge ordering (think psychology) and information pro-cessing (think computer science), but have increasingly been applied to other humanistic and scientific domains, including (you guessed it) nanotech [4]. Top-down is often used as a synonym for decomposition of a complex system into its component building blocks (from the top), whereas bottom-up refers to di-rect synthesis of a complex system from smaller pieces (from the bottom). In a top-down approach, a broad-brushed conceptual view of the complex system is first formulated, specifying but not detailing high-level subsystems. Each layer of the subsystem is then further refined in greater detail, and so forth, until the entire system-level functional specification is reduced to its most basic elements. A top-down model is typically specified by using "black boxes" that make the decomposition process easier to manipulate and understand. That is, here is what a given subsystem must actually do, but I don't need to understand precisely what's inside it to enable it to do this or that function – it's opaque, a black box. Clearly, however, black boxes, by definition, aren't detailed enough to actually validate and build the system, and at some point they will themselves need to be specified and made into "clear boxes." But that comes later. Practicing engineers will immediately recognize this top-down process – all complex pieces of modern technology (a jet, or a car, or a computer, or a cell phone) are designed and built in this top-down manner. It's efficient and works well. Tried and true.

In a bottom-up approach, on the other hand, the individual basic building blocks of the complex system are first specified in great detail. These building blocks are then linked together to form increasingly larger subsystems, which are in turn linked together, until a complete complex system is realized (this may require many layers of synthesis). This bottom-up strategy often resembles an organic "seed-to-tree" model; it begins small and simple, but rapidly grows in complexity and completeness, hopefully to the correct end result. Clearly nature favors this bottom-up assembly process – it is how we are made after all. It truly is miraculous that a human adult containing roughly 10 trillion cells can begin life as a single cell with an encoded set of construction blueprints – aka DNA.[2] There can easily be pitfalls in such organic bottom-up assembly strategies, resulting in a tangled mess of elements from overdevelopment or underdevelopment of isolated subsystems, or perhaps because a given subsystem was subjected to local optimization instead of being properly directed toward a global end construction goal. Nature has obviously figured this problem out amazingly well, astoundingly well, but it has proven to be a far more challenging problem for technologists to tame.

Fine. Back to nanotech. Top-down and bottom-up are used as two different construction paradigms for producing nanoscale materials, devices, and systems. Top-down is what micro/nanoelectronics is all about, and it does it exceptionally well. We use increasingly smaller dimensions in our fabricated patterns and layers to build more highly integrated and complex electronic systems. For instance, a thumbnail-sized modern microprocessor chip might have 1-billion transistors, with miles of internal wiring to connect them all up. Each transistor itself is 10s of nanometers in size, with nanometer-thin layers inside them. Traditional micro/nanoelectronics is a classical top-down assembly strategy and uses the traditional microfabrication methods, in which externally controlled tools are used to cut, mill, and shape materials into the desired shape and order. Micro/nanopatterning techniques, such as classical photolithography and ink-jet printing are obviously tools of the trade.

Bottom-up nanotech approaches, however, use nanoscale components to arrange themselves into increasingly complex assemblies and ultimately systems. Such bottom-up approaches typically exploit the unique chemical properties of particular molecules to induce single-molecule components to automatically arrange themselves (this is called "self-assembly" in nanospeak) into some useful configuration, and often utilizes the concepts of molecular self-assembly or molecular recognition known from biochemistry. At least on paper, bottom-up nanotech synthesis approaches should be able to produce end products, whatever they might be, in a highly parallelized fashion, hopefully resulting in far cheaper solution than that for a top-down synthesis approach. This seems logically compelling given that bottom-up nanotech self-assemblies are essentially trying to mimic what nature does so well. Seed-to-tree; and now grow a complete forest at a time. Nanomaterial synthesis, and especially bioapplications of nanotech, typically embody some level of bottom-up self-assembly strategy. At present, however, top-down nanotech is far more advanced than bottom-up nanotech. We are still a long way away from synthesizing a microprocessor in a beaker!

---

[2] Geek Trivia: 10,000,000,000,000 cells. Placed end to end, those cells would stretch around the Earth about 47 times. If you could count them at a rate of one cell per second, it would take you over 2,600 years. A 10-year-old has only about 1/2 that many cells. Talk about a growth spurt during puberty!

Evolutionary nanotech? Revolutionary nanotech? The goal of evolutionary (read: more straightforward, a natural progression) nanotechnology is to improve existing processes, materials, and applications by simply scaling component sizes down into the nanorealm in order to exploit the unique quantum and surface phenomena that matter exhibits at nanoscale dimensions [5]. Top-down. This nanotrend is driven by industry's relentless quest to improve existing products by creating smaller components with better performance, all at a lower cost (the smaller, cheaper, faster axiom). Micro/nanoelectronics and the resultant computer industry (Moore's law) are the poster children for such evolutionary nanotech, and they have been extraordinarily successful. Reminder: electronic "stuff" is a $1T global market. Evolutionary nanotech also covers the product improvements being currently driven by nanoadditives and nanocoatings and is a rapidly emerging commercial juggernaut because of the plethora of potential everyday products it touches. Think of this as nanomaterials meets low-tech. Evolutionary nanotech.

By contrast, truly revolutionary nanotech envisages a newer (and of course tougher) bottom-up synthesis paradigm, in which functional nanomaterials, nanodevices, and perhaps entire nanosystems could be built literally atom by atom (grow a computer?). This nanotech synthesis idea, as intellectually appealing as it is, remains in its very early infancy, but is receiving increasingly focused attention (and fortunately funding), especially in the biological nanotech arena. There is clearly no accurate means to crystal-ball it and project a market value on such potentially visionary technology and the likely myriad of hypothetical downstream products. The promise of a truly bottom-up revolutionary nanotechnology is fun to think about, though, and is the latest rage in the global research community (and popular press). The potential payout to civilization is incalculable. But remember. Early infancy. We just don't have the answers as yet as to how practical, and within what application contexts, this might represent a viable path forward. Enough said.

12.3 Darwinian evolution in microelectronics: The end of the silicon road

I have waxed poetic about the heavy and powerful footprint of evolutionary silicon-based micro/nanoelectronics on modern life. Micro/nanoelectronics very much defines our world, and will for the foreseeable future. Business-as-usual Moore's law-driven microelectronics has marched steadily downward in scale from 100 μm to 10 μm to 1 μm and now to 0.1 μm (100 nm) and below [8]. Microelectronics meets nanotech. Evolved into would actually be more apt. Nanoelectronics is currently yielding billion-transistor-count ICs (Fig. 12.6). Microelectronics has very naturally morphed into nanoelectronics via a top-down evolutionary nanotech paradigm. I say business as usual, but I should actually qualify that. To sustain Moore's law growth into the nanotech regime, a plethora of new fabrication processes have had to be developed, including, roughly in time

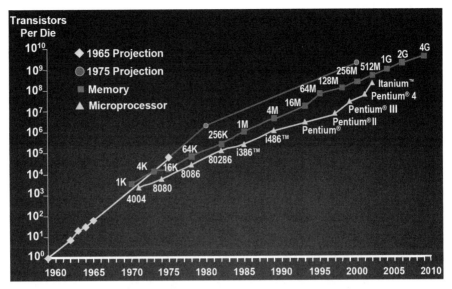

Evolution of the number of transistors per die for both microprocessors and memory chips, as well as the minimum feature size used to fabricate those dies. Moore's 1965 extrapolation is also shown (after G. Moore).

sequence, polysilicon gates; silicon nitride dielectrics; titanium silicide; tungsten "studs"; titanium nitride diffusion barriers; cobalt silicide; and copper metalization. Not to mention all sorts of clever doping design strategies (aka "drain engineering" in MOSFETs) to ensure adequate transistor reliability (Fig. 12.7). In the past couple of years, so-called "strain engineering" has been successfully engaged to enable the penetration of the nanoscale barrier in electronics, and is currently in active use in 90-nm nanoelectronics, and below (the next immediate target nodes for CMOS are 65 nm and 45 nm – all engage strain engineering in a myriad of forms). Strain engineering? Yep. We put the silicon crystal lattice either under compressive strain (pushing inward) or tensile strain (pulling outward) to improve the carrier transport properties (higher mobility, hence faster).

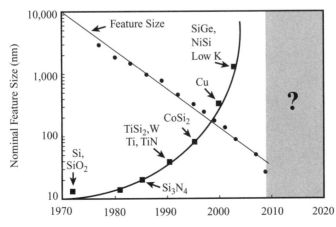

MOSFET feature size evolution, showing the number of new materials and processes that must be engaged to produce a viable IC design platform.

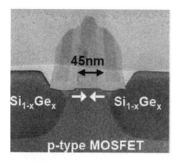
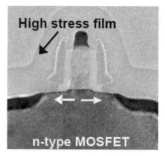

Figure 12.8   SEM cross-sectional views of 65-nm *n*-channel and *p*-channel MOSFETs devices embodying varying levels of bandgap and strain engineering to enhance performance (courtesy of Intel Corporation).

Figure 12.8 shows a reminder of what Intel's 90-nm CMOS transistors look like. The *p*-channel MOSFET is compressively strained by SiGe alloys embedded in the source and drain regions, and the *n*-channel MOSFET is tensilely strained by high-stress films wrapped around the transistor's gate. Sound sophisticated? It is. Sound costly to develop? It is. Why do it? Well, basically to continue to improve the transistor performance (e.g., improved switching speed at acceptable off-state leakage) as we further scale Mr. Transistor into the sub-100-nm regime. Strain engineering quickly becomes mandatory at these dimensions; otherwise you just spent a billion dollars developing a new IC technology that gives no better performance than the previous technology generation. Yep, that will get you fired quickly!

A logical question suggests itself. Gulp... exactly how long can we sustain Moore's law size reductions into the nanometer regime by using "conventional" approaches? Said another way, when will the party end? And when it does end, what comes next? Well, there are two divergent paths currently being pursued, both somewhat speculative and the arena of much present-day research (and hand-wringing!). In the first path, we attempt to continue to push conventional "planar" MOSFETs (transport direction is along the silicon surface from source to drain and modulated by the the gate electrode above the channel – Fig. 12.9) into the deep-nanometer regime. That is, take the age-old vanilla MOSFET structure and pull out all of the stops in strain/doping engineering to continue to improve performance with scaling. Most (but not all) people would say that 65-nm and

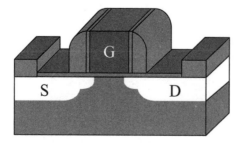

Figure 12.9   The cross-sectional structure of a conventional "planar" MOSFET.

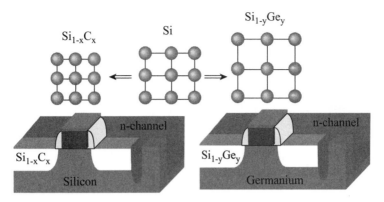

Possible paths forward for continued planar MOSFET scaling: (left) with SiC alloy source and drains and a Si channel, and (right) SiGe source and drains and a Ge channel (inspired by S. Thompson, University of Florida).

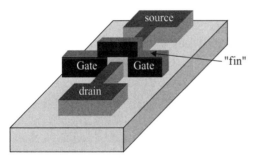

The cross-sectional structure of "Fin-FETs."

45-nm CMOS nodes will probably be built this way (multiple 65-nm examples already exist), and perhaps even 32 nm (the jury is still out). Beyond that, one probably will need to engage yet more radical strain engineering techniques inside the standard planar MOSFET; things like using Si_yC_{1-y} alloys in the source and drain regions, and perhaps resorting to using pure Ge (or even III-V) materials for the transport channels instead of Si (Fig. 12.10). Not cake, but it may allow us to get to viable solutions for 32-nm and 20-nm nodes.

In path two, we instead step back and say, forget the sophisticated strain engineering, let's instead change the fundamental structure of the MOSFET itself to our scaling advantage. "Enter the realm of the Matrix"[3] ... no wait; enter the realm of the 3D, multigate MOSFET. These speculative MOSFETs go by many names, including "double-gate MOSFETs," "Fin-FETs" (Figure 12.11), "trigate MOSFETs" (Figure 12.12), Omega Fin-FETs, and "gate-all-around" MOSFETs [9–13]. In multigate, nonplanar MOSFETs, the channel region is surrounded by several gates on multiple surfaces, allowing more effective suppression of parasitic "off-state" leakage current, enhanced current drive, and faster speeds. Nonplanar devices are also much more compact than conventional planar

[3] You know; Morpheus, Neo, Trinity, and the gang, in the sci-fi classic *Matrix*.

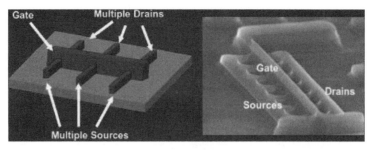

Figure 12.12 The cross-sectional structure of a "trigate" FET (left), and a SEM micrograph of a fabricated trigate device (right) (courtesy of Intel Corporation).

transistors, enabling higher transistor density and thus higher integration levels. Clearly a good thing. Moving from planar (flat) to vertical (3D) is not simple and requires significant innovation on the fabrication front to reach "gigascale" (billion-transistor) integration levels [14]. Still, I think most (not all) would say that nonplanar, multigate MOSFETs, in whatever form we eventually settle on, are likely to play an important role in the CMOS end game, say at 32-nm and 20-nm nodes, and below.

Both radical strain engineering and nonplanar MOSFETs lie at the center of current CMOS research and will clearly occupy us for some years to come. Stay tuned, much is at stake. But what comes beyond 20 nm? Well, most (not all) would say that the end of the CMOS road probably lies in that 10–20-ish-nm-scale range. Said another way, CMOS is likely to meet its scaling demise at 10–20-ish nm [15]. Kaput. End of the road. "Game over, man."[4]

Why out of gas at 10–20 nm? Well, there are actually two sets of fundamental limits to be faced: one theoretical (physical limits) and one practical (economic limits). Physical limits first. The lattice constant of silicon is 0.543 nm, so a 10-nm transistor would be roughly only 18 atoms across. Now imagine accurately controlling the properties of 10–100 billion such transistors in a microprocessor on a centimeter-sized piece of silicon and getting them to behave according to plan. Likely? Extremely challenging at best. In addition, the gate dielectric thicknesses needed to build a functional 10-nm MOSFET are daunting; likely 1.0–1.5 nm at the 10-nm CMOS node. Even with alternative (high-K) gate dielectrics it will prove extremely challenging to maintain acceptably low off-state leakages in a 10-nm MOSFET. Remember, if you have 1 nA (one billionth of an amp) of leakage when the transistor is "off," multiply that by 1 billion transistors on the IC and you now have 1 A of leakage current! Congratulations, major problem. AND, you need to create some means for interconnection of all of these 10–100 billion transistors on a single silicon die. The infamous "interconnect nightmare" [15]. These are three biggies to solve. More lurk close by. It has long been recognized that operating MOSFETs at very cold temperatures (say −196°C, the temperature

---

[4] You know; the wild-eyed, jittery space marine with a slight morale problem in the sci-fi classic *Aliens* – don't laugh, you'd be scared too!

at which liquid nitrogen boils = 77.3 K) can help partially mitigate some of these identified scaling problems [15], and even contemplating operation at 20 nm is likely to mandate some level of IC cooling. Still, carrying around a liquid nitrogen bath for your laptop is likely to prove highly problematic at the 10-nm CMOS node!

Economic limits? You bet. Think about it; let's say we could actually build a 10-nm CMOS technology node with 1 trillion transistors on a centimeter-sized microprocessor. The sophistication level of the fabrication processes required for doing this at high yield is staggering to contemplate. Suddenly your $5B 90-nm CMOS fab becomes a $100B 10-nm CMOS fab. It is not obvious how anyone could recoup the GDP-sized investment needed to produce such 10-nm ICs. One thing is certain – if there is no money to be made, no company will pursue the path. A very practical, but just as rock-solid, limit for scaling.

So what comes next? Well, it's time to think "out of the box." It seems pretty clear that the days are numbered for business-as-usual micro/nanoelectronics. 2015? 2020? Soon. Clearly, however, humankind's appetite for computing power and memory storage is insatiable and likely to remain so for the foreseeable future. So what are our options for further progress? Hummm . . . Read on.

12.4 Buckyballs, nanotubes, and graphene

The future, in short, is all about carbon! BUT, not the carbon you and I are made of, or the carbon in a diamond ring. Same atom, yes, but VERY different arrangement. Until recently, only two forms of pure carbon were known to us: graphite (think pencil lead) and diamond (as in ring). This all changed in 1985, when a totally unexpected third form of carbon was stumbled on – a hollow cluster of 60 carbon atoms (written C_{60}) shaped like a soccer ball (see sidebar discussion). Really! Got to have a cool name, right? How about Buckminsterfullerene, or more commonly, and quite affectionately, "buckyballs." Buckyballs were named for the American architect R. Buckminster Fuller, the pioneer of the geodesic domes – think 1960s. The buckyball is a "truncated icosahedron" – a polygon with 60 vertices and 32 faces, 12 of which are pentagons and 20 of which are hexagons. The cluster geometry is identical to the pattern of seams on a soccer ball (Fig. 12.13).

Buckyballs are definitely the coolest large molecules in town. Exceedingly stable despite their size (a shared virtue with geodesic domes), buckyballs can survive extreme temperatures and environmental conditions without deflating. Very rugged. And they turn out to have rather extraordinary chemical and physical properties. For instance, buckyballs react with chemical species known as "free radicals," a key to the polymerization processes used in industry (think plastics). Buckyballs rapidly have become one of the new loves of organic chemists, and buckyball chemistry is all the rage. Even a whole family of related carbon-cluster molecules has since been discovered, including C_{28}, C_{70} (looks kind of like a rugby ball!), C_{84}, and even perhaps C_{240} (talk about big!).

Mr. Buckyball

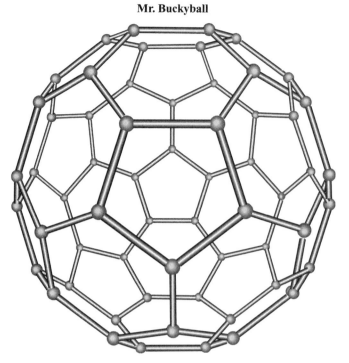

Figure 12.13 A 3D model of the now-famous Buckminsterfullerene C_{60}, aka the "buckyball," the simplest of the carbon nanostructures known as fullerenes.

Some Buckyball History

Richard E. Smalley and Robert F. Curl, Jr. of Rice University, and Harold Kroto of the University of Sussex, in Brighton, England, joined forces to investigate carbon "clusters" in the early 1980s [16]. Why carbon clusters? Well, Kroto, a microwave spectroscopist, was especially interested in carbon-rich red giant stars. Curl, a microwave and infrared spectroscopist, was a friend of Kroto's who often collaborated with Smalley. Smalley had designed and built a contraption at Rice for creating and then characterizing clusters formed from almost any element. These researchers carried out a series of carbon-cluster experiments in September of 1985 [17]. They generated carbon clusters by laser vaporization of graphite into a pulsed stream of helium and then analyzed the clusters by mass spectrometry. A variety of polyacetylenes (chem-speak for burnt carbon) were formed, but under certain conditions, a mass peak corresponding to C_{60} (60 atoms of carbon) totally dominated the spectrum. Hummm ... The controversial inference was that C_{60} must be a giant spherical carbon molecule (Smalley's insight). All three gentlemen shared the 1996 Nobel Prize in Chemistry for their discovery. Curl famously remarked, "This is the dream of every kid who's ever owned a chemistry set." You gotta love it. The rest is history.

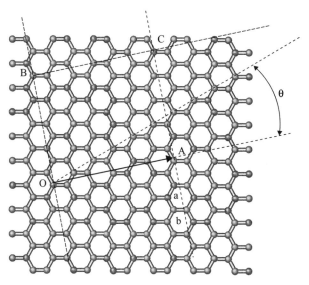

Figure 12.14 A construction guide for going from flat sheets of graphene to the various single-walled CNTs (© 2007 IEEE).

Buckyballs definitely have cachet. But for the next radical step in transistors, they are of limited utility because of their cumbersome (albeit beautiful) shape. Far more interesting for the future of electronics are carbon nanotubes (CNTs), a related fullerene [18,19]. Refer to the sidebar discussion on the contentious history of CNTs. CNTs are essentially long sheets of "graphene" (a 2D sheet of hexagonal carbon bonds; looks like chicken wire – Fig. 12.14) rolled up into neat little cylinders (tubes) with a diameter of only 1–2 nm but a length as long as several centimeters! There are single-walled CNTs and multiwalled (cylinder within a cylinder) CNTs. The length-to-diameter ratio of a CNT can thus be huge – 10,000,000:1. Impressive and VERY unique. A CNT can be synthesized in a variety a ways, but CVD (refer to Chap. 7) has rapidly emerged as the go-to growth technique for CNTs (Fig. 12.15) [20].

Figure 12.15 A plasma-enhanced chemical vapor deposition (PECVD) chamber modified to grow CNTs. (See color plate 40.)

The Contentious History of Carbon Nanotubes

Most everyone who knows a bit about CNTs will tell you they were discovered by Sumio Iijima of NEC in Japan, and point to the now-classic 1991 paper in *Nature (London)*, "Helical microtubules of graphitic carbon" [21]. In that paper Iijima identified single-walled CNTs among the remains of the insoluable material left over after burning graphite rods (you know – as in those pencil-thin, brilliant carbon arc lamps illuminating German bombers during the London Blitz, or perhaps the blazing arc lamps at the Oscars in Hollywood, CA). Iijima's paper created a major buzz in the global scientific community and was a clear catalyst for much that was to follow in the nanotube world. The arc-discharge technique was previously known to produce the famed buckyball, but this was pretty much a hit-or-miss venture. Iijima found that adding transition-metal catalysts to the carbon in the arc discharge yielded a much more controllable synthesis process. Case closed, Iijima wins, right? Well, not so fast.

    M. Monthioux and V. Kuznetsov pointed out in a recent editorial in the journal *Carbon* that nanotube history was a bit more complex than generally supposed [22]. For example, in 1952 Radushkevich and Lukyanovich published clear images of 50-nm-diameter tubes made of carbon in the Soviet (no, I can't read Russian either) *Journal of Physical Chemistry*, which for obvious reasons remained largely unnoticed by the scientific community. A 1976 paper by Oberlin and co-workers clearly showed hollow carbon fibers with nanoscale diameters, as well as a TEM image (transmission electron microscopy – the tool of choice, because it offers enough magnification to see those little buggers) of a nanotube consisting of a single wall of graphene. In addition, in 1979, Abrahamson presented evidence of CNTs at the *14th Biennial Conference of Carbon*, in which he described CNTs as "carbon fibers" that were produced during arc discharge. In 1981 a group of Soviet scientists published the results of an extensive chemical and structural characterization of synthesized carbon nanoparticles by using a combination of TEM images and x-ray diffraction. The authors even offered that their "carbon multi-layer tubular crystals" were formed by rolling graphene layers into cylinders and further suggested two possibilities of such arrangements: a circular arrangement (armchair nanotubes) and a spiral, helical arrangement (chiral nanotubes). Wow! Now enter IP. Go figure. In 1987, H.G. Tennent of Hyperion Catalysis was issued a U.S. patent for the production of "cylindrical discrete carbon fibrils" with a "constant diameter between about 3.5 and about 70 nanometers . . ." [4].

    The net? Well, clearly the "discovery" of CNTs remains a debated topic. It seems likely that there were synthesized CNTs hanging around long before before people developed the tools to actually be able to see them (after all, the carbon arc lamp was invented by Sir Humphry Davy during the first decade of the 19th century!). That said, most people would fairly credit Iijima's 1991 paper as seminal, because it brought CNTs directly into the limelight[5] for the scientific community for the first time.

    CNTs are intriguing because of their truly remarkable properties [20]. For instance, depending on how we roll the graphene sheet into a tube, we can either obtain metallic or semiconducting behavior (Fig. 12.16). Reminder: The usefulness of semiconductors for building electronic devices fundamentally rests

---

[5]You asked for it: "Limelight"?! Literally. Scottish surveyor and politician Thomas Drummond invented limelight in 1825 by burning calcium oxide (aka lime) in a hot hydrogen–oxygen flame to create a brilliant white light bright enough to be used for surveying or even as a warning beacon to ships (think lighthouse).

Zigzag

Armchair

Chiral

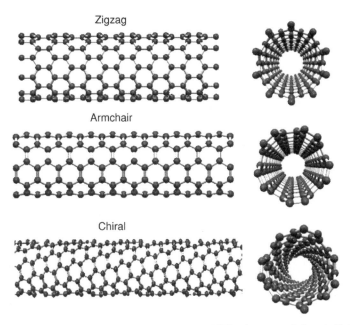

Figure 12.16   Another view of the three families of single-walled CNTs: zigzag, armchair, and chiral (©2007 IEEE).

on our ability to change its resistivity. Same idea here, but we do this by structural means, not by doping. Electronically speaking, CNTs are interesting because they can carry exceptionally large current densities, have both high carrier mobility and high saturation velocity, have excellent thermal conductivity, and are tiny (1–2-nm diameter); all highly desirable features for any potential electronic device. But the fun doesn't end there. CNTs also have remarkable mechanical properties: a Young's modulus (formally the ratio of stress to strain – a measure of material "stiffness") of over 1 TPa and a tensile strength (how much force it takes to pull it apart) of over 200 GPa. Both are HUGE numbers in the materials world. CNT as a nanomaterial is pound-for-pound the strongest material around, by a long shot! CNTs also have a very high surface-to-volume ratio, making them uniquely suited for a host of biological sensor applications in which you want to expose the surface to various biological or chemical agents and look for changes. The virtues of CNTs don't end there. The ends of pristine CNTs look like buckyballs, and can be tricked into all sorts of clever self-assembly arrangements (read: revolutionary nanotech in action) by altering the chemistry of the ends of the CNTs as they are grown (e.g., Fig. 12.17, CNT "towers"; and Fig. 12.18, CNT parallel arrays) [20].

Although the technology surrounding CNTs is still very immature (they have been around less than 20 years after all), MANY clever ideas abound, including transistors (CNT transistors and even simple CNT circuits have already been demonstrated); interconnects for 3D packaging (very useful in mitigating the interconnect bottleneck due to their superior electrical properties); improved chemical and biological sensors (e.g., NO_2 gas sensors and detectors for lipoproteins, and various other cancer biomarkers, are already in use); transparent CNT

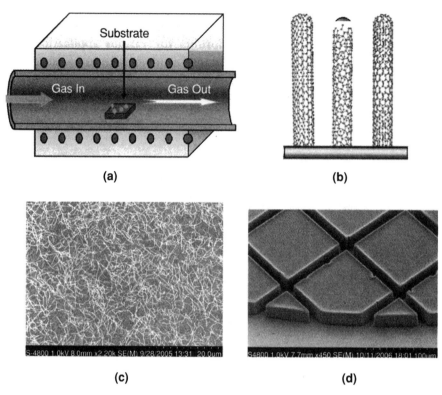

(a)   (b)

(c)   (d)

Figure 12.17   From nanotube growth to nanotube towers: **(a)** schematic view of the CVD synthesis process; **(b)** nanotube base growth vs. tip growth; **(c)** a SEM image of single-walled CNTs on top of a silicon subtrate coated with silicon dioxide; and **(d)** a SEM image of a CVD-synthesized multiwalled carbon nanotube tower (©2007 IEEE).

"paper" for nanofiltering and improved electrodes for energy storage (this should lead to better batteries and capacitors).

The newest kid on the fullerene block is graphene itself, the building block of CNT, and importantly with perhaps the highest overall potential for breaking the scaling limits of conventional silicon-based electronics [23]. As transistor components CNTs have some serious disadvantages, because, in order to embed them into a MOSFET, for instance, you have to literally figure out a way to stretch hundreds or thousands of CNTs from source to drain. Not easy. And even if you could do it, trying to do that within a conventional high-volume IC manufacturing

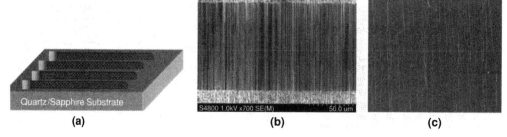

(a)   (b)   (c)

Figure 12.18   A schematic representation **(a)** of how one synthesizes aligned CNTs and SEM images of aligned single-walled CNTs on **(b)** quartz and **(c)** sapphire (©2007 IEEE).

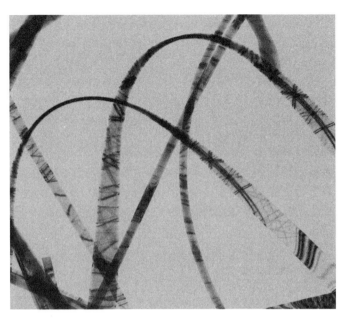

Figure 12.19 Nanobelts of zinc oxide (ZnO). Such nanobelts define a new class of nanomaterials for applications in nano-electronics, nanosensors, nanoactuators, and biosensors (courtesy of Z. L. Wang, Georgia Tech).

environment on a billion transistors per die on 300-mm wafers . . . is a stretch. What if instead we could simply unroll CNTs and make them flat and then extend their shapes over macroscopic distances. Presto – graphene – a single 2D sheet of carbon atoms (Fig. 12.14) with some of the same remarkable properties as CNTs. The C–C bond length in a graphene sheet is only 0.142 nm. Importantly, however, one can fairly easily then envision a seamless integration path of classical microfabrication techniques with graphene sheets, in which the graphene would form the conductive channel of a fairly conventional planar MOSFET. Result? The graphene FET. With its superior electronic transport properties compared with those of silicon (the mobility of electrons in undefected graphene is thought to be potentially as high as 200,000 at 300 K, HUGE by silicon standards), graphene FETs would be faster at much larger dimensions, effectively extending the transistor evolutionary timeline. Are we there yet? Nope, still an idea, but it has attracted serious interest and significant funding. Initial prototypes are just beginning to emerge, so stay tuned.

12.5 Emerging nanoapps

One thing is certain. Our ability to synthesize broad new classes of nanomaterials is accelerating rapidly, and those nanomaterials make for some very cool images. Can you say nanoart? One of the more exciting nanomaterials presently coming on line is zinc oxide (ZnO), the synthesis of which has been pioneered by Professor Z. L. Wang's group at Georgia Tech [24–31]. Some ZnO nanoexamples include nanobelts (Fig. 12.19), nanowire arrays (Figs. 12.20 and 12.21), nanohelixes (Fig. 12.22), nanopropellers?! (Fig. 12.23), and nanosensors (Fig. 12.24).

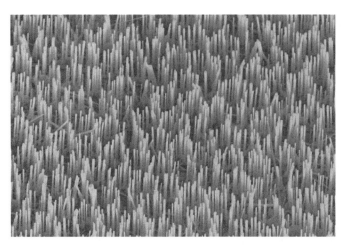

Figure 12.20   Aligned arrays of ZnO nanowires for applications in field-emission displays, vertical FETs, and biofluid transport (courtesy of Z. L. Wang, Georgia Tech).

A nanograb-bag of cool new nanoapps? Sure, why not!

- A nanowire-based dc nanogenerator using aligned ZnO nanowire arrays that can scavenge tiny amounts of energy from local vibrations for use in powering other nanomachines (Figs. 12.25–12.27).
- Polymers-turned-metals: Traditional polymers can be strengthened by nanoparticles, resulting in novel materials that can be used as lightweight but very strong replacements for very expensive metals.

Figure 12.21   SEM image of aligned ZnO nanowire arrays synthesized by a new vapor–solid process. The source materials used for the synthesis were commercial ZnO, SnO_2, and graphite powders, which were placed in a tube furnace. By heating the source materials to a high temperature, reduction of Sn from SnO_2 occurred, serving as the catalyst for the growth of the ZnO nanowires. This chemical approach has also been applied to grow aligned nanowire arrays, which usually occurs at low temperatures ($70°C$) and on a larger scale. Growth of aligned arrays of nanowires is of great importance in nanobiotechnology, because they are a fundamental structure needed for biosensing, manipulation of cells, electron field emission, and converting mechanical energy into electricity for powering nanodevices (courtesy of Z. L. Wang, Georgia Tech). (See color plate 41.)

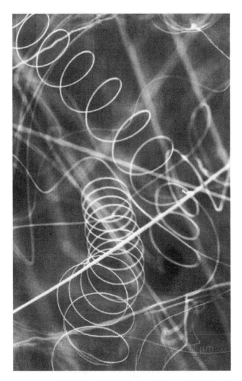

Figure 12.22 Nanoscale helixes of ZnO that exhibit superior piezoelectric properties, an important effect for converting a mechanical signal into an electrical signal, and vice versa (courtesy of Z. L. Wang, Georgia Tech). (See color plate 42.)

- "Quantum dots" (nanoparticles with quantum confinement properties, such as size-tunable light-emission wavelength), when used in conjunction with conventional MRI (magnetic resonance imaging), can produce exceptional images of tumors for improved success in cancer diagnosis and treatment.

Figure 12.23 Aligned propeller arrays of ZnO, which have applications in nanoscale sensors and transducers (courtesy of Z. L. Wang, Georgia Tech). (See color plate 43.)

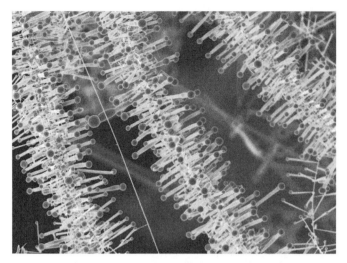

Figure 12.24 Hierarchical nanostructures of ZnO for chemical and biomedical sensor applications. These types of nanomaterials have the potential of measuring nanoscale fluid pressure and flow rate inside the human body (courtesy of Z. L. Wang, Georgia Tech). (See color plate 44.)

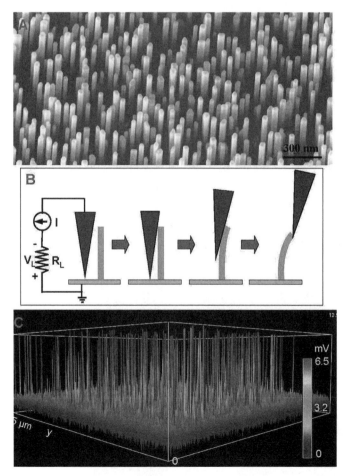

Figure 12.25 Piezoelectric nanogenerators based on aligned ZnO nanowires: **(A)** SEM images of as-grown ZnO nanowires on a sapphire substrate, **(B)** schematic experimental procedure for generating electricity from a nanowire by using a conductive AFM, **(C)** piezoelectric discharge voltage measured at an external resistor when the AFM tip is scanned across the nanowire arrays (courtesy of Z. L. Wang, Georgia Tech). (See color plate 45.)

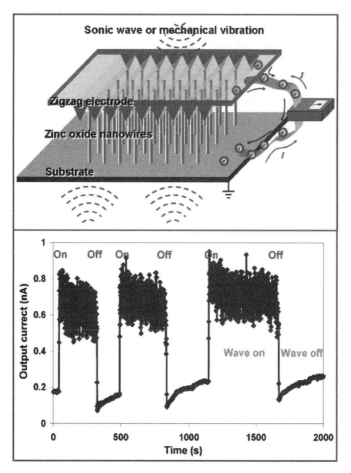

Figure 12.26

Schematic diagram showing the dc nanogenerator built with aligned ZnO nanowire arrays with a zig-zag top electrode. The nanogenerator is driven by an external ultrasonic wave or mechanical vibration, and the output current is continuous. The lower plot is the output from a nanogenerator when the ultrasonic wave is on and off (courtesy of Z. L. Wang, Georgia Tech).

These quantum dots are much brighter than conventional organic dyes used today and need only one light source for excitation. This means that the use of fluorescent quantum dots should be able to produce a higher-contrast image at a much lower cost.

- On a related front, the high surface-area-to-volume ratio of nanoparticles allows biologically functional chemical groups (say an enzyme or a drug molecule) to be attached to a nanoparticle, which can then seek out and bind to specific tumor cells, delivering drug therapy locally, with minimal impact to unintended (non-tumor) cells.

- The ultimate dream in nano-oncology would be to cheaply mass-produce multifunctional nanoparticles that could first detect, then image, and then proceed to directly destroy a tumor inside the body, in one step, without the systemic impact of conventional toxic cocktails used in chemotherapy. The potential for using nanotech in revolutionary cancer treatments is truly exciting.

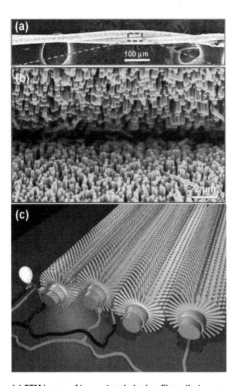

Figure 12.27   (a) SEM image of two entangled microfibers that were covered radially with piezoelectric ZnO nanowires, with one of them coated with gold. The relative scrubbing of the two "brushes" generates electricity. (b) A magnified SEM image at the area where the two "brushes" meet teeth-to-teeth, with the top one coated with gold and the bottom one with as-synthesized ZnO nanowires. (c) A schematic illustration of the microfiber–nanowire hybrid nanogenerator, which is the basis of using fabrics for generating electricity (courtesy of Z. L. Wang and X. D. Wang, Georgia Tech). (See color plate 46.)

- A potential world-changing nanotech innovation surrounds the (cheap) production of potable water by use of low-cost nanofiltration techniques. In this case, nanofilters remove contaminants quickly and effortlessly, because they are far too large to pass though the tiny pores of a nanofilter membrane. Similarly, magnetic nanoparticles offer an effective method to remove heavy-metal contaminants from waste water by using magnetic separation techniques. Using nanoscale particles increases the efficiency to absorb the contaminants and is comparatively inexpensive compared with traditional precipitation and filtration methods. Nanofiltration could be a very big deal in third-world countries. For reference, UNICEF estimates that lack of safe drinking sources kills roughly 1.5 million children every year (2006 statistic).
- The popular lab-on-a-chip concept is rapidly emerging for use in medical diagnostics and gene sequencing and combines conventional microelectronics fabrication with nanotech. Biological tests measuring the presence of selected substances become quicker and more sensitive when nanoscale particles are used as chemical tags or labels. For example, magnetic nanoparticles, when bound to a suitable antibody, can be used to label specific molecules or even microorganisms. Gold nanoparticles tagged with short segments of DNA can

be used for detection of the genetic sequence in a given sample. Multicolor optical coding for biological assays has been achieved by embedding different-sized quantum dots into polymeric microbeads. And nanopore technology for analysis of nucleic acids converts strings of nucleotides directly into electronic signatures. Medical diagnostics is changing as we speak, in no small measure because of ongoing nanotech innovations.

- Nanoenabled "tissue engineering" uses nanomaterial-based scaffolds and growth factors to generate artificial tissue, with one end goal being the synthesis of artificial organs (clearly still a long way off).

- On the electronics front, novel nanodevices exploiting various magnetic phenomena fall under the heading of "spintronics" and are generating significant interest. The dependence of the resistance of a given material (because of the spin of the electrons – hence the name spintronics) on an external magnetic field is called "magnetoresistance." This effect can be significantly amplified, however, when nanosized objects are used, and is then called "giant magnetoresistance" (GMR). GMR is already used in the hard disk drives in your computer. A new and related effect, however, called tunneling magnetoresistance (TMR), is based on the spin-dependent tunneling of electrons through adjacent ferromagnetic layers, and are envisioned as creating improved non-volatile memories.

- On the photonics front, nanoengineered "photonic crystals" and quantum dots are proving to be quite exciting. Photonic crystals are materials made with periodic variations in their refractive index, with a "lattice constant" that is half the wavelength of the light that you shine on them (hence nanoscale). Because of some complex physics, photonic crystals exhibit a selectable "optical bandgap" prohibiting the propagation of light in the crystal at a certain wavelength. This is actually analogous to what happens for electrons in a semiconductor (an electronic bandgap), and look at how we have exploited that fact to our advantage. Photonic crystals could potentially allow photonic components to mimic electronic components, but at far higher operating frequencies. Quantum dots are already finding their way into semiconductor laser design as a means of tuning the laser wavelength (by varying the quantum dot geometry) and greatly improving efficiency.

This nanolist could go on for miles. One final question of practical interest: Just how safe and ecofriendly is all this nanostuff? Hummm... excellent question. We'll visit the (considerable) environmental implications of micro/nanotech in the next chapter.

Whew! Done. Nanofinito. Witness the amazing world of nanotech. Although we have scratched only the surface, hopefully your nanogeek-speak skills have improved dramatically, and you can now converse with all manner of nanonerds you may come across. Despite the considerable hype, nanotechnology is clearly one of the most exciting scientific developments to come along in quite some time. Funding is being aggressively applied, and the global tool-up sequence has been engaged. Those who pretend to possess the proverbial crystal ball and offer any level of certainty of where nanotech will actually lead us . . . should be taken with several grains of salt. It is very tough to imagine what the nano-end-game will look like. However, the one certainty you can always take to the bank is that, when lots of clever humans get involved with new science, engineering soon follows, and then technological fruit will start to emerge. It has always been so, and in nanotech this process is now well underway. It will be an immensely interesting next decade for nanotech R&D. On to our last stop – societal transformations and ethical issues surrounding our microelectronics and nanotechnology revolution.

References and notes

1. I am indebted to my friend, Dr. Guofu Niu, of Auburn University for translating Confucius into a beautiful Mandarin Chinese script.
2. For a cute (and informative) on-line video introduction to nanotechnology, see "The Twinkie Guide to Nanotechnology," by Andrew Maynard. Visit *http://penmedia.org/video/maynard.html*.
3. J. Kahn, "Nano's big future," *National Geographic*, June 2006.
4. See, for instance, Wikipedia: *http://en.wikipedia.org*.
5. See, for instance, *http://www.nanowerk.com/spotlight/spotid=1792.php*.
6. See, for instance, *http://www.nanotechproject.org/inventories/consumer/* for a compiled listing of known nanotechnologies in consumer products.
7. R. P. Feynman, "Plenty of Room at the Bottom," the text of which can be found at *http://www.its.caltech.edu/ feynman/plenty.html*.
8. Internal Roadmap for Semiconductors (ITRS), *http://public.itrs.net/*.
9. L. Risch, "Pushing CMOS beyond the roadmap," *Proceedings of IEEE European Solid-State Circuits Conference*, p. 63, 2005.
10. H.-S. Wong *et al.*, "Self-aligned (top and bottom) double-gate MOSFET with a 25 nm thick silicon channel," *Technical Digest of the IEEE International Electron Devices Meeting*, p. 427, 1997.
11. X. Huang *et al.*, "Sub 50-nm FinFET: pMOS," *Technical Digest of the IEEE International Electron Devices Meeting*, p. 67, 1999.
12. D. Hisamoto *et al.*, "Impact of the vertical SOI 'delta' structure on planar device technology," *IEEE Transactions on Electron Devices*, Vol. 41, p. 745, 1991.
13. N. Singh *et al.*, "High-performance fully depleted silicon nanowire gate-all-around CMOS devices," *IEEE Electron Device Letters*, Vol. 27, p. 383, 2006.
14. J. D. Meindl, J. A. Davis, P. Zarkesh-Ha, C. S. Patel, K. P. Martin, and P. A. Kohl, "Interconnect opportunities for gigascale integration," *IBM Journal of Research and Development*, Vol. 46, pp. 245–263, 2002.
15. *IBM Journal of Research and Development*, Special issue on "Scaling CMOS to the Limits," Vol. 46, nos. 2,3, 2002.
16. See, for instance, *http://www.3rd1000.com/bucky/bucky.htm*.

17. H. W. Kroto, J. R. Heath, S. C. O'Brien, R. F. Curl and R. E. Smalley, "C60: Buckminsterfullerene," *Nature (London)*, Vol. 318, pp. 162–163, 1985.

18. M. Meyyappn (Editor), *Carbon Nanotubes: Science and Applications*, CRC Press, Boca Raton, FL, 2004.

19. V. Mitrin, V. Kochelap, and M. Stroscio, *Introduction to Nanoelectronics*, Cambridge University Press, New York, 2008.

20. C. Zhou *et al.*, "Small wonder: The exciting world of carbon nanotubes," *IEEE Nanotechnology Magazine*, Vol. 1, pp. 13–17, September 2007.

21. S. Iijima, "Helical microtubules of graphitic carbon," *Nature (London)*, Vol. 354, pp. 56–58, 1991.

22. M. Monthioux and V. Kuznetsov, "Who should be given the credit for the discovery of carbon nanotubes?" *Carbon*, Vol. 44, 2006.

23. A. K. Geim and K. S. Novoselov, "The rise of Graphene," *Nature Materials*, Vol. 6, pp. 183–191, 2007.

24. Z. W. Pan, Z. R. Dai, and Z. L. Wang, "Nanobelts of semiconducting oxides," *Science*, Vol. 291, pp. 1947–1949, 2001.

25. X. Y. Kong, Y. Ding, R.S. Yang, and Z. L. Wang, "Single-crystal nanorings formed by epitaxial self-coiling of polar-nanobelts," *Science*, Vol. 303, pp. 1348–1351, 2004.

26. P. Gao and Z. L. Wang, "Nanopropeller arrays of zinc oxide," *Applied Physics Letters*, Vol. 84, pp. 2883–2885, 2004.

27. X. Wang, C. J. Summers, and Z. L. Wang, "Large-scale hexagonal-patterned growth of aligned ZnO nanorods for nano-optoelectronics and nanosensor arrays," *Nano Letters*, Vol. 3, pp. 423–426, 2004.

28. P. Gao and Z. L. Wang, "Self-assembled nanowire–nanoribbon junction arrays of ZnO," *Journal of Physical Chemistry B*, Vol. 106, pp. 12,653–12,658, 2002.

29. Z. L. Wang and J. Song, "Piezoelectric nanogenerators based on zinc oxide nanowire arrays," *Science*, Vol. 312, pp. 242–246, 2006.

30. X. Wang, J. Song, J. Liu, and Z. L. Wang, "Direct current nanogenerator driven by ultrasonic wave," *Science*, Vol. 316, pp. 102–105, 2007.

31. Y. Qin, X. Wang, and Z. L. Wang, "Microfiber–nanowire hybrid structure for energy scavenging," *Nature (London)*, Vol. 451, pp. 809–813, 2008.

13 The Gathering Storm: Societal Transformations and Some Food for Thought

The integrated circuit meets Mother Nature. (Courtesy of International Business Machines Corporation. Unauthorized use is not permitted.) (See color plate 47.)

A Garden Patch,
Or a Redeemed Social Condition;
To Know That Even One Life
Has Breathed Easier
Because You Have Lived.
This is to Have Succeeded.

Ralph Waldo Emerson

Ah yes . . . success. Clearly something we all can sink our teeth into. I've always loved Emerson's take on what it means to truly succeed during the very brief span of years we inhabit this planet. What is success? To make money? To drive a nice car? Have a big house? Attain a position of power? This is certainly is a message emphatically trumpeted in most of the world these days. But is that really the path to success? To fulfillment? To happiness? Emerson didn't think so; neither do I. Rather, success should be measured with a different ruler: "To leave the world a bit better . . . to know that even one life has breathed easier because you have lived. This is to have succeeded." Amen.

Microelectronics and nanotechnology, as they become increasingly pervasive in the global infrastructure of day-to-day existence, have the power to do much good for our world. And, alas, much harm. There is tremendous historical precedent for developing impressive new technologies "just because we can," without pausing to reflect much about the potential good and harm that could come from such powerful innovation. Or else we rush to demonstrate our technical prowess by building a slick new toy, a clever new tool, and table all thoughts on the downstream implications with a "let's worry about that later." History teaches that this can be a dangerous course to follow, especially with technology.

We have come to the technical end of our micro/nano journey. But not the end end. Assuming you have paid close attention and read this book slowly, cover to cover (did I tell you I'm a naive optimist?!), you now have a background deep and broad enough to entertain some serious discussion on virtually any technical topic in this fascinating micro/nano field. You'll be able to follow the often-ridiculous claims made in the media with an informed awareness, and will experience that wonderful swelling of confidence that empowers you to knowingly whisper to your friend "They don't know what they are talking about – here's the real deal." Congratulations, you've earned it!

I began this book by "throwing down the gauntlet." I challenged the legions of all you bright young people of our world, you students on a learning curve, to become the next practitioners, the future technological movers-and-shakers, to first understand what microelectronics and nanotechnology are really all about; and then gleefully revel in the glamour and excitement, appreciating the incredible myriad of future applications awaiting only an inventive mind. But that's not all. Now I challenge you to "take up the gauntlet." Knowing what you now know, I invite you to step back, reflect, muse a bit, and then take a stand regarding HOW the development of these remarkable micro/nano inventions can be put to best use in serving our global community. For the greater good. The many evolving societal transformations and the myriad issues swirling around the ensuing microelectronics and nanotechnology revolution is what this final chapter probes.

Yes, yes, I know I have to be careful here. My intent is not to preach; nobody likes that. And I certainly don't want to be a total downer – I told you I am intrinsically an optimist! Most, if not all, of these micro/nano issues are quite gray; fuzzy; frustratingly murky. There are no clearcut answers, and bright and reasonable people argue from both sides – that's the point of this exercise. Furthermore, I am admittedly no expert in the topical areas I am about to engage. No worries. These issues I raise can be viewed as simply representative of the many things that need be thought through. They need to be well researched and considered carefully. These are certainly things that I reflect on; consider; get excited about; worry over; and generally occupy a substantial amount of my waking hours. Some are just plain fun to think about; some are filled with a wonderful potential for changing our planet in positive ways; others possess a subtle dark side; still others really mandate VERY careful consideration because of their serious implications and potential to do harm.

By the way, I do not intend to answer any of the various questions that I raise. Sorry! Rather, in my view these are topics that should be thought about by all budding practitioners of the art of microelectronics and nanotechnology. Carefully considered; mulled over from time to time; debated in class; talked about with friends over a cup of coffee or, better yet, a cool pint of Guiness. These issues matter. To you; to me; to the future of our civilization and planet Earth. So here goes . . . don't stop reading now!

13.1 The Internet: Killer app . . . with a dark side

Arguably the most important innovation the microelectronics and nanotechnology revolution has spawned in terms of its direct impact on our day-to-day lives is . . . drumroll please . . . the Internet (capital "I" – hey, it's a big deal and deserves some respect). The ultimate "killer app." [Needless to say, subtract transistors and there are no computers ("servers"); no servers, no Internet.] If you are reading this book, I am 99.9% sure that you are an Internet user. But here's a good question. What exactly IS the Internet anyway?! Let's keep it simple. The Internet is a global, publicly accessible, massively interconnected, enormous computer network that allows us to transmit data from point A to point B in the network. The Internet is thus a "network of networks of networks" that consists of literally millions of smaller home, academic, business, and government networks, all of which together provide various information and services. Prevalent examples of such useful services would include e-mail, on-line chatting, blogs, file transfer, and the interlinked Web pages of the World Wide Web (WWW). Clearly it is a highly distributed and highly redundant system; part of its inherent robustness. And it's a beast! Care to have a glimpse? Sure! Figures 13.1 and 13.2 show visualizations

The Internet

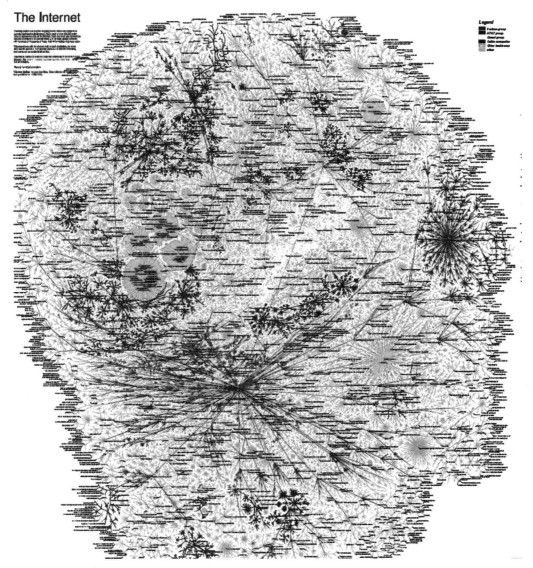

Figure 13.1 A cool visual image of the Internet backbone in North America (courtesy of Lumeta Corporation). (See color plate 48.)

of the Internet [3]. A veritable rat's-nest of millions of interconnected networks. So just how big is the Internet, and how rapidly is it growing? Unbelievably big and unbelievably fast – and getting bigger and faster each day (refer to Figs. 13.3 and 13.4). As of late 2007, 20% of the world population use the Internet (over 1.3 billion and counting), with a 265% growth in usage from 2000 to 2007. Just imagine what it will look like in 7 more years! For a brief history of the Internet and the WWW, refer to the sidebar discussion.

By the way, do appreciate that the Internet (aka the Net) and the WWW (aka the Web) are not the same thing (a common mistake). The Net is a collection of highly interconnected networks, whereas the Web is a collection of highly interconnected documents (web pages), linked together by so-called "hyperlinks" and "universal

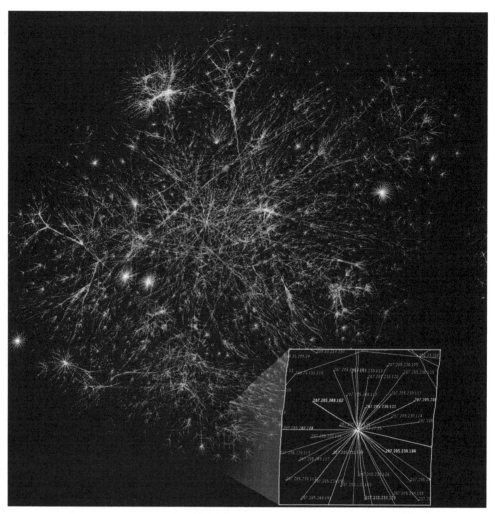

Figure 13.2   A higher-level view of a portion of the global Internet backbone. Each line is drawn between two Internet nodes, representing two IP addresses. The lengths of the lines are indicative of the delay between those two nodes. This graph represents less than 30% of the Class C networks reachable by the data collection program in early 2005.

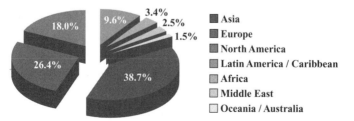

Figure 13.3   A regional breakdown of global Internet usage by geographical region, as of December 2007 (courtesy of Miniwatts Marketing Gorup).

WORLD INTERNET USAGE AND POPULATION STATISTICS				
World Regions	**Population (2007 Est.)**	**Internet Users**	**% of Population**	**Usage Growth 2000-2007**
Africa	941,249,130	**44,361,940**	4.7%	882.7%
Asia	3,733,783,474	**510,478,743**	13.7%	346.6%
Europe	801,821,187	**348,125,847**	43.4%	231.2%
Middle East	192,755,045	**33,510,500**	14.4%	920.2%
North America	334,659,631	**238,015,529**	71.1%	120.2%
Latin America/Caribbean	569,133,474	**126,203,714**	22.2%	598.5%
Oceania / Australia	33,569,718	**19,175,836**	57.1%	151.6%
WORLD TOTAL	6,606,971,659	**1,319,872,109**	20.0%	265.6%

Figure 13.4 A regional breakdown of global Internet usage by geographical region, as of December 2007, compared with region population data (courtesy of Miniwatts Marketing Group).

resource locators" (URLs). The Web is thus only one of the many services accessible to all via the Net. Other services enabled by the Net would include e-mail, on-line gaming, music file sharing, etc. In short, the Net is much more than the Web. Fine, but HOW does the Internet actually work? Good question. Read on.

A Brief History of the Internet and the World Wide Web

Following the Soviet Union's launch of Sputnik in October of 1957, the United States created the Advanced Research Projects Agency (ARPA) (later to become DARPA, the Defense Advanced Research Projects Agency) in February 1958, nominally to regain a technological lead over the Soviets (you know, the cold war) [1]. ARPA then created the Information Processing Technology Office (IPTO) to explore the utility of computer networks in national defense. Network champion and visionary, J.C.R. Licklider, was selected to head the IPTO, and Licklider recruited L. Roberts to head a project to actually implement a real computer network. After several starts and stops, the first two nodes of what was to become ARPANET were interconnected between UCLA and SRI International in Menlo Park, California, on October 29, 1969. The ARPANET is considered one of the "Eve" networks (as in Adam and Eve) of today's Internet. Importantly, ARPANET was built on the then-controversial notion of "packet switching" of information around the network. Packet switching is a communications paradigm in which packets (discrete blocks of digital data) are routed between network nodes over data links that are shared with other packet data traffic. In each network node, packets are "queued" (buffered) before being sent, resulting in a variable delay in their point-to-point transit time. Packet switching is in stark contrast with the other principal communications paradigm, circuit switching, which sets up a limited number of constant-bit-rate (constant-delay) connections between network nodes for their exclusive use for the duration of the information exchange. Packet switching has proven to be the optimal approach to creating more robust and stable networks. A fortuitous choice for ARPANET.

In 1973 Vinton Cerf and Robert Kahn developed the first description of "TCP protocols" for packet transfer on interconnected networks. Use of the term "Internet" to describe a single global TCP/IP (Transmission Control Protocol/Internet Protocol) network originated in December 1974 with the publication of RFC 674, the first full specification of TCP, written by Vinton Cerf, Yogen Dalal, and Carl Sunshine. During the next 9 years, work proceeded to refine the IPs and to implement them on a wide range of computer operating systems.

The first TCP/IP wide-area network became operational by January 1, 1983, when all hosts on the ARPANET were switched over from the older protocols to TCP/IP. In 1985, the U.S. National Science Foundation (NSF) commissioned the construction of a university 56-kbit/s network backbone using computers called "fuzzballs" by their inventor, David L. Mills. The following year, NSF sponsored the development of a higher-speed 1.5-Mbit/s backbone, soon to be called NSFNET. A key decision to use the DARPA TCP/IP protocols in NSFNET was made by Dennis Jennings, then in charge of the supercomputer program at NSF.

The opening of NSFNET to commercial interests began in earnest in 1988. The U.S. Federal Networking Council approved the interconnection of the NSFNET to the commercial MCI mail system, and other commercial electronic e-mail services were soon connected, including OnTyme, Telemail, and Compuserve. Also in 1988, three commercial Internet service providers (ISPs) were launched: UUNET, PSINET, and CERFNET. Various other commercial and educational networks, including Telenet, Tymnet, and JANET were soon interconnected with the now rapidly growing Internet. Next in line was SKYNET.[1] The ability of TCP/IP to work well over virtually any preexisting communication network was clearly key in the Internet's rapid growth, and with the addition of powerful network data routers for seamless network-to-network connections and "Ethernet" for easy installation of dime-a-dozen local-area networks, the Internet was soon growing by leaps and bounds.

So how about the WWW? A killer app for the killer app. Even though the basic applications and guidelines that make the Internet possible had existed for almost a decade, the network did not gain real public interest until the 1990s. Then, on August 6, 1991, CERN, the European particle physics consortium, located on the border between France and Switzerland, announced its new "World Wide Web" project, invented by British scientist Tim Berners-Lee back in 1989 as the means to process and then display the vast amounts of information collected in giant particle accelerator experiments (really!) [1]. The WWW is a system of interlinked hypertext documents accessed via the Internet. With a "Web browser" (think Microsoft Explorer or Firefox), a user views Web pages that may contain text, images, videos, and other multimedia, and, importantly, can navigate between them by using "hyperlinks" (basically page-to-page linkages).

A simplified end-to-end visualization of the Internet, from you, the end user, to "stuff" that people put onto the Internet that you want to access (pictures, blogs, web pages, Map Quest, Google, etc.) is shown in Fig. 13.5 [2]. From you, the user, to the content on the Internet you aim to use, a (simplified) ordered path would go something like this [2].

[1] Joking! You'll recall that SKYNET was the self-aware, ultimately machine-friendly but human-averse, evil network from *Terminator 2: Judgement Day*.

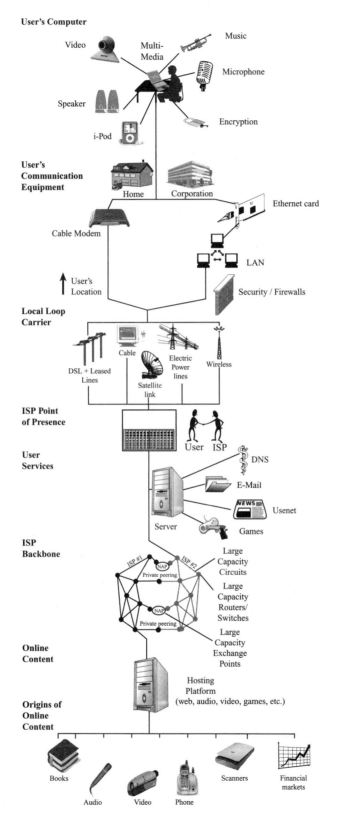

Figure 13.5   A 50,000-ft view of the end-to-end structure of the Internet (inspired by R. Haynal).

- *User PC*: This is you in your home office using the Internet. You have a multimedia-enabled PC equipped to send and receive all manner of files to and from the Internet.
- *User Communication Equipment*: This hardware connects your User PC to what is known as the "local loop" and is likely located at the user's physical location. Think: cable or digital subscriber line (DSL) modem, a wireless router, an ethernet card, Company X's local-area network (LAN) that you are plugged into, etc. This communication equipment hopefully also has some level of Internet security built into it (aka "firewalls") to keep your neighbors from snooping on you.
- *Local Loop Carrier*: The local loop connects the user location to the ISP's point of presence (POP). There are obviously various local communication links involved, including phone lines, cables lines, satellite uplinks, and the ever-popular wireless "hot-spots." Sigh . . . so much jargon!
- *ISP POP*: This is where connections between you the user and your local Internal service provider (ISP – just what it says; the people you pay your monthly access charge to!) are accepted and authenticated – in short, you are now at the edge of the ISP's network.
- *User Services*: Resident here are the various services most users would want to use while connected to the Internet: e-mail, newsgroups, web hosting, file sharing, ftp, etc.
- *ISP Backbone*: The ISP backbone interconnects the various ISP POPs with each other and interconnects the ISPs to other ISPs and various on-line content. Networks within networks within networks. The Internet backbone starts to embody some serious machinery with very high speeds and data transfer bandwidth, and might include fiber data-pipe connections, monster routers, ATM switches, SONET, and various broadband access points. Here lies the famous "T1 line" for Internet access. T1 is a widely used telecommunications standard in North America and Japan for transmitting voice and data over the Internet. A T1 line is made up of 24 channels, each with 1.54-Mbps transmission bandwidth.
- *On-line Content*: These are the "host" sites that you as an Internet user will directly interact with. Here lie the Web "servers" – very powerful computers with LOTS of memory and very fast communications capability (Fig. 13.6). Sometimes these servers are ganged up to construct monster "server farms" for massive web sites.
- *Origins of On-line Content*: Here are the original resources for the on-line information you'd like to access via the Internet, and include music, pictures, scanned images, word files, powerpoint presentations, etc., that have been uploaded to the Internet by others . . . using another computer at some other location and also attached to the Internet!

Pretty complicated, huh? You'd do well to appreciate that complexity each time you enter your favorite URL and press "go."

Figure 13.6   A prototypical Web server.

And how about the Web? How does that work? Well, "viewing" a Web page normally begins either by typing the URL of the page into your Web browser, or by following (double click on) a hyperlink to that page [1]. The Web browser then initiates a series of communication messages, hidden to you, to fetch and display the Web page. First, the server-name portion of the URL is resolved into an IP address using the global, distributed Internet database known as the domain name system, or DNS (see Fig. 13.7). This IP address is necessary to contact and send data packets to the Web server. The browser then requests the Web page you're after by sending an HTTP request to the Web server at that particular IP address. In the case of a typical Web page, the HTML text of the page (HTML – Hyper-Text Markup Language – just computer language used to construct Web pages) is requested first and parsed immediately by the Web browser, which will then make additional requests for images and any other files that form a part of the Web page. Statistics measuring a web site's popularity are measured by the number of "page views" or associated server "hits" (file requests). Having received the required files from the Web server, the browser then renders the page onto the screen as specified by its HTML code. Any images and other resources are incorporated to produce the on-screen Web page that the user sees. Cool, huh?! Now you are officially ready to "surf the Web." The amazing thing is how fast that entire Web page load process happens; almost instantaneously on a good computer with a decent high-speed link. Never forget, though, that this speed ultimately boils down to how fast Mr. Transistor can switch from a "0" to a "1" and how fast electrons can be moved from point A to point B.

Figure 13.7 A part of a single floor of this office building in Marina Del Rey, California, is home to the Internet Corporation for Assigned Names and Numbers (ICANN), where unique identifiers on the Internet, including domain names, IP addresses, and protocol port and parameter numbers are assigned.

Clearly civilization, on a global scale, is being rapidly transformed by the Internet. The way we live our daily lives has been disruptively altered – how we communicate, shop, play, socialize, entertain ourselves, job hunt, create, educate. You name it – the Internet has changed it. Permanently. Consider MapQuest; Google Earth; YouTube; on-line banking; gaming; IM; Facebook; e-mail. Many of these are great things (in my view the jury is still out on e-mail; more on this later). New doors have been opened. Boundless new possibilities emerge, and some offer promise for a better world. But, alas, the Internet has brought with it some not-so-great things, too. A smattering of current issues are presented, each followed by some logical questions for your further reflection and discussion.

Internet Regulation: The Internet architects have always vigorously insisted on a total lack of content–subscriber regulation, with a mantra of "hands off my Internet, buster" . . . aka an "inter-free-for-all-net." Ergo, anybody, anywhere, anytime can pretty much put anything up on the Web. Move it around via the Net at will, with no easy way to trace my global footsteps. Anonymity in its purest form. Admittedly, this content–subscriber–path freedom is the basis for much of the appeal and wild success enjoyed by the Internet . . . but it comes with a dark side. Predictable secondary result? As more people with questionable motives get involved with using the Internet, WWW has slowly but surely morphed from benign geek-playground into the Wild, Wild West. Think Hollywood. You know,

Tombstone. The O.K. Corral. Anything goes, and you best watch your back. As N. Allard states in "Frontier Justice," "The relatively unfettered frontier of cyberspace is showing the strains of a commercial gold rush. It often resembles Wild West boomtowns, populated with earnest PC pioneers and homestead users, internet preachers, copyright rustlers, perverts, scam artists, and plain old crooks" [4]. Apt. Some remarkably recent, less-than-savory Internet happenings? You bet. Cyberporn (THE big $ business of the Internet); on-line gambling (all it takes is a credit card to be in Vegas); child exploitation; illegal or prescription drug sales; virus construction and dissemination (worms, trojan horses and their ilk, just for kicks); hackers (they think of themselves as benign); crackers (the criminal angle of hacking); scams, pfishing; identity theft; bait-and-switch; boldfaced fraud. The list is long and troubling.

> **Question 1**: Is it time for the Internet to be actively regulated? If so, how? By whom?
>
> **Question 2**: Can a government, any government, get involved in such regulation without crashing the whole system?
>
> **Question 3**: Should we require a (traceable) "driver's license" to surf the Web?

Child Exploitation: What better idea than a global playground for kids, right? Meet people from other cultures; visit educational web sites; find your house on Google Earth; do your homework with your buddies; instant access of virtual encyclopedias; meet new friends in virtual chat rooms. All sounds good on paper – but not nearly so benign in practice. Let's look for moment at a particularly disturbing Internet-abuse trend: on-line exploitation of children. Some disturbing facts await [5]:

- Approximately one in seven young people aged 10–17 using the Internet have received a sexual solicitation or approach on-line.
- Four percent (4%) of those received an "aggressive" sexual solicitation (e.g., a solicitor who asked to meet them somewhere; called them on the telephone after connecting on-line; sent them an off-line hardmail, money, or gifts).
- Thirty-four percent (34%) experienced unwanted exposure to on-line sexually explicit material – pictures of naked people or people having sex.
- Only 27% of the youth who encountered such unwanted sexual material told a parent or guardian. If the encounter was defined as "distressing" – episodes that made them feel very or extremely upset or afraid – only 42% told a parent or guardian [6].

Humm … Gives one pause for thought. Thankfully, organizations have been specifically launched to address this travesty and are beginning to gain traction, ironically in many cases using the Internet itself as an effective weapon against Internet-based perpetrators. For instance, *CyberTipline* was established in March 1998 by the National Center for Missing and Exploited Chidren (NCMEC) [5] to anonymously report instances of on-line child pornography and exploitation. To date, CyberTipline has received more than 546,600 reports involving the possession, manufacture, and distribution of child pornography, the on-line enticement of children for sex acts, child prostitution, child sex tourism, child molestation, unsolicited obscene material sent to a child, and misleading domain names. Still, the problem is enormous in scope and growing rapidly, transcending geographical borders. Transcultural. Useful resources for parents interested in educating their young people on safe ways to engage the Internet might consult *NetSmartz* [7] and *Inhope*, the International Association of Internet Hotlines [8].

> **Question 1**: *What is best way to confront and then deal with Internet-based exploitation of children and other at-risk individuals?*
>
> **Question 2**: *How can parents most effectively limit what their children are exposed to on the Internet?*
>
> **Question 3**: *How can law enforcement agencies effectively function across country borders rendered instantly porous on the Internet, and where they likely have no legal jurisdiction? Should there be a global Internet police force?*
>
> **Question 4**: *Should some active "tracing protocols" be put in place to more easily track down perpetrators?*

E-Commerce and Taxation: The Internet has rapidly evolved into THE major driver of the global economic engine. Electronic commerce (e-commerce) refers to the buying and selling of products, services, or both, over electronic systems connected to the Internet, and a number of innovations have grown up to support the emerging dominance of e-commerce, including electronic funds transfer (EFT), supply chain management, Internet marketing, on-line transaction processing, electronic data interchange (EDI), inventory management systems, and automated data collection systems [1]. E-commerce typically uses the Web at least at some point in any given transaction's lifecycle, and your company is a nobody in today's business world if you don't have a sophisticated web site to support this. E-commerce can be conducted entirely electronically for virtual items such as access to web site content, but most e-commerce involves old-fashioned transportation of physical items purchased via on-line retailers (e-tailers!). E-commerce conducted between businesses is also obviously huge, often referred to as Business-to-Business (B2B) transactions. So … ummm … does one pay tax on Internet purchases? Resounding … NO! The 1998 Internet Tax Freedom Act was a United States law authored by Representative Chris Cox and Senator Ron Wyden, and signed into law on October 21, 1998 by President Bill Clinton in an effort to preserve, promote, and ultimately grow the commercial, educational, and informational potential of the Internet [1]. This law bars federal, state and local governments from taxing Internet access and from imposing discriminatory Internet-only taxes such as bit taxes, bandwidth taxes, and e-mail taxes. The law also bars multiple taxes on e-commerce. It has been extended four times by the United States Congress since its original enactment, the most recent of which was the Internet Tax Freedom Act Amendments Act of 2007, which added seven years to the life of the tax moratorium, prohibiting state and local governments from taxing Internet access and e-commerce until November 1, 2014. But is this the right idea?

> ***Question 1***: *How has the emergence of virtual e-commerce giants such as Amazon.com changed the retail landscape? For the better? For the worse?*
>
> ***Question 2***: *Should Internet access be tax-free to everyone, everywhere, at all times?*
>
> ***Question 3***: *How should states or countries best deal with the potential (a debate in itself) loss of tax revenue that is due to increasingly prevalent on-line retail purchases, which effectively bypass local sales taxes typically used to support many public needs?*
>
> ***Question 4***: *Should we consider some sort of taxation for on-line retail purchases that could be directed to those globally in need? If so, for what items? At what percentage? How would we collect and distribute the proceeds locally? Globally?*

I'll stop here, but you get the idea. The list of Internet-enabled transformational topics is nearly endless. Not a bad idea to jot down some of your own Internet-enabled "issues."

13.2 E-addictions: E-mail, cell phones, and PDAs

Ah, yes, electronic addictions (e-addictions). Okay, so let me admit this right now. Yea verily, I too am an e-mail addict. There, that's not so hard. Try it. Very liberating. E-mail is the first thing I do in the morning when I get to work, and it often stretches far into my evenings. I am rarely out of e-mail contact for more than an hour at a time, if that. Want to reach me anytime, anywhere? Drop me an e-mail. I am frighteningly diligent about replying to non-SPAM e-mails. Sigh . . . the resultant flood of e-mail traffic in my daily life can be simply overwhelming. Dark clouds. Seriously depressing. Major bummer. Really take some time off and shut down my e-mail for a few days of well-earned vacation, and it takes me a full morning (often a full day) just to wade through the 300+ e-mails awaiting my return (post-SPAM filter). So, yes, I grind my teeth, wondering whether that vacation was really worth the major pain of catch-up. Welcome to the 21st century.

To my mind, e-mail is a wonderful concept, but with a subtle, sneaky, dark side when actually put into practice. In my dealings with my 20-odd graduate students in my research team, and with my many colleagues literally scattered around the world, e-mail is absolutely essential. MAJOR productivity boost. I couldn't even contemplate conducting 20% of what I do daily by old-fashioned means (landline; snail-mail). E-mail is definitely my preferred means of communicating. I often joke with my friends that it could be a full-time job just keeping up with my e-mail. While I'm confessing, let me also admit to something I'm not especially proud of. I can derive TANGIBLE PLEASURE by pressing delete on that last finally answered e-mail, kicking my feet back, and beholding that now empty inbox with a satisfied grin. P.S. – it's fleeting; that nice warm feeling lasts only 30 s until the next herd of e-mails is forwarded from the server. E-mail. Seemingly indispensable productivity tool ... suddenly morphed into an agent of evil. Inevitable consequence? You begin to feel that queasy pang in your gut when you are away from your Internet connection, out of e-mail contact. Who needs me? What am I missing? What fires need to be put out? Interestingly, over time e-mail often produces a weird false sense of importance and exaggerated proportions. You know what I mean – "If I don't answer this e-mail RIGHT NOW, the world may literally end." Poof! How crazy.

In case you are wondering, I'm not alone. Examples? Sure! A survey of more than 7,800 business managers indicated that 25% felt routinely overwhelmed by their e-mail. A survey of "busy people" (who work 60+ hours a week in high-stress jobs) found that 59% said technology lengthens, not shortens, their workday; 64% said it encroaches on their family time. A survey of 1,500 executives found that 82% want to keep their e-mail close at hand when they're away from their desk but still in the office. Another 84% said e-mail is the application they need access to most when traveling on business or commuting. 4,012 Americans aged 18 and older were queried about their home e-mail habits: 41% check e-mail first thing in the morning on rising; 23% check e-mail in bed; 4% check e-mail in the bathroom. Wow ... "Houston, we have a problem."[2]

Can't live without it. Irresistible craving. Distorted perceptions of reality. Sound familiar? Yep, e-addiction. Try this test sometime. Shut off your e-mail when you get to work, and begin to go about your normal routine, all the while objectively observing your state of mind over the next few e-mail-less hours. Result? Hint: It won't be pretty. Try it.

Forget computers. How about those other Internet-enabled gadgets? You know: cell phones, PDAs, and their ilk (e.g., the now-famous BlackBerry; aka "CrackBerry").[3] Thank goodness. NOW I can have Internet access on a PDA or a decent 3G cell phone so that I don't even need my computer to check my e-mail. It's with me 24/7 – in the car; at the airport; at the dinner table; at the PTA meeting; at the restaurant; in church; at the movies. Anywhere. Anytime. And I get this cute little inch-square keyboard to enter text. How lucky am I?![4] Talk about dropping a free fifth of vodka into the lap of an alcoholic! The results shouldn't surprise anyone. Inevitably we become enslaved by the very tools we created to serve us. Is there hope? Definitely! See sidebar discussion on breaking e-addictions.

---

[2] You know, Captain Jim Lovell, the Apollo 13 Mission Commander's classic understatement while-all-hell-was-breaking-loose on the way to the Moon.

[3] I am pleased to say that I have drawn the proverbial line-in-the-sand and do not carry a cell phone or a PDA. Hooray for me!

[4] I find it truly dazzling to observe teenage finger dexterity during a mile-a-minute text messaging "exchange" (too mild a descriptor) on a cell phone. My daughter, Joanna, can text faster than I can compose on my laptop ... with only her thumb, and with a blindfold on!

Here are some logical questions to ponder in this context:

Question 1: *How comfortable are you with going cold-turkey on your e-mail, for even a day? Your cell phone? Your PDA? Does your answer qualify you as an e-addict?*

Question 2: *Imagine that e-mail was erased from the planet starting . . . now. How would your life change for the better? For the worse? Repeat this for cell phones.*

Question 3: *When is e-mail appropriate, and when is it not, as a human communications venue?*

Question 4: *Can global businesses function effectively using e-managers? ("virtual" managers directing their people at remote locations via e-mail, e-teleconferences, etc.)*

Question 5: *What toll has e-addiction exacted on families? How has it reshaped family time and family relational dynamics? Has there been any impact on family structure and family values? For better or for worse?*

Breaking E-Addictions

Fortunately, Jean Chatzky of *Money Magazine* has created a useful "5-Step Program" for kicking e-addiction [9]. Give it a try!

1. *Admit Your Problem*: Think hard, revisit your work routine with respect to e-mail, and decide if you have a problem. 'Fessing up is always the first step to recovery.

2. *Repeat This Mantra: There Are No E-Mail Emergencies*. If it really were an emergency requiring your immediate action, people will ALWAYS find a way to reach you. They'll call. Hey, they might even knock on your office door!

3. *Tally Up The Cost*: Checking e-mail is inevitably disruptive to your concentration on the task at hand. You're working away, and then stop to answer that one pinged e-mail that just arrived and MUST be important. Thirty minutes later (after revisiting your whole inbox), you resume your work. It takes time to refocus, bring your concentration back to bare, ultimately harming your productivity rather than helping it.

4. *Fight Tech With Tech*: First, unsubscribe to all that useless electronic news-junk you receive. Second, if your PDA or computer buzzes or dings or flashes an envelope when you have new e-mail, turn off that feature. Silence the beast! You'll check messages when you want to, not because you've been conditioned to do it. Third, use an away-message that's specific. An autoreply for your e-mail (or instant-messaging program) that says you're writing a report or on a conference call, letting people know that you're truly busy and not just blowing them off.

5. *Hide the Gadgets*: Put the PDA in the drawer when you get home. Turn it off first! Don't keep your laptop sitting open and begging for your attention while at home as you pass by it. Power it down. Close its lid and let it rest. Keep it in your briefcase. *Choose* to power off your cell when you're at the family dinner table. They WILL call back or leave a message. Remind yourself – real life did exist before cell phones. Try leaving your cell at home, or at least in the car, when you go to the movies or out to dinner. You'll be better for it; we'll all be better for it. And, hey, then we can eliminate that 5-minute dressdown prior to the feature film when the theater workers come in and lecture the audience about keeping cells and PDAs off during the show, or else.

13.3 Gaming and aggressive behavior: A causal link?

So let me begin by freely admitting I'm not a gamer. Pong? Yep. Pacman? Sure. Space Invaders? Definitely. Doom? Nope. Halo? Nope. Myst? Nope. Grand Theft Auto? Thankfully, no. Sorry. Truthfully, my game of choice has always been pinball – call me old-fashioned! The Moore's law-driven performance improvements in both microprocessor speed and memory capacity have enabled frighteningly realistic games to emerge. At present, it's pretty much "if you can imagine it you can build it" in the game-creation universe. Not surprisingly, gaming is big business. In 2007 the computer game industry was valued at $9.5B. Big bucks. So what are people playing? See the sidebar for the gamers' Top 25.

On-line gaming, in which you can simultaneously go head-to-head in combat with a guy in Sweden, a gal in Argentina, and a guy in Singapore, is now all the rage, powered by "Flash" and "Java." "Flash" has become THE popular method for adding animation and interactivity to Web pages. Result? "Rich" Internet for a true-life gaming experience. Flash can manipulate vector and raster graphics and support bidirectional streaming of both audio and video, all prerequisites for lifelike advanced gaming. Java was originally "just" a new programming language, but importantly it also allows you to run mini-applications (aka Java applets) WITHIN a Web page; again, commonly needed for a slick on-line gaming experience. Gamers the world over are finally pleased with what the Communications Revolution hath wrought. Let the games begin!

The resultant jaw-dropping realism of both computer and on-line game experiences presents some interesting dilemmas worth thinking about. If I were to spend 8 h a day playing ultraviolent, shockingly realistic "shooter" games, for instance, would it make me a more aggressive person? Would I be more prone to committing a violent crime? Given that we as a global people presumably value peace as at least a desirable end goal for our planet, and the fact that even a casual glance at the evening news presents a depressing picture of the amount of violence and crime in our world, this is a pretty important question. Said another way, isn't the world violent enough without using the fruits of our transistors and computers to make it still more violent, even though it might be on the surface "cool" or "fun" to play them?! Parents have a vested interest, of course, because young kids are clearly highly impressionable. In fairness, a rating system (see sidebar) is presently in place to help parents with their game purchasing decisions, but how often that is actually followed is debatable. An interesting and highly relevant question: Is there a causal link between computer gaming and aggressiveness?

Top-25 Computer Games (April 2008) [10]

1. Black & White 2
2. Grand Theft Auto III
3. Perfect Dark Zero

    4. Halo 3
    5. NBA 2K6
    6. Age of Empires III
    7. Dragon Ball Z: Budokai Tenkaichi
    8. Grand Theft Auto: San Andreas
    9. F.E.A.R.
   10. The Movies
   11. Battlefield 2
   12. Madden NFL 06
   13. Castlevania: Dawn of Sorrow
   14. Metal Gear Solid 4: Guns of the Patriots
   15. Mass Effect
   16. SOCOM 3:
   17. Doom 3: Resurrection of Evil
   18. Far Cry Instincts
   19. Peter Jackson's King Kong
   20. Sniper Elite
   21. Indigo Prophecy
   22. Civilization IV
   23. Phoenix Wright: Ace Attorney
   24. City of Villains
   25. Day of Defeat

A highly controversial topic to be sure. The gaming industry obviously has a strong opinion on the subject! Recall – it's a $9.5B question. "Objective" studies inevitably suggest, however, that a link does exist, albeit usually a minor one. Recent research at Villanova University is typical [12]. One-hundred and sixty-seven university undergraduates participated in a study designed to assess the impact of gaming on aggressive behavior. After completing background and personality surveys, the undergrads were randomly chosen to play selected computer games, and then complete a questionnaire designed to quantify their resultant level of aggression. Two groups separately played either "violent" games (*Mortal Kombat*, *Deadly Alliance*, and *Doom 3* – some serious gore) and "nonviolent" games (*Return to Castle Wolfenstein Against Tetris Worlds*, *Top Spin Tennis*, and *Project Gotham Racing* – pretty tame). Result? Overall, players of the violent video games indeed produced significantly more aggressive responses than the non-violent games players. Ah-ha! BUT, it turned out that initially "mild-mannered" students were the least affected by gaming, whereas initially "angry" students were the most affected by gaming. That is, the starting personality of the gamer largely defined the impact of the gaming experience. Initial conditions matter. Humm ... so much for clearcut answers! This topic is worth thinking about, though, because so much is riding on the answer. And, truthfully, the matter is far from settled. Consider.

Question 1: Time to weigh in: Do you believe there is a link between computer gaming and aggressive or violent behavior? Defend your answer.

Question 2: Is the influence of the game dependent on the nature of the violence indulged? For example, is there a difference in impact between "shooting" people and battling alien creatures to the death?

Question 3: Is the current game-rating system sufficient? If not, how would you change it?

Question 4: Should stricter controls be in place limiting who has access to especially graphic or violent computer games? How could this be made effective for controlling on-line gaming?

Game Ratings [11]

The Entertainment Software Rating Board (ESRB) rates computer games to provide parents with ammunition to hopefully make better-informed decisions on game purchases for their kids. The rating consists of a symbol, followed by some description on the suitability for a given audience. Game Ratings include these:

- Early Childhood (EC): Suitable for ages 3 and older. Contains no material that parents would find inappropriate.
- Everyone (E): Suitable for ages 6 and older. May contain minimal cartoon, fantasy, or mild violence, and/or infrequent use of mild language.
- Everyone 10+ (E10+): Suitable for ages 10 and older. May contain more cartoon, fantasy, or mild violence, mild language, and/or minimal suggestive themes.
- Teen (T): Suitable for ages 13 and older. May contain violence, suggestive themes, crude humor, minimal blood, simulated gambling, and/or infrequent use of strong language.
- Mature (M): Suitable for persons ages 17 and older. May contain intense violence, blood and gore, sexual content, and/or strong language.
- Adults Only (AO): Should be played only by persons 18 years and older. May include prolonged scenes of intense violence and/or graphic sexual content and nudity.

Much like the ever-sliding "R" rating system for movies, however, the proverbial end-around (e.g., a 14-year old purchasing an M-rated game) is likely trivially accomplished for experienced gamers (kids will be kids).

13.4   The human genome, cloning, and bioethics

Arguably the most exciting development of the 21st century is the seemingly out-of-thin-air successful sequencing of human DNA. Ah, DNA – our inherited genetic erector set (Fig. 13.8). What makes you you, and me me; each of us fundamentally unique. What makes us human and a fruit fly a fruit fly.

The complete genome of *Homo sapiens* is stored on 24 distinct chromosomes (22 autosomal + an X and a Y for determining sex), and contains an estimated 20,000–25,000 genes, with just over 3,000,000,000 DNA base pairs.

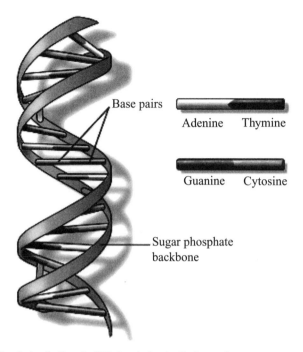

Base pairs

Adenine   Thymine

Guanine   Cytosine

Sugar phosphate
backbone

Figure 13.8 A visualization of a DNA strand, showing the base pairs.

The $3B Human Genome Project (HGP) was launched in 1990 to determine the sequence of chemical base pairs that make up the genome and to specifically map our 20,000-ish genes, from both a physical standpoint (where they are located on the various chromosomes), and importantly, also from a functional standpoint (i.e., which gene does what). Initially directed by James Watson at the U.S. National Institutes of Health (you know, Watson and Crick, co-discoverers of the structure of DNA), HGP released a "working draft" of the human genome in 2000 (announced jointly by then U.S. President Bill Clinton and British Prime Minister Tony Blair on June 26, 2000), with an essentially complete sequence released in April of 2003. In May 2006, the last chromosome was published. Importantly, the results of the HGP have been made freely available to the public for use by all interested parties. A shared global resource. Very nice. Perhaps not surprisingly in a capitalist world, a parallel project using a different sequencing technique was also conducted by the private company Celera Genomics (can you say gene patents?!). Those results are NOT for public consumption, as you might imagine. Most of the HGP sequencing was performed in universities and research centers from the United States, Canada, and Great Britain.

Curious how they did it? Alas, it's complicated; but in short, combine genetic engineering techniques with MANY POWERFUL COMPUTERS. Ergo, subtract transistors, no computers; no computers, no genome. Case closed. Genetic dark ages resume. Here's a mini-taste of the process [1]: First the DNA strand in question is broken into small, manageable pieces, each approximately 150,000 base pairs in length. These pieces are then spliced into the DNA of genetically engineered "bacterial artificial chromosomes" (BACs), which can then be inserted into legions of bacterial factories where they are mass produced by the bacteria's own DNA replication machinery (clever!). Each of these small pieces is then painstakingly "sequenced" (the base pair ordering identified) via sophisticated computer algorithms using a random "shotgun" approach. The now-sequenced 150,000 base pairs are then fit together end to end to eventually construct a fully sequenced chromosome. Now work your way through all 24. Sound easy? It's not. Remember, it cost $3B to do it! Read: Lots of computers required.

Some interesting genomic tidbits? Sure! The human genome has fewer genes than originally expected, and surprisingly, only about 1.5% of our genes code for proteins (our basic building blocks), whereas the rest comprise various RNA genes, complicated regulatory sequences involve how genes are expressed, and a WHOLE BUNCH of so-called "junk DNA" that doesn't seem to do much of anything (Hint: a dose of humility might be in order here; nature tends to be pretty subtle and remarkably clever – my own guess is that we have lots to learn still about what constitutes junk). The actual number of human genes is only 2× or so greater than that of FAR SIMPLER organisms (roundworms, fruit flies, and their ilk); rather remarkable given the resultant differences in complexity. However, human DNA instead makes extensive use of "alternative splicing" to produce several different proteins from a single gene. Clever. Human genes are also surprisingly distributed quite unevenly across the various chromosomes, with numerous gene-rich and gene-poor regions. The significance of these nonrandom patterns of gene density remains poorly understood. Last one. So ... like ... ummm, whose DNA actually got sequenced? John Doe? Jane Smith? Who's the lucky person? HGP researchers collected blood (female) or sperm (male) samples from a large cadre of donors, whose identities were then carefully protected to preserve anonymity. Supposedly (Urban Legend alert!), however, much of the DNA for the public HGP came from a single anonymous male donor living in Buffalo, New York (code name RP11) [1].

So why is it such a huge deal to have sequenced the human genome? Well, scientifically, knowledge of our genome is a treasure trove for understanding how life evolved, and hopefully will help solve the numerous riddles concerning how the human body actually does its magical thing – growth, healing, aging, you name it. But that is only the tip of the iceberg. Think health care. Most aspects of human biology are determined by both genetic (inherited) and complex environmental factors. Some inherited genetic variations determine relatively benign aspects of who we are (e.g., height, eye color, shoe size, gulp ... hair loss; perhaps not so benign!). But there are whole classes of genetic miswirings that are not nearly so benign; ironic given that genetic mutation fuels life's evolutionary engine. Some genetic disorders cause disease only when in combination with the appropriate environmental factors (e.g., diet, sunburn, exposure to secondhand cigarette smoke), especially troubling given the plethora of manmade chemicals we now routinely release into our ecosystem without batting an eye. Other genetic disorders, however, are defined by unintended genomic DNA sequence variations – miswirings. In the most straightforward cases, such genetic disorders can even be associated with variations of base pairs within a single gene, For example, cystic fibrosis is caused by one of 1,300 known mutations in the CFTR gene and is the most common recessive genetic disorder in Caucasian populations [1]. Disease-causing mutations within specific genes usually have severe consequences for gene function, and hence their impact on health is strongly felt. Fortunately these tend to be statistically rare. However, given the sheer number of genes potentially involved in mutations, in aggregate they comprise a significant percentage of the known medical problems in the general population, especially in children. There are at present about 2,200 genetic disorders for which the underlying causal gene(s) has been identified, and this is growing rapidly thanks to our knowledge of the genome. How would you use the genome to combat this?

Step 1: Identify the gene(s) involved in said disorder.
Step 2: Develop genetic testing techniques to determine the predisposition of various at-risk individuals.
Step 3: Develop an effective therapy to treat those in need.

Thus our new knowledge of the human genome can be leveraged as a powerful tool in the development of more effective medicines, and even may eventually help us eradicate the myriad of tragic diseases associated with genetic defects. I think it is fair to say that most scientists believe that our best hope for ultimately conquering cancer, for instance, rests here, with the development of various "gene therapies." Time will tell, of course, but our medical future looks substantially brighter with the sequenced genome in place. Exciting times. Go transistors!

This all seems rather altruistic, right? Use the human genome for the betterment of humankind. But there are inevitably some troubling logical downstream consequences that should give you pause for thought:

Question 1: Should any company have a right to file a patent (invoke ownership) on a specific gene or gene sequence that they have identified as potentially useful? Pros? Cons?

Question 2: With the genome in place, one can more easily envision a future in which designer children could be tailormade by manipulations of specific gene expressions to have certain desirable attributes (e.g., hair color, skin tone, athletic prowess, good looks, smarts). Would parents ever be justified in such tampering with the random fate of nature? What are the unforeseen implications?

Question 3: With our rapidly increasing ability to identify gene function and develop viable genetic testing for various maladies, our ability to foresee our own individual medical future is getting clearer by the day. However, such testing is extremely expensive, and likely to remain so. Who should get these tests, and under what circumstances? Who pays for them? What impact would this have on already skyrocketing health care costs?

Figure 13.9   Dolly, the famous cloned sheep. (Courtesy of Gusto Images/Science Photo Library.)

Question 4: The pioneering genetic engineering techniques used to sequence the genome have also led to a veritable arms race in animal cloning. At present we have tadpoles (1952), carp (1963), mice (1986), sheep (the famous Dolly – Fig. 13.9 – 1997), rhesus monkeys (2000), cows (2001), cats (nicknamed "copycat"! – 2001), mules (2003), and horses (2003); and counting. The genomes of many such animals are currently being decoded. Soon we will be able to replicate ad infinitum a prized stallion, an especially tasty cow, a near-extinct species, or even beloved Fido who got run over by a car. Under what circumstances should animal cloning be allowed? Should animal cloning be subject to (global) regulation? How? By whom? Should cloned food animals be allowed into the global food supply? Why or why not?

Question 5: It seems only a matter a time before a human is cloned. Ratchet up the ethical issues by a mile or three! Would human cloning under any circumstance ever be morally justifiable? Example: One might imagine cloning an individual to then "harvest" organs (stem cells, DNA?) before that clone was "born" in order to treat another person suffering from a devastating terminal disease. Do the ends ever justify the means in such scenarios? How do we best regulate human cloning? Locally? Globally? Can human cloning be prevented even if we choose to stop it? How?

Question 6: Given that evolution operates by random genetic variations and mutations (read: gene manipulations), are there potential long-term consequences (over thousands or even hundreds of thousands of years) for unintentionally altering human evolution by (artificial) genetic engineering, say, ending in a "Planet of the Apes" scenario? What are the potential consequences? Any upsides? Downsides?

The questions could go on for days. Remember: The sequencing of the human genome and its plethora of logical consequences represent an classic unforeseen result directly emanating from the development of the transistor and computer technology that it has enabled.

13.5 The changing face of education

Computers are clearly an essential part of modern life. Given the fact that virtually no 21st-century profession can be transacted without using computers at some level (at least in the first world), it seems blindingly obvious that computer-based learning and instruction is the way to best educate our young people, right? Banish those blackboards; put a laptop on every desk! Create slick computer "games" that will instantly engage student's attention; empower them to learn at their own pace; think outside the box; nurture their creativity. Heck, school might even be fun! Let's get kids rolling on computers from minute one, establish some computer literacy for comfort, and then turn them loose. The sooner the better. Pre-K? Why not! K–12? Definitely! College? Duh!

It's not hard to imagine that we teachers have begun to feel a bit under attack. Teacher obsolescence seems only a heartbeat away. "Teachers? Hah! Outdated. Who needs them?!" But not so fast ... Numerous recent studies of test scores and student achievement, particularly at the elementary school level, present (at best) a murky view of the net effectiveness of computer-based education. Test scores simply do not bear witness that computer-based education is superior to "old-fashioned" teaching. And even if test scores did improve, clearly there is far more to education that simply being able to make a high mark on the PSAT, the SAT, or the GRE. Education is clearly a communal process, and social skills can be just as important to long-term success as book smarts.

One opinion, admittedly extreme [13]: "A basic truth in education is that a child must be literate before he or she is computer literate. The best teacher has always been a person, not a machine. Which is a better learning environment: reality or virtual reality? In the three dimensional, real world, kids encounter the unexpected. On the two dimensional screen, children see only the choices a programmer has developed for them. While presenting the illusion of options, a computer can only deliver a limited number. The keyboard and mouse constrain a child's option to reach out and touch the world. In the real world we can teach, explore and learn the patterns of connection which link different people, plants, animals and places. If education software even attempts to deal with these crucial concepts, the limits of the media may make the presentation inflexible, superficial, and inadequate."

Pundits, of course, will rush to point out that if only computers were used the RIGHT WAY, the results would be very different. Perhaps. Perhaps not. And the argument that computer software is not yet good enough would seem to suggest that someday it might be, given Moore's law of evolutionary influence on computational power. Clearly, of course, the effectiveness of learning within ANY classroom environment depends on the quality of the teacher.

In my own experience, admittedly at the university level (Fig. 13.10), the role of computers in effective teaching is actually rather limited. Tons of professors have embraced the now-pervasive trend of teaching via powerpoint "slides." Are they slick to look at? Yep. Do they minimize prep time for lecture once

Figure 13.10 The Georgia Tech Tower (courtesy of The Georgia Institute of Technology).

developed? Sure. Are they easier to modify as you hone your course year after year? Definitely. Can you cover more ground during lecture? Absolutely. Much more. Is it easier to distribute your material to the students so they can follow along during lecture? Yep. Less hand cramp from writing? You bet; for you and them! But ... do students actually learn as much from a powerpoint lecture as from a slower-paced, traditional at-the-board lecture style? In my experience ... no, they do not. I have repeatedly queried my classes over the past 17 odd years on this matter. The results point to an *overwhelming* preference on the part of students to old-fashioned at-the-board lecturing. Why? Well, it is slower paced, for one. Easier to absorb. Writing also tends to keep students awake. There are natural pause-points on the board where questions can naturally arise and nuances gleaned. Questions asked and answered, and re-asked. For most people, writing instead of just listening also helps to solidify material; helping to "carve it" into their memory. Understanding and retention are improved. And of course, a bit selfishly, an at-the-board lecture style clearly better indulges a professor's natural showmanship tendencies. Guilty as charged! Could note writing be effectively incorporated into powerpoint slides? No doubt. But in my admittedly anecdotal experience, computers should be left in the backpack during class. Have them open, and the tendency to surf or IM or even game can get awfully tempting (remember those e-addictions!).

And that is not to say that students don't need to engage their computers. They do. That is where they will compose; research; run programs to calculate answers to problems; and even design things. BUT, that happens outside of the classroom, not inside.

A complicated topic for sure, but one we teachers (and you students) need to think hard about given the profound downstream implications of good vs. bad education. Some follow-up questions:

> ***Question 1****: Which lecture style to you prefer to receive, powerpoint or at-the-board? Why? What are some of the most annoying features of powerpoint-based lectures?*
>
> ***Question 2****: Do Internet-based classes work well in a modern university environment? Weigh in on taking on-line exams; using chat rooms for communication with profs/TAs; wading through self-study lectures.*
>
> ***Question 3****: Do on-line degree programs produce a similar caliber of graduate as conventional universities? Why? Why not?*
>
> ***Question 4****: What components of the traditional university "experience" translate well to a virtual e-education? Which do not?*
>
> ***Question 5****: Is there a viable path to on-line graduate study? In which majors might that work? Not work?*

13.6 The evolution of social media

"*Social media* is an umbrella term that defines the various activities that integrate technology, social interaction, and the construction of words and pictures. This interaction, and the manner in which information is presented, depends on the varied perspectives and 'building' of shared meaning, as people share their stories, and understandings." Ironically, this statement was taken from Social Media Platform 101 – Wikipedia [1]. Social media represent a very new and quite fascinating intersection between micro/nanoenabled high technology and global society. Given its newness, social media tend to be a land for younger folk, although this is rapidly changing. Social media venues assume a staggering array of forms, including Internet forums, message boards, blogs, podcasts, picture–video–music sharing, wall-postings, e-mail, instant messaging, etc. Not surprisingly, social media "applications" (places to see and be seen) have sprung up to corral this herd of media into more interface-friendly venues: things like MySpace (social networking, Fig. 13.11), Wikipedia (reference, Fig. 13.12), Google Groups (reference, social networking), Facebook (social networking); Last.fm (personal music); YouTube (social networking and video sharing); Second Life (virtual reality); Flickr (photo sharing); Twitter (social networking and microblogging), etc.

Social media differ fundamentally from traditional media (newspapers, television, books, radio, etc.) because they function via interactions between people from all walks of life, at all locations, from all cultures, at a very rapid pace, and its ever-changing and highly dynamic form rests on building a collective,

Figure 13.11 The MySpace logo (courtesy of MySpace.com).

editable-as-we-go, "shared-meaning" via an enabling technological conduit (aka the Internet) [14]. There is no queen bee controlling things; no stuffy editorial board – the hive rules. Get an Internet connection and you too can play. Let's face it, social media are fun; addictive even. And time consuming! No Facebook page? Shame on you! I personally find Wikipedia, for instance, to be a remarkably useful resource (see sidebar discussion). Need some general information very quickly on a topic you know little about? Go to Google Search in your Web browser and type in *social media AND wiki*. Press go. The world likely knows something about it. Of course, the wiki content, by definition, is only as a good as the article's author(s) (and the various downstream editors), but if one maintains a healthy skepticism of all Web-based "free" information (students need to be constantly reminded of this!) then no worries. In my own experience, "most" of the content that I see on wiki (that I actually know something about) is generally pretty darn accurate. You'll note wiki references sprinkled throughout this book (thanks for your hard work all you wiki authors!).

Wikipedia

Wikipedia is a free, multilingual, open-content encyclopedia project operated by the nonprofit Wikimedia Foundation [1]. The name Wikipedia is a cojoining of *wiki* (a social media for creating collaborative web sites) and *encyclopedia*. Launched in January of 2001 by Jimmy Wales and Larry Sanger, it is the largest, fastest-growing, and most popular Internet-based general reference work available. As of April 2008, Wikipedia has attracted over 684 million visitors, accessing over 10 million articles, in 253 languages, and comprising a total of over 1.74 billion words (and counting). Wikipedia's articles are written collaboratively by volunteers all around the world, and importantly, its articles can be edited by anyone with access to the Internet. The intent is that Wikipedia content naturally grows over time, with a built-in, self-regulating mechanism for improving overall accuracy during that growth. You certainly can't argue with its success. Wikipedia ranks among the top-10 most-visited web sites worldwide. There are critics, of course (can you say Britannica? – joking!). Having content accuracy defined by consensus rather than by experts will always be suspect at some level. Wiki-vandalism (intentional misinformation in Wikipedia articles) is also a concern, although the thought is that such perversions should be self-correcting (with a billion people looking over your shoulder).

Social media are so very new, and growing so very rapidly, that predicting how they will change civilization is an interesting topic for discussion all on its own. Several logical questions for your consideration:

Figure 13.12 The Wikipedia logo (courtesy of the Wikimedia Foundation).

> *Question 1: Do social media "water down" the truth by allowing everyone to be an expert and weigh in on topics, regardless of their credentials?*
>
> *Question 2: How do we ensure that Wikipedia entries get it right without destroying the compelling original concept? Is there a viable way to vet entries before being posted, or would "Wikipedia police" quash the party?*
>
> *Question 3: Should social networking venues (Facebook, MySpace, etc.) check ID for admission? Should someone (if so, who?) monitor activity and/or troll for less-than-savory members not playing by the rules? At what age are such social media venues appropriate places to "play" for young people?*
>
> *Question 4: Does the proliferation of social media risk diluting the very notion of photography–video–music as valid "art forms"? Does this matter?*
>
> *Question 5: How have social media affected our understanding and use of written language? For example, does an all-lowercase, punctuation-free, mile-a-minute, spelling-error-prone, social media writing style matter in the grand scheme of things? Why or why not?*
>
> *Question 6: Do social media change the way we think? Or, in recent parlance, "Is Google making us dumber?" Why or why not?*

13.7 E-activism and e-politics

Internet-based activism (aka e-activism or cyberactivism) centers on the use of Internet "tools" (e-mail, web sites, podcasts, etc.) to promote various forms of political activism. Why use the Internet? Simple: To speed up communications links between various movement constituents, deliver messages faster and cheaper to target audiences, and to reach far more people than knock-on-door campaigning or snail-mailing can ever hope to reach. Related activities also include e-fundraising, e-lobbying, and e-volunteering in conventional political campaigns. In short, Internet meets politics – electronic politics (e-politics) [15,16].

Internet activism can be broken into three basic categories [1]: Awareness–Advocacy, Organization–Mobilization, and Action–Reaction. The Internet is often a vital resource for activists in totalitarian countries, particularly when the intended advocacy message runs counter to the party line, effectively preventing information dissemination. Examples might include getting the word out on human-rights violations in a regime that controls the media sources, the reporting of atrocities, political or religious suppression, or election fraud. The Internet also enables cash-strapped advocacy organizations to communicate with their members (and recruits) in an inexpensive and timely way. Gatherings and protests can be easily e-organized, mass e-announcements sent, and even live-coverage feeds of the event e-distributed to all members in real time. Action–Reaction is a more radical category consisting of "denial-of-service" attacks (spam a target server until it cries "uncle" and has to reboot), taking over and vandalizing someone's web site (drawing a mustache on your opponents web site pic), uploading "trojan horses" to disable an opponents political e-machinery. Such radical activism is sometimes known as hacktivism and can obviously get you in trouble! It should be obvious how the anonymity afforded by the Internet would prove indispensable.

The Internet has also played an increasingly important role in political campaigns, and e-politics has seemingly reached a crescendo in the 2008 U.S. presidential election (Fig. 13.13). "The Internet is tailor-made for a populist, insurgent movement" . . . having its "roots in the open-source ARPANET, its hacker culture, and its decentralized, scattered architecture make it difficult for big Establishment candidates, companies and media to gain control of it. And the establishment loathes what it can't control. This independence is by design, and the Internet community values above almost anything the distance it has from the slow, homogenous stream of American commerce and culture. Progressive candidates and companies with forward-looking vision have an advantage on the Internet, too. Television is, by its nature, a nostalgic medium. Look at Ronald Reagan's campaign ads in the 1980s – they were masterpieces of nostalgia, promising a return to America's past glory and prosperity. The Internet, on the other hand, is a forward-thinking and forward-moving medium, embracing change and pushing the envelope of technology and communication" [17]. One significant e-politics-induced change has been the emerging role of "small donors" in helping finance mega political campaigns [Then Senator Obama was very successful at this.]. In pre-Internet days, small-donor fundraising was

Good ole Uncle Sam (courtesy U.S. Library of Congress).

prohibitively expensive because of printing and postage costs. Not so for Internet-based fundraising. Go to the candidate's web site, enter your credit card, press go, and now you too can be a financial supporter. Grass-roots here we come. This is clearly empowering to young people who may want to have a political say. Now they can.

Are there limitations to the overall effectiveness of e-activism and e-politics? Most definitely. For example, the term "digital divide," popularized by then-President Bill Clinton in a 1996 speech in Knoxville, Tenneesee, refers to the gap existing between people with Internet access and those without. This divide is typically either driven by individual's socioeconomic status (they can't afford the requisite computer or money to pay the ISP for a connection), their location (no Internet access is available), or some other practical barrier. The "global digital divide" extends that notion to the world scene, where third-world countries may be at a major disadvantage because of the lack of a basic communications infrastructure (no Internet backbone). The concern here is that e-politics potentially gives disproportionate representation to those with disproportionate Internet access or technological means – read: an unfair advantage; undemocratic. Clearly, for e-politics to be successful in the long run, the divide separating the various target audiences must be closed (globally), and this will not be easy. Remember, only 1.3 billion people currently have Internet access, out of a world population of 6.5 billion.

Some food for thought related to e-activism and e-politics? Sure!

> *Question 1: How can the Internet best be used to "take back" politics from big business and special interest groups and put it in the hands of the people?*
>
> *Question 2: Does the Internet have a role to play in human-rights monitoring? How would this best be implemented in third-world countries without a viable communications infrastructure?*
>
> *Question 3: Is it ever justifiable to use the Internet as a weapon to overthrow a totalitarian regime? How would such a cyberwar actually be fought?*
>
> *Question 4: How do we best close the global digital divide? Should we try?*

13.8 The web we weave: IC environmental impact

Okay, blood pressure check. Deeeepppp breath in. Now slowly let it out. Feeling relaxed? Good. Now – repeat after me: "GLOBAL WARMING." Any temple throbbing? Heart palpitations? Facial tics? Cold sweat? Welcome to Contentious Issues 101. Global warming is a fact. No one argues this. The world is getting demonstrably hotter (by perhaps 3°C this century). Very bad things are quite likely to happen as a result (think melting ice caps, poor Venice, shifting weather patterns, etc.). However, people, lots of people, do argue, heatedly (pun intended, sorry), about the role humans are actually playing in creating this observed warming, and whether we have a responsibility as a global community in trying to help Ma Earth out by cutting back on our greenhouse gas emissions. This is a big deal and highly contentious because there is a strong coupling between the world's economic engine and greenhouse gas production – you can't have it both ways. Al Gore has an opinion. So does the United Nations Intergovernmental Panel on Climate Change (IPCC), a collection of the world's leading climate scientists, which concluded (February 2007) that "major advances in climate modelling and the collection and analysis of data now give scientists 'very high confidence' – at least a 9 out of 10 chance of being correct – in their understanding of how human activities are causing the world to warm" [18]. But clearly the debate continues . . . while the planet warms. As Voltaire famously said, "Men argue, nature acts." Intuitively, the inertia of the Earth's climate is not small (major understatement alert), so if it turns out in the end that we are indeed a Titanic headed right for the iceberg, our options for last-minute sidestepping are depressingly limited. It's an important issue to wrestle with by any reckoning.

Climate change aside, something that all can agree on is an increasingly acute need for resource conservation. Oil, titanium, ironwood, fresh water, you name it – these are all limited resources. Use them up and they are gone. Of course, 100 years ago no one really cared, because the world was such a huge place. But at 6.5 billion souls and counting, coupled with the emergence of capitalism and its sidekick consumerism, things are rapidly getting out of hand. Rain forests are being desecrated, pollution-intensive electricity production using, say, coal-fired power plants to support manufacturing energy needs, are rapidly proliferating, and fresh-water acquifers are being drained or polluted or both. Not good. Yep, resource conservation has emerged as a major issue of concern for the 21st century. It has been refreshing to see just how quickly the terms "Green" and "Carbon Footprint" as the "new cool" have tiptoed into the global lexicon. No longer the parade ground of hippies, tree-huggers, and Euell Gibbons,[5] in only the past couple of years the world has jumped on board with all-things-green. Recycling, reusable grocery bags, turning the water off when you brush your teeth, driving a fuel-efficient car (better yet a hybrid), planting trees, tracking and then minimizing your carbon footprint. Encouraging signs. Alas, pragmatism often wins the day, at least for the time being. Can third-world countries worried about feeding their people afford to worry about greenhouse gas production? Can the first-world juggernauts afford to dial down their economic engines with "carbon taxes" and sustainable resource allocations if that growth decrease literally ripples around the world affecting everyone's prosperity? Complicated issues to be sure. Tough decisions to be made.

One might logically wonder in this context just how "green" micro/nano fabrication actually is. Fortunately silicon is a resource we have plenty of, so no worries there. BUT, we have already seen just how complex it is going from beach sand to functional IC. And let's not forget about the requisite toxic dopants

---

[5] You know, the famous Grape Nuts commercial – "Ever eat a pine tree? Many parts are edible."

Environmental Impact of ICs

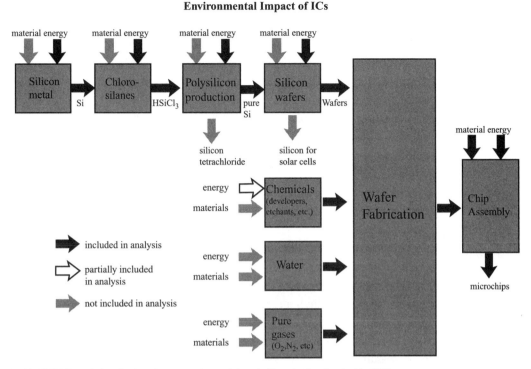

Figure 13.14 A visualization of energy and material use in IC production (inspired by [20]).

(boron, arsenic), nonbiodegradable organic polymers, precious metals (tungsten, copper, aluminum, gold), gobs of pure water, and of course the massive amounts of energy required for powering the IC fab beast. And EVERYTHING in question must be the absolute ultimate in purity to ensure high yield (99.99999% pure). Net result? The environmental footprint of IC manufacturing is not especially pretty to behold [19].

Some numbers? Sure! The total fossil fuel + chemical mass needed to produce a 2-g IC die? About 1.7 kg (Fig. 13.14). The total energy required for producing an IC and using it? About 56 MJ (Fig. 13.15) [20]. To put this in perspective, we define a "secondary materialization index" (SMI) to measure the material and energy "intensity" of building Widget X (a smaller SMI is generally better for environmental impact). The SMI is defined to be the total weight of all secondary materials consumed during production, divided by the total weight of the final product. The SMI of an automobile? 1–2. Refrigerator? 2. IC chip? 640! Gulp. An interesting explanatory thought on the origins of the extreme size of this number for ICs: Because ICs are highly organized, very low-entropy objects, the materials and energy required for building them are especially large [20]. Worth pondering.

Given this rather bleak picture for building the micro/nano "stuff" needed to fuel our burgeoning Communications Revolution, what is one to do, stop using computers? Doubt it. Shut down the Internet? Not likely. Some ideas:

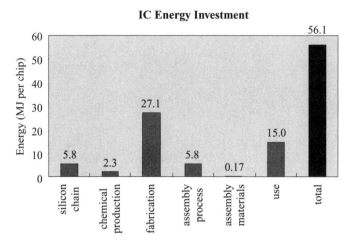

Figure 13.15   Energy breakdown for production and use of a single IC.

- Establish national (global?) systems for recycling computers and electronic widgets to either reclaim and reuse the resources resident in them, or at least to keep them out of the environment and food/water chain.
- Ban the use of especially environmentally toxic chemicals historically used in IC manufacturing. For example, mandate 100% lead-free IC packaging (already common in many places).
- Use eco-friendly labels to reward companies for particularly green IC manufacturing approaches.
- Extend the lifecycle of computers and electronic widgets. The computer practice of built-in obsolescence in the computer industry is troubling in this context. Can't we figure out a creative use for the slightly aged but still functional laptop?
- Carefully look over the shoulders of IC manufacturing companies, especially those outsourcing IC production to non-first-world countries where governmental oversight on environmental impact generally is lacking.

All this said, there are other potentially very positive roles for micro/nanoelectronics to play in greening-up the planet. Some are obvious, some are buried under the covers. For instance, photovoltaics (PV) as a viable energy production means is a highly visible semiconductor-driven renewable energy source that will clearly play an increasingly important role now that oil prices are skyrocketing and fossil-fuel-based electricity production is frowned on as non-eco-friendly. And let's not forget the role that electronics can play in energy use "optimization." That is, in complex energy-intensive systems (cars, homes, buildings, cell phones), micro/nanoelectronics can be used to first sense, and then electronically optimize, in real time, the system's own energy usage. For instance, the electronic systems in your car presently adjust the engine operational parameters dynamically to maximize fuel efficiency. GPS optimizes truckers' travel routes to minimize distance and maximize fuel efficiency. Your laptop and cell phone both have varying levels of sleep modes and sophisticated power-down circuitry to extend battery life. In your home, electronic sensors will soon turn off lights when they are not in use or adjust the thermostat in real time to maintain comfort while maximizing energy efficiency. Opportunities for creativity abound.

So, yes, micro/nano is not necessarily a bright shade of green at first glance, but possibilities do exist for brightening this picture up and are receiving increasing attention. Some questions on the eco-side of the IC world include these:

> *Question 1: Time to weigh in. Just how important a role do you think humans are playing in global warming? Do some research. Find evidence both pro and con.*
>
> *Question 2: Should first-world, cash-rich countries to be able to buy "carbon offsets" from cash-starved, third-world countries to minimize their own carbon footprints? That is, should they be able to trade money to maintain their status quo in greenhouse gas production? In 2006, for instance, about $5.5 billion of carbon offsets were purchased, representing about 1.6 billion metric tons of CO_2 [1].*
>
> *Question 3: The Catch-22 of photovoltaics: If it actually takes massive amounts of electricity to produce the silicon solar cells that I then plan to use to produce "green" renewable electricity, where is the "real" break-even point? Meaning, if it takes 100 kWh of electricity from a coal-fired power plant to build that slick solar array that generates 100 kWh from a renewable source (the Sun), have I really gained anything?*
>
> *Question 4: What are some new ideas for greening up our world via energy use optimization with micro/nano widgets?*

Whew! And so the end finally has arrived. I hope you have enjoyed the journey. Questions abound, yes; but many, many promises and possibilities too. My final challenge?

Ad Eundum Quo Nemo Ante Iit.

Take up the gauntlet.

References and notes

1. See, for instance, Wikipedia: *http://en.wikipedia.org*.
2. For a nice visual introduction to the structure of the Internet, see R. Haynal, "Internet, the Big Picture," at *http://navigators.com/internet_architecture.html*.
3. See, for instance, *http://www.saschameinrath.com/*.
4. See, for instance, *http://dragon.ep.usm.edu/ it/capstone/tonette/ethics1.html*.
5. See, for instance, *http://www.missingkids.com*.
6. D. Finkelhor *et al.*, "Online victimization of youth: Five years later," *National Center for Missing and Exploited Children*, Vol. 33, pp 7–8, 2006.

7. See *http://www.netsmartz.org/*.

8. See *http://www.inhope.org/*.

9. J. Chatzky, *Money Magazine* editor-at-large, "Kick your e-mail addiction," *Money Magazine*, February 22, 2007, *http://money.cnn.com/2007/02/22/magazines/moneymag/chatzky.moneymag/index.htm*.

10. See, for instance, *http://www.gamespot.com/misc/top100_pop.html*.

11. See, for instance, *http://www.esrb.org/ratings/ratings_guide.jsp*.

12. C. Campbell, "Games do cause violent behavior (but not much)," *Business Week*, April 25, 2007, *http://www.businessweek.com/innovate/content/apr2007/id20070425_615390.htm*.

13. "Learning in the Real World," *http://www.realworld.org/*.

14. D. Tapscott, *Growing Up Digital: The Rise of the Net Generation*, McGraw-Hill, New York, 1999.

15. S.F. Hick and J.G. McNutt, *Advocacy, Activism, and the Internet: Community Organization and Social Policy*," Lyceum Books, 2002.

16. M. Dartnell *Insurgency Online: Web Activism and Global Conflict*, University of Toronto Press, 2006.

17. J. Trippi, *The revolution will not be televised: Democracy, the Internet, and the overthrow of everything*, Regan Books, 2004.

18. See, for instance, *http://www.ipcc.ch/*.

19. R. Kuehr and E. Williams (Editors), *Computers and the Environment: Understanding and Managing Their Impact*, Kluwer, 2003.

20. E. Williams, "Environmental impacts of microchip and computer production," 2003, *http://www.environmentalfutures.org/Images/williams.pdf*.

Properties of Silicon

Table 1.1 contains the bulk structural, mechanical, optical, and electrical properties of the semiconductor silicon, driver of the Communications Revolution [1–3].

Table 1.1 Properties of bulk silicon

Parameter	Units	Silicon
Atomic number	–	14
Atomic density	(atoms/cm^3)	5.02×10^{22}
Atomic weight	(g/mole)	28.09
Density	(g/cm^3)	2.329
Electronic orbital configuration	–	(Ne)$3s^2 3p^2$
Crystal structure	–	diamond
Lattice constant (298 K)	(Å)	5.43107
Dielectric constant	–	11.7
Breakdown strength	(V/cm)	3×10^5
Specific heat	(J/g °C)	0.7
Melting point	(°C)	1412
Index of refraction	–	3.42
EM transparency region	(μm)	1.1–6.5
Thermal conductivity (300 K)	(W/cm °C)	1.31
Thermal expansion coeff. (300 K)	(°C^{-1})	2.6×10^{-6}
Young's modulus	(dyn/cm^2)	1.9×10^{12}
Energy bandgap (low doping)	(eV)	1.12 (300 K)
Effective electron mass (300 K)	($\times m_0$)	1.18
Effective hole mass (300 K)	($\times m_0$)	0.81
Intrinsic carrier density (300 K)	(cm^{-3})	1.02×10^{10}
Electron mobility (300 K)	(cm^2/V s)	1450
Hole mobility (300 K)	(cm^2/V s)	500
Electron diffusivity (300 K)	(cm^2/s)	37.5
Hole diffusivity (300 K)	(cm^2/s)	13
Intrinsic resistivity (300 K)	(Ω cm)	3.16×10^5

References

1. J. D. Cressler and G. Niu, *Silicon-Germanium Heterojunction Bipolar Transistors*, Artech House, Boston, MA, 2003.
2. M. Shur, *Physics of Semiconductor Devices,* Prentice-Hall, Englewood Cliffs, NJ, 1990.
3. R. Hull (Editor), "Properties of crystalline silicon," EMIS Datareviews Series, Number 20, INPSEC, 1999.

Some Basic Concepts from Physics and Electrical Engineering

In case it's been awhile since you have studied high school or college physics, and feel a little rusty, have no fear! In case you have NEVER studied electrical engineering, no worries! Here is a bare-bones background on some very basic concepts necessary for our introductory discussion of microelectronics and nanotechnology [1].

A reminder: In its very broadest sense, physics is the science of nature. Physics deals with matter and energy and the fundamental forces that nature uses to govern the interactions between all things. Simply put, the grand challenge of physics is to understand how the universe does what it does, across dimensions ranging from the nanoscale to millions of light years. Physics does not deal with questions of "why," but rather of "how." Clearly, physics is the basic underpinning of all of engineering. Read: Pretty important in the grand scheme of things. Electrical engineering, or more commonly today in most universities, the bundle of electrical AND computer engineering, is the practical application of electrons, and the current flow generated from them, to cool new technology. It ranges from the electrical wiring in your house, to the global power grid, to the design of your cell phone or laptop, to the Internet. It most definitely spans the understanding, development, and creative use of semiconductors to ultimately yield the fields of microelectronics and nanotechnology. So, let's now review a few very basic ideas from both fields.

2.1 Energy

Energy is about as fundamental a concept as we have. Everything that exists, everything, has energy, and it occupies space, and it evolves in time. Energy (E) and mass (m) are fundamentally linked by the speed of light (c), according to[1]

$$E = mc^2. \tag{2.1}$$

---

[1] Einstein's famous result, and arguably our best-known mathematical equation. Regarding it, Einstein stated: "Politics is for the moment, an equation is for eternity."

Mass and energy are one and the same thing. This nonobvious result led inexorably to the nuclear age.

Physical reality, our experience of the universe, is simply a manifestation of the various forms of energy and matter. Not surprisingly, if you stop and think about it, energy is actually at the crux of human civilization. Consider: To manipulate nature in accordance with our basic needs as a society (to eat, to drink, to talk on your cell phone, to surf the Net), we must expend energy. Hence, engineering and technology, at their most basic level, are concerned with first identifying and then relentlessly taming the various forms of energy that exist in our environment. All things in nature, whether from a perspective of physics, or chemistry, or biology, or electrical engineering, operate by regulating or transforming energy from one form into another, and this transformation is the basis for the amazingly complex array phenomena we refer to as "nature": you, me, horses, transistors, apple trees, Fido, the cosmos. Importantly, energy is conserved in isolated systems, meaning energy is neither created nor destroyed; rather it is only changed from one form into another. Examples of energy? Sure! Kinetic, potential, gravitational, nuclear, electrical, solar, etc. The units of energy are joules ($J = kg\ m^2/s^2$), the same as "work." In fact, energy is often defined as the "capacity to do work"; latent work. In the micro/nano world, energy is usually expressed in electron volts, eV, but that is merely for convenience in calculations. 1 eV is simply 1.602×10^{-19} J. Big difference in magnitude, but the same thing – it's all energy.

2.2 Force and the theory of everything

In general terms, a "force" in physics is a "push" or a "pull" that can cause an object with mass to accelerate. According to Newton's famous second law, an object will experience acceleration (a) in proportion to the net force (F) acting on it, and in inverse proportion to the object's mass (m); mathematically,

$$F = ma. \tag{2.2}$$

Surprisingly, there are only four fundamental forces found in nature. In order of decreasing strength, they are (1) the strong nuclear force (governs how nuclei inside atoms stick to together); (2) the electromagnetic force (electricity and magnetism – our land of electrons); (3) weak nuclear force (governs radioactive decay); and (4) gravity (you know, Newton and his apple). One of the major themes in physics is to prove that all forces are fundamentally one and the same. Physicists have cleverly used monstrous particle accelerators (atom smashers) to probe these linkages, and at present a unified theory of forces 2–3 exists (electroweak), and substantial progress has been made in unifying the electroweak and strong nuclear forces (1, 2, and 3). All the craziness associated with "string theory," with its murky, multidimensional notions, centers on bringing gravity (general relativity) into the fold, for a complete "theory of everything" (physicists are not timid in their terminology!).

2.3 Electrons, quantum mechanics, and Coulomb's law

Discovered in 1897 by J. J. Thomson, the electron is a fundamental (can't be broken down into further constituents) subatomic particle that carries a negative electric charge ($q = -1.602 \times 10^{-19}$ C, first measured by R. A. Millikin in 1909). Hold on tight – the electron is a fermion, and has a spin quantum number of 1/2, importantly allowing it to participate in electromagnetic force interactions. Hooray for us! The electron has a mass (m_0) approximately 1/1,836 that of a proton ($m_0 = 9.11 \times 10^{-31}$ kg – pretty darn light!). Together with atomic nuclei, consisting of protons and neutrons, electrons make up atoms and are largely responsible for the way they chemically interact (bond), ultimately yielding all of the matter of the universe. There are lots of electrons to go around (as close to a real infinity as you'll ever come), and electrical engineering evolved as a discipline largely in response to the increasingly clever ways people found to move electrons around to harness their resultant energy. It's fun too!

Do bear in mind that the electron, as an elementary particle of minuscule size and mass, obeys the laws of quantum mechanics. And those laws can be strange indeed! At a fundamental level the electron is actually a "particle wave," exhibiting particlelike (think billiard ball) properties in some instances, but wavelike (think sound or water waves) properties in other instances. All the while obeying the governing physics of quantum mechanics via Schrödinger's equation:

$$-\frac{\hbar^2}{2m} \, \nabla^2 \, \psi(x, y, z, t) + U(x, y, z)\psi(x, y, z, t) = i\hbar \, \frac{\partial \psi(x, y, z, t)}{\partial t}. \quad (2.3)$$

Don't even ask! The quantum world is a world of probabilities, where I cannot simultaneously know both Mr. Electron's position and velocity (the famous Heisenberg uncertainty relations). The quantum world is indeed a strange one, highly counterintuitive at every bend in the road; and yet, demonstrably TRUE.

Let Mr. Electron step out into the macroscopic world (fling him across the room or let him travel down a copper wire from light switch to lightbulb), and he will respond to the applied force via Newton's second law. BUT, importantly, electrons confined inside crystals, and moving on a nanometer scale, exhibit their full-blown quantum nature. This is semiconductor land. Micro/nanoelectronics. Thus every time you switch on your laptop or answer that cell phone call, you are in fact holding and engaging a living, breathing example of the quantum world.

Fine. So what forces act on electrons? The "electrostatic force." In 1784 Coulomb showed that the electrostatic force varies as an inverse square law acting in the radial direction (r) between the two charges (q_1 and q_2), and can be either attractive (pull the charges together) if the charges are different in sign, or repulsive (push them apart) if they have the same charge (e.g., two negatively charged electrons), independent of the mass of the charged objects. Mathematically,

$$F = \frac{1}{4\pi \epsilon_0} \frac{q_1 \, q_2}{r^2}, \quad (2.4)$$

where ϵ_0 is a fundamental constant known as the permittivity of free space $= 8.854 \times 10^{-14}$ F/cm. A useful companion concept is the "electric field" generated by a single point charge q, and is given by

$$\mathcal{E} = \frac{1}{4\pi\epsilon_0}\,\frac{q}{r^2}. \tag{2.5}$$

Thus the force acting between two electrons and the field generated by a single electron are related by

$$F = q\,\mathcal{E}. \tag{2.6}$$

Think of it this way. An applied electric field (we'll see how in a moment) will produce the electrostatic force, which will then cause Mr. Electron to move.[2] This is good, because a moving charge generates an electrical "current," and that current can be easily harnessed to do an amazingly diverse set of useful things. It would not be overkill to say that virtually ALL technology, at some basic level, taps into this current-harnessing capability associated with Mr. Electron.

2.4 Voltage, circuits, field, current, and Ohm's law

Fine. So what exactly is "voltage," about as fundamental a concept as one can imagine from electrical engineering, and how does it relate to electric field and currents? Simply put, voltage is the difference in "electrical potential" between two points of an electrical "circuit," and is expressed in volts (V). The term "circuit" in electrical engineering means something specific. An electrical network is a general interconnection of electrical elements such as resistors, inductors, capacitors, transmission lines, voltage sources, current sources, and switches. An electrical circuit, however, is an electrical network that has a closed loop, giving a return path for the current to flow through.

Voltage measures the potential energy of an electric field to produce an electric current in an electrical conductor. Between two points in an electric field, such as exists in a circuit, the difference in their electrical potentials is directly proportional to the force that tends to push electrons from one point to the other. The potential difference can be thought of as the ability to move an electron through a resistance. Voltage is thus a property of an electric field, not of individual electrons. An electron moving across a voltage difference experiences a net change in energy, often measured in electron volts (eV). This effect is analogous to a mass falling through a given height difference in a gravitational field (e.g., Galileo dropping two cannonballs of different mass off of the Leaning Tower of Pisa).

When using the term voltage, one must be clear about the two points between which the voltage is specified. When specifying the voltage of a point ("node") in a circuit, it is implicitly understood that the voltage is being specified with

---

[2] The magnetic force also obviously is in play with electrons, but is not important to the context of this discussion.

respect to a stable and unchanging reference point in the circuit that is known as "ground." Often this is just what it says. Take the reference point in the circuit and connect it with a conductor (wire) to a stake literally driven into the ground (aka "Earth ground"). The Earth is an infinite source of electrons and thus serves as a stable reference. Voltage, then, is really a voltage *difference*, with one of the two points being ground. A voltage (with respect to ground) can thus be either positive or negative.

So how do we relate voltage to electric field? Simple. Electric field is just voltage per unit distance (l), or

$$\mathcal{E} = \frac{V}{l}. \tag{2.7}$$

Finally, what is electrical current (I)? Well, voltage can be thought of as the ability to move an electron through an electrical "resistance." That is, resistance (R) is simply a measure of the degree to which an object opposes an electric current moving through it. Moving electrons generate an electric current, measured in units of amperes (A = C/s). Sooo ... apply a voltage to a resistance, and some amount of electrical current flows, as mathematically described by Ohm's law (1827):

$$V = IR. \tag{2.8}$$

Obviously that flowing current is physically composed of a large number of electrons (N) racing from point A to point B with some average velocity v. If we imagine, then, applying a voltage V across the ends of a slab of material of cross-sectional area A and volume V, we have

$$I = q \, A \, (N/V)v. \tag{2.9}$$

Check the dimensions (C/s $= I$). Done.

Reference

1. See, for instance, *http://en.wikipedia.org*.

Electronic Circuits, Boolean Algebra, and Digital Logic

As I have argued at length, transistors make-the-technological-world-go-round. Why? Well, they are the fundamental building blocks of ALL electronic circuits, and these electronic circuits are the fundamental building blocks of ALL of the infrastructure surrounding the Communications Revolution; cell phone, to digital camera, to laptop, to Internet. There is no such thing as a piece of modern technology without electronic "stuff" in it. Why transistors for building electronic circuits? Transistors exhibit gain. That is, they can amplify signals. And ... transistors make beautiful regenerative (signal-restoring) switches, enabling complex (millions, billions) of electronic circuits to be hooked together to do VERY complex things (*Playstation, Wii*, etc.).

I thought it would be useful to explicitly show some examples of this transistor-to-circuit connection. Electronic circuits exist for performing an almost infinite variety of electronic functions. One studies electrical and computer engineering in large measure to first understand, and then learn to design, and ultimately learn to use such electronic circuits to build complex systems; pieces of modern technology. Yep, it takes years of hard work. Some examples of common electronic circuits? Sure! Figure 3.1 shows a precision voltage reference – literally an electronic circuit that supplies a highly stable (e.g., to temperature variations) voltage for use by other circuits. This particular voltage reference is a BiCMOS circuit, meaning that it contains both MOSFETs and BJTs, two types of transistors. Figure 3.2 shows a low-noise amplifier (LNA), which is designed to amplify (make larger) tiny RF input signals received by your cell phone and then pass them on to the rest of the cell phone's radio and DSP circuitry. Figure 3.3 is known as a operational amplifier (aka op amp), a general-purpose amplifier. Figure 3.4 shows a common interface circuit [aka an input–output (I/O) circuit] that enables one circuit to send data to another circuit very quickly without distorting its signal shape. Finally, Figure 3.5 is called a comparator circuit, which compares an input signal with some known reference value and then makes an electronic "decision" based on this information. Very quickly. This particular comparator is used in the front end of a high-speed analog-to-digital converter (ADC), which changes an input analog signal [$V(t)$] into a digital signal ("1s" and "0s") at a rate of billions of times per second. These electronic examples are known as "analog"

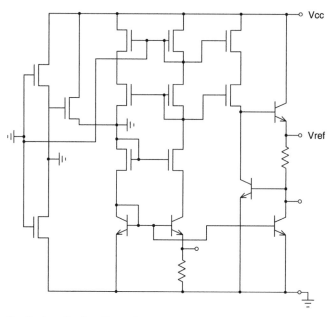

Figure 3.1   Circuit schematic of a voltage reference.

circuitry, because they do not directly manipulate digital information, but rather manipulate voltages and currents that are continuous in time (analog).

As we saw in Chapter 8, however, transistors are also uniquely suited for dealing with signals that are discrete in time (digital). Why? Well, transistors are either "on" or "off"; think "1s" and "0s." Binary code. We also need digital circuitry to implement a vast array of electronic functionality supporting the Communications Revolution. For instance, a microprocessor is essentially a highly complex (read: millions of circuits) decision-making system (logic) with an ability to remember what it decided (memory). If the input is X, then the output should be Y. Electronic cause and effect. The on–off binary nature of transistor switches can be conveniently translated into "true" or "false" logical statements for electronic decision making, PROVIDED we possess binary mathematics with which

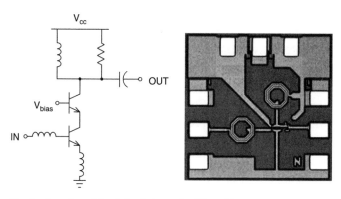

Figure 3.2   Circuit schematic and die photo of a low-noise RF amplifier.

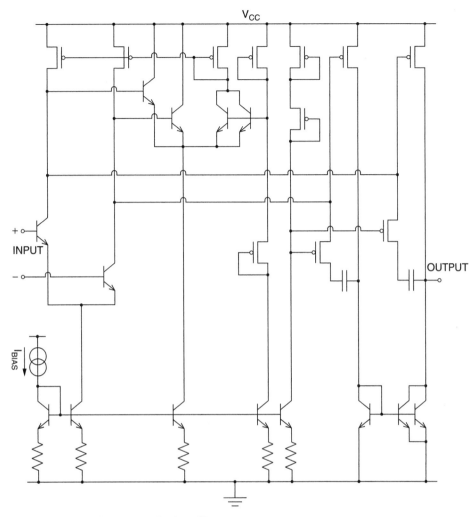

Figure 3.3   Circuit schematic of an operational amplifier.

to build these decision-making systems. We thus need a new binary mathematics, not our common base-10, number-based, mathematics. Problem? Nope.

In 1854, George Boole, in his book *An Investigation of the Laws of Thought*, introduced what is now known as Boolean algebra, a variant of ordinary taught-in-high-school algebra. Boolean algebra differs from ordinary algebra in three major ways: (1) the values that variables may assume are of a logical nature (true or false, 1 or 0), instead of a numeric character; (2) the operations applicable to those binary values differ; and (3) the properties of those mathematical operations differ. For example, certain operations from ordinary algebra (addition: $A + B$; multiplication: AB; and negation: $-A$) have their counterparts in Boolean algebra (Boolean operations: OR, $A + B$, read "A or B"; AND, AB, read "A and B"; and NOT, $\overline{A}$, read "not A," respectively). For instance, the Boolean AND function (AB) behaves on 0 and 1 exactly as multiplication does for ordinary algebra: If either A or B is 0 then so is AB, but if both A and B are 1 then so is AB. The Boolean

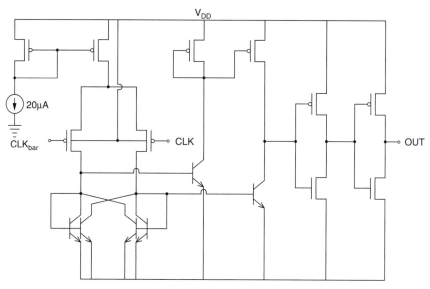

Figure 3.4   Circuit schematic of an I/O circuit.

NOT function (aka complement) resembles ordinary negation in that it exchanges values, but whereas in ordinary algebra negation interchanges 1 and -1, 2 and -2, etc., while leaving 0 fixed, in Boolean algebra, NOT ($\overline{A}$) interchanges 0 and 1 and vice versa. Just as with normal algebra, we can also combine (chain together) Boolean functions to build complex logic systems: e.g., $Y = \overline{A + B}$ [first OR A and B and then complement (NOT), the result]. You get the idea. Some prominent Boolean operations, their logical symbols, and some useful Boolean identities (just as in traditional algebra) are given in Fig. 3.6.

Seem complicated? It definitely can be. Why do it? Well, we can use transistors to quite easily implement highly complex Boolean logical functions. Because those transistors act as regenerative switches, we can hook together literally millions of such logical electronic circuits. This is called "digital logic." Ergo, we can now in principle build a decision-making "engine" from millions of

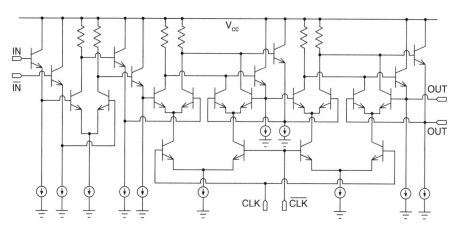

Figure 3.5   Circuit schematic of comparator circuit used in a high-speed ADC.

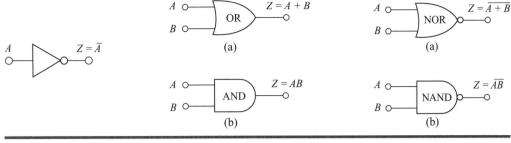

$$A + 0 = A \qquad\qquad A \cdot 1 = A \qquad\qquad \text{Identity operation}$$
$$A + B = B + A \qquad\qquad AB = BA \qquad\qquad \text{Commutative law}$$
$$A + (B + C) = (A + B) + C \qquad\qquad A(BC) = (AB)C \qquad\qquad \text{Associative law}$$
$$A + BC = (A + B)(A + C) \qquad\qquad A(B + C) = AB + AC \qquad\qquad \text{Distributive law}$$
$$A + \overline{A} = 1 \qquad\qquad A \cdot \overline{A} = 0 \qquad\qquad \text{Complements}$$
$$A + A = A \qquad\qquad A \cdot A = A \qquad\qquad \text{Idempotency}$$
$$A + 1 = 1 \qquad\qquad A \cdot 0 = 0 \qquad\qquad \text{Null elements}$$
$$\overline{A} + \overline{B} = \overline{AB} \qquad\qquad \overline{AB} = \overline{A} + \overline{B} \qquad\qquad \text{DeMorgan's theorem}$$

Figure 3.6   Basic Boolean logic symbols (top), and some useful Boolean identities (bottom).

individual transistor logic circuits. Presto – a microprocessor, the electronic brain of all computers, is born.

For example, consider Fig. 3.7, which shows a transistor-based (in this case CMOS – nMOSFET + pMOSFET; refer to Chap. 8) implementation of a Boolean NOT logic "gate" (a logic gate is just a primitive building block) [1]. You can construct the so-called "truth table" to verify its logical function, and then trace out which transistor is "on" or "off" to correctly implement that logical

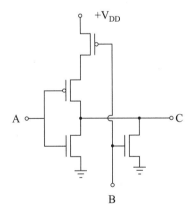

A	B	$Y = \overline{A+B}$	nFET-A	nFET-B	pFET-A	pFET-B
0	0	1	off	off	on	on
0	1	0	off	off	on	off
1	0	0	on	on	off	on
1	1	0	on	on	off	off

Figure 3.7   A transistor-level CMOS implementation of a digital NOR gate, together with its truth table and the operational state of the various transistors used.

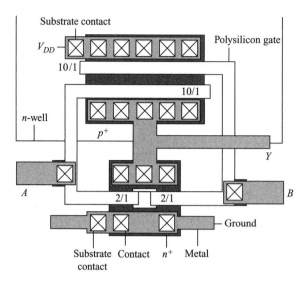

Figure 3.8 An example transistor-level "layout" (the shapes that get placed in the silicon wafer surface) of the CMOS NOR gate (inspired by [1]).

function (try it, it's not hard). This logical-to-physical (Boolean logic-to-transistor) mapping is completed by actually designing the silicon implementation of the requisite transistors and their interconnections (Fig. 3.8). You're now well on your way to microprocessor design! Concerned about how you would extend this construction path to the 100,000,000 logic gates you'll need to build a modern microprocessor? You should be! Currently, powerful computers are required for very rapidly "synthesizing" complex digital logic systems into their transistor circuit equivalents, and then of course logically verifying that its logic is correct. A classic sci-fi-like example of computers begetting computers. Sounds a little scary, but that is the only way to do it.

Reference

1. R. C. Jaeger and T. N. Blalock, *Microelectronic Circuit Design*, McGraw-Hill, New York, 2008.

A Grab-Bag Glossary of Useful Techno-Geek Terms and Acronyms

One of the more serious challenges in tackling any new technological field is figuring out exactly what all the jargon is. So much new vocabulary to contend with! I have included here a highly downselected list of some particularly common terms and acronyms that you will come across and be forced to contend with [1].

- **3G**: Third-generation wireless mobile communications networks (also 2G, 4G, etc.).
- **ADC**: Analog-to-digital converter.
- **Angstrom**: One ten-billionth of a meter (1×10^{-10} m = 0.1 nm).
- **ASIC**: Application-specific integrated circuit. Designed to suit a customer's particular requirement (application), as opposed to DRAMs or microprocessors, which are general-purpose ICs. Customized.
- **Bandwidth**: A communications figure-of-merit measure derived from the difference between the highest and lowest frequency of a given EM wave used to transmit data. Bandwidth is also used to describe the amount of data that can be sent through a given communications channel (e.g., 1 Mbps = 1,000,000 bits of data per second).
- **Binary**: Characteristic of having only two states, conveniently represented by "on" and "off." The binary number system uses only 1s and 0s.
- **Bipolar**: A type of transistor noted for its speed. The current is composed of BOTH electrons and holes ("bi" = two; hence the name).
- **Bit**: A binary digit, and the basic unit of all digital communications. A bit is a "1" or '0" in a binary language.
- **BJT**: Bipolar junction transistor (see bipolar).
- **Bluetooth™**: A wireless personal-area network (PAN) technology from the Bluetooth Special Interest Group, founded in 1998 by Ericsson, TBM, Intel, Nokia, and Toshiba. Bluetooth is an "open standard" for short-range transmission of digital voice and data between mobile and desktop machines.
- **Broadband**: Loosely specifies the bandwidth (data capacity) of a communications link (e.g., 10 Mbps = broadband).
- **Buckyball**: A sphere of 60 carbon atoms in an arrangement that looks like a soccer ball. Buckyball is a popular name for the C_{60} buckminsterfullerene molecule (named after Buckminster Fuller, inventor of the geodesic dome).

- **Byte**: A unit of binary data consisting of 8 bits.
- **CAD**: Computer-aided design. Sophisticated, computerized workstations and software used to design integrated circuits (among other things).
- **CDMA**: Code-division multiple access, also called "spread spectrum," a term for a form of digital spread-spectrum cellular phone service that assigns a code to all speech bits, sends a scrambled transmission of the encoded speech wirelessly, and then reassembles the speech into its original format. CDMA has up to 20 times the capacity of traditional analog cellular service and is known for its superior call quality and long battery life.
- **Chip**: Or microchip. Or die. An individual IC built in a tiny, layered rectangle or square on a silicon wafer.
- **Clean room**: The sterile rooms in a microelectronics fabrication facility where microchips are built. The air in these rooms is tens of thousands of times cleaner than that in a typical hospital operating room.
- **CMOS**: Complementary metal-oxide semiconductor. Combines both p-channel and n-channel field-effect transistors in the same circuit (CMOS = nFET + pFET).
- **DSL**: Digital subscriber line. A telecommuncations technology that increases the bandwidth communications over ordinary phone lines. DSL has a distance limitation of around 2 or 3 miles between the telephone company's central office and the customer site.
- **DSP**: Digital signal processing. DSP is used to enhance, analyze, filter, modulate, or otherwise manipulate standard analog functions, such as images, sounds, radar pulses, and other such signals by analyzing and transforming waveforms (e.g., transmitting data over phone lines via modem).
- **Diode**: A device that allows current to flow in one direction but blocks it from flowing in the opposite direction. Typically made from a pn junction inside a semiconductor.
- **DRAM**: Dynamic random-access memory. A type of RAM.
- **Doping**: A wafer fabrication process in which exposed areas of silicon are bombarded with chemical impurities to alter the way the silicon conducts electricity in those areas.
- **E-Beam**: Electron beam. Refers to a rather large machine that produces a highly energetic stream of electrons that can be used to expose photoresists directly on a wafer or on a mask. Electron-beam lithography is essentially a micro/nanoprinting technique.
- **EEPROM**: Electronically erasable programmable read-only memory. A type of memory.
- **Ethernet**: A local-area network (LAN) used for connecting computer, printers, workstations, terminals, servers and other computer hardware. Ethernet operates over twisted pair wires or coaxial cables at speeds up to 10 million bits per second (10 Mbps).
- **Fab**: The micro/nano fabrication facility, or fab, is the manufacturing plant where the front-end process of making semiconductors on silicon wafers is

completed. The package and assembly (aka "back-end") stages are typically completed at other facilities.

- **Fabless**: A semiconductor company with no internal wafer fabrication capability.
- **Flash**: A faster form of EEPROM memory that permits more erase/write cycles. Goes in a thumb drive.
- **Foundry**: A wafer production and processing plant. Usually used to denote a facility that is available on a contract basis to companies that do not have wafer fab capability of their own, or that wish to supplement their own fab capabilities.
- **FPGA**: Field-programmable gate array. A FPGA is a gate array that can be programmed at the end-use location (in the field; hence the name).
- **Gate array**: A semicustomized chip. The IC is preprocessed to the first metal interconnect level. The remainder of the interconnections are then customized to meet specific requirements of the customer (often used in ASICs).
- **GSI**: Gigascale integration. The art of putting billions of transistors onto a single IC.
- **HDTV**: High-definition TV.
- **IC**: Integrated circuit. A chip etched or imprinted with network or electronic components such as transistors, diodes, and resistors, along with their interconnections.
- **IEEE**: Institute of Electrical and Electronics Engineers. A worldwide engineering, publishing, and standards-making body for the electronics industry. The IEEE is the world's largest professional organization.
- **ISP**: Internet service provider. Any of a number of companies that sell Internet access to you and me.
- **LAN**: Local-area network. A communications network that serves users within a confined geographical area, and is made up of servers, workstations, a network operating system, together with some sort of wired or wireless communications link. See also WLAN (wireless LAN).
- **Lithography**: The transfer of a pattern or image from one medium to another, (e.g., from a lithographic mask to a wafer). If light is used to effect the transfer, the term "photolithography" applies.
- **Microcontroller**: A microcontroller is a stand-alone device that performs computer functions within an electronic system without the need of other support circuits. Think of it as a stripped-down specialized microprocessor. Microcontrollers are used in TVs, VCRs, microwave ovens, and automobile engines. Everywhere.
- **Micron**: One millionth of a meter (1×10^{-6} m = 1,000 nm).
- **Microprocessor**: A central processing unit (CPU) fabricated on one or more chips, containing the basic arithmetic, logic, and control elements required by a computer for processing data. A microprocessor accepts coded instructions, executes those instructions, and then delivers signals that describe its internal status.

- **Mixed-Signal**: A class of ICs combining analog, RF, and digital circuitry.
- **Modem**: A piece of communications equipment that links computers via telephone lines (or coaxial cables) and enables the transmission of digital data. Derived from the words "modulate" and "demodulate" because a modem converts, or modulates, transmission signals from digital to analog for transmission over analog telecommunications lines, and then converts them back, or demodulates the signals, from analog to digital.
- **MOSFET**: Metal-oxide semiconductor field-effect transistor. The guts of a CMOS.
- **Nanometer**: One billionth of a meter (1×10^{-9} m).
- **Nanotube**: A cylindrical tube of nanoscale dimensions, most frequently made of carbon (carbon nanotubes, CNTs). These nanotubes, which are stronger and lighter than steel, can have varying electrical properties, with applications in computing, materials manufacturing, and biotechnology.
- **Network**: An arrangement of objects that are interconnected. In communications, the network consists of the transmission channels interconnecting all clients and server stations, as well as all supporting hardware and software to make it all work (e.g., routers).
- **Optoelectronics**: An electronic device that is responsive to or that emits or modifies light waves. Examples are LEDs, optical couplers, laser diodes, and photodetectors.
- **PCB**: Printed circuit board. A flat material (thin board) on which electronic components are mounted. Also provides embedded wires that act as electrical pathways connecting components.
- **PDA**: Personal digital assistant. Think Blackberry.
- **PLD**: Programmable logic device. A digital IC that can be programmed by the user to perform a wide variety of logical operations. FPGAs are PLDs.
- **Quantum dot**: A class of nanocrystals that emit varying colors of light depending on their size. They can be used to label different biological structures.
- **RAM**: Random-access memory. May be written to, or read from, any address location in any sequence. Also called a read–write memory, RAM stores digital bits temporarily and can be changed rapidly as required. RAM constitutes the basic read–write storage element in computers.
- **RF**: Radio frequency. The range of electromagnetic (EM) frequencies above the audio range and below that of the visible light. All broadcast transmissions, from AM radio to satellites, fall into this range, between roughly 30 kHz and 300 GHz.
- **ROM**: Read-only memory. Permanently stores information used repeatedly by a widget, such as tables of data, characters for electronic displays, etc. A popular type known as PROM is programmable in the field with the aid of special equipment. Programmed data stored in a ROM are often called "firmware."
- **Roadmap**: An international reference document of requirements, potential solutions, and their timing for the semiconductor industry (e.g., SIA Roadmap).

It identifies needs and encourages innovative solutions to meet future technical challenges, and also provides an ongoing emphasis on obtaining consensus industry drivers, requirements, and technology timelines.

- **Self-Assembly**: The ability of objects to assemble themselves into an orderly structure. Routinely seen in living cells, this is a property that nanotechnology may someday extend to inanimate matter.
- **SRAM**: Static random-access memory. A type of RAM.
- **Standard cell**: Predefined circuit elements that may be selected and arranged to create a custom or semicustom IC more easily than through design. Designers build ASICs using standard cells.
- **System-on-a-chip (SoC)**: A chip that is a self-contained minisystem, including processing, memory, and input–output functions.
- **Transistor**: A solid-state electronic device capable of amplifying electronic signals similar to the vacuum tube but made from a semiconductor material such as silicon or germanium.
- **VLSI**: Very large-scale integration. The art of putting hundreds of thousands of transistors onto a single IC.
- **Wafer**: A thin slice, typically less than 1 mm thick, sawed from a cylindrical ingot (boule) of extremely pure, crystalline silicon, typically 200–300 mm in diameter. Arrays of ICs or discrete devices are then fabricated on the wafers during the IC manufacturing process.
- **Yield**: Yield refers to the percentage or absolute number of defect-free (fully functional) die on a silicon wafer or of packaged units that pass all performance specifications. Because it costs the same to process a wafer with 10% "good" die and 90% "good" die, eliminating defects and improving yield become the critical variables in determining the final cost per chip. Higher yield is definitely better.

Reference

1. See, for instance, *http://www.sia-online.org/ind_glossary.cfm* or *http://www.icknowledge.com/glossary/glossary.html*.

Index